Microbial Biomass Process Technologies and Management

Basanta Kumara Behera • Ajit Varma

Microbial Biomass Process Technologies and Management

Springer

Basanta Kumara Behera
Amity Institute of Microbial Technology
Amity University Uttar Pradesh
Noida, Uttar Pradesh
India

Ajit Varma
Amity Institute of Microbial Technology
Amity University Uttar Pradesh
Noida, Uttar Pradesh
India

ISBN 978-3-319-53912-6 ISBN 978-3-319-53913-3 (eBook)
DOI 10.1007/978-3-319-53913-3

Library of Congress Control Number: 2017942789

Printed on acid-free paper

This Springer imprint is published by Springer Nature
The registered company is Springer International Publishing AG
The registered company address is: Gewerbestrasse 11, 6330 Cham, Switzerland

Foreword

Industry based on microbial processing technology is enormous, second only in gross revenue to the information industry. It runs at breakneck speed in efforts to keep pace with the growing global population, now in excess of 7.4 billion people. Advances in all fields of microbial processing are providing a never-before-seen level of health care products. The world's population demands not only safe health care products, but also high quality and convenient products that meet their latest requirements.

The biotechnological, biopharmaceutical, and biological industries are among the most advanced and sophisticated as a result of their major investments in research and development (with a delay in results), use of the complex metabolism of microbes in biocatalysis, and understanding of complex human biology. Earlier manufacturing processes took place in batches, with some traditional touches. Thanks to modern microchip-based upstream and downstream processes, continuous manufacturing has been developed and is on the verge of transforming the biotechnological value chain. This is having an impact on every biotechnology-based company, from giant multinationals to the third-party manufacturers that small start-ups hire to make their products. The process of continuous manufacture with completely modernized supply chain management has increased the scope of the biotechnological product market at the international level. The biotechnology industry now wants to be less expensive, more eco-friendly, and provide a timely supply of goods on a quality basis.

Keeping the above facts in mind, the authors present the concept of microbial processing technology, using impressive models and the most current data to explain the social and industrial significance of microbes. This work will be immensely helpful to graduate scientists, engineers, and biochemical engineers, enabling them to step into the art of commercialization of microbial processing technology.

This book focuses on the core biochemical engineering principles that enable the translation of advances in the life sciences into real processes or products. Readers will develop an understanding of advanced engineering skills (bioprocess design, bioreactor engineering, downstream processing, biopharmaceutical supply chain management, material balance), state-of-the-art life science techniques (cell disruption, cell organelle separation, biomolecule polishing and stabilization), and essential business and regulatory knowledge (quality control, biopharmaceutical supply chain management, material balance).

I am highly impressed by the way the authors have presented updated knowledge on the science and technology of microbial process engineering to reveal the impact that transferring laboratory data has on sustained development of the biotechnology industry.

I sincerely believe that the book will be of high value to students, research scholars, and microbial technology professionals and provide a major boost to the microbial process technology industry worldwide.

Amity University Gurgaon P.B. Sharma
Manesar, India

Delhi Technological University
Delhi, India
17 May 2017

Preface

Microbial biomass process technology refers to the systematic study of the basic principles of engineering and technology and their application to the large-scale isolation and purification of microbial metabolites obtained through fermentation processes. These metabolites are widely used in biopharmaceuticals, biochemicals, foods, nutraceuticals, and diagnostics. Knowledge of microbial engineering enables scientists and practitioners to conduct microbial processes with innovation and high productivity. Thus, microbial biomass process technology is of academic interest and also has tremendous industrial use. It entered academia at the undergraduate and postgraduate levels as a taught course, as usually happens with any new topic. It is now taught as a specialized course in higher studies in order to involve biotechnology and biochemical engineering students in industrial activities.

Currently, numerous microorganisms are used in industrial microbiology, including naturally occurring organisms, laboratory-selected mutants, and genetically modified organisms (GMOs). Debate over the use of GMOs in food sources is gaining momentum, with more and more supporters on both sides. Keeping this in mind, the first chapter highlights the salient features of industrial microbial processes, including safety, mixing, environmental conditions, and the use of selected microorganisms to obtain higher productivity at lower cost. In addition, there is a section on microbial products in the global market and a brief discussion on possible improvements in the market share. The discussion is intended to induce the reader to think about how to solve the problem of inadequate production of microbial products at the international scale.

The next two chapters are devoted to significant aspects of upstream and downstream processes, and suggest how innovative research is urgently required to establish economically viable technologies that operate on the spectrum of opportunities offered by microorganisms. The authors describe how to monitor microbial process design in order to solve specific biological and technological constraints and minimize the cost of product during production and downstream operations.

Innovative microbial processing management techniques are therefore urgently needed to increase the efficacy of process control systems at all stages of the production system. Bioengineering is under the constraint of eco-ethics, sustainability, and public acceptance. Another option for industrialists is an effective microbial product supply chain to enable sustainability in the biotechnology and biopharmaceutical industries. Therefore, the fourth chapter is fully dedicated to innovative management technology for maintaining the quality and quantity of microbial products, and their safe and timely supply to the end user. The main aim of the chapter is to introduce the idea of integrated knowledge of microbial resources and bioengineering, which is necessary for faster development of industrial microbial processes and generation of the appropriate manpower.

The last chapter of this book describes application of the basic principles of thermodynamics and mass balance to help understand the ongoing status of a specific fermentation process and enable optimum production in the shortest possible time. The authors describe material helpful in understanding material balances, the concept of stoichiometric coefficients for the overall reaction representing the biological system, and how to obtain information about yield factors, substrate consumption, and product formation rates.

We are greatly indebted to Mrs Asha Sharma and Katyayanee for their help in computing and in drawing models and computer-generated graphics. We would also like to thank our respective families, who spent many lonely hours while we were busy preparing the manuscript.

Support from the publisher during preparation of the manuscript is also gratefully acknowledged.

Noida, UP, India

Basanta Kumara Behera
Ajit Varma

About the Book

This work describes a rational and conceptual framework for appraising how the diversified range of microbes can be used as effective and sustainable resources to meet current challenges in finding suitable economic solutions for the biopharmaceutical, enzyme, food additive, nutraceutical, biochemical, and microbial fuel industries. In addition, various aspects of microbial regulatory acts and their effective application have been highlighted. To make these subjects more applicable, special attention has been given to the *stoichiometry of material balance, with interesting examples*. The book also discusses the management processes involved in effective production of microbe-based products and in utilization of eco-friendly biotechnology industries for green living. The prime aim is to provide a ready reference for technologists and policymakers who need an instant and pragmatic guide to the fast emerging field of microbial process technology development and implementation.

Contents

About the Authors

Basanta Kumara Behera held the position of professor of biotechnology at three distinguished Indian universities, where, since 1978, he was regularly involved in research and taught postgraduate courses on topics in bio-energy management and biomass processing technology. In 2009, he joined a multinational company as adviser for specialty chemical production and design. Dr. Behera has authored or co-authored three books, including one on English literature. His works on hydrogen fuel cells and microbial transformation of regiospecific complex botanical molecules have appeared in reputed international journals.

Ajit Varma is a distinguished scientist and Professor of Eminence at Amity Institute of Microbial Technology, and vice chairman of Amity Science, Technology & Innovation Foundation (ASTIF), Amity University, Uttar Pradesh, India. Before this, until 2004 he served as Professor of Microbial Technology, School of Life Sciences, Jawaharlal University. Dr. Varma has contributed over 304 research articles to national and international journals. He has also published 78 books in the area of microbial technology through Academic Press, CRC Press (USA), IDRC (Canada) and Springer Verlag (Germany). He is the editor-in-chief for *Soil Biology*, published by Springer-Verlag (Germany).

Abbreviations

AD	Anaerobic digestion
ADP	Adenosine diphosphate
ALF	Airlift fermenter
ATCC	American Type Culture Collection
ATP	Adenosine triphosphate
ATPS	Aqueous two-phase system
BG Medium	Blue-green medium
BIO	Biotechnology Industry Organisation
BLA	Biological license application
CAGR	Compound annual growth rate
CAPEX	Capital expenditure
CC	Column chromatography
cGMP	Current good manufacturing practice
CMMi	Capability maturity model integration
CoGs	Cost of goods
COGS	Cost of goods sold
CRM	Customer relationship management
CRP	Continuous replacement process
CSTR	Continuous stirred tank reactor
DEAE	Diethylaminoethyl
DMSO	Dimethyl sulfoxide
DNA	Deoxyribonucleic acid
dNS	2′-Deoxyribonucleoside
DOE	Department of Energy
DSP	Downstream processing
EBT	Earn before tax
ecoABC	eco-Agriculture Biofuel Capital Initiative
EDTA	Ethylenediaminetetraacetic acid
EOR	Enhanced oil recovery
ERP	Enterprise resource planning
FDA	Food and Drug Administration

FFDCA	Federal Food, Drug, and Cosmetic Act
FIFO	First in, first out
FIPCO	Fully integrated pharmaceutical company
FP	Framework programme
GC	Gas chromatography
GLP	Good laboratory practice
GSCM	Green supply chain management
HDFE	High-density polyethylene
HPH	High-pressure homogenizer
INDA	Investigational drug application
ISO	International Organization for Standardization
IT	Information technology
LDFE	Low-density polyethylene
LIFO	Last in, first out
LLE	Liquid–liquid extraction
LTV	Long tube vertical evaporator
M&A	Merger and acquisition
MAE	Microwave-assisted extraction
MF	Membrane filter
MNC	Multinational company
MPE	Microbial processes economics
MPN	Most popular number
MSA	Manitol salt agar
NBE	New biological entity
NDA	New drug application
NPV	Net process value
OMEGA	Offshore membrane enclosure for growing algae
OPEX	Operating expense
PAA	Polyacetic acid
PCTE	Polycarbonate track etch
PEG	Polyethylene glycol
PHA	Polyhydroxyalkaneate
PHSA	Public Health Service Act
PPQA	Process and product quality assurance
PTFE	Polytetrafluoroethylene
PVC	Polyvinyl chloride
QA	Quality assurance
QbD	Quality by design
QC	Quality control
QR	Quick response
R&D	Research and development
RCE	Relative centrifugal field
RFFC	Rising-falling film evaporator
RMS	Reverse micellar system

RNA	Ribonucleic acid
RRFC	Recirculating rising evaporator
SAMM	Safety assessment of marketed medicine
SAP	System application product
SC	Supply chain
SCF	Supercritical fluid
SCM	Supply chain management
SOP	Standard operation protocol
TB	Terrific Broth
TLC	Thin-layer chromatography
TQM	Total quality management
UHPH	Ultrahigh-pressure homogenizer
UV	Ultraviolet
XLD	Xylose lysine deoxycholate
YM	Yeast and mold

Chapter 1
Concept of Microbiological Processes

"Microbiological processes" refers to processes in which microorganisms (or their parts) are used to make or modify products, or in which new microorganisms are developed for specific uses. Consequently, the concept of "microbe-based products" encompasses products that are made or modified by microorganisms, as well as new microorganisms. In this connection, the term "microorganism" means not only bacteria and yeasts, but also fungi, algae, protozoa and human, animal, and plant cells, that is, all generally unicellular organisms with dimensions not visible to the naked eye that can be propagated and manipulated in a laboratory, including plasmids and viruses.

Nonpathogenic microbes are mostly sustainable natural sources of interesting biocompounds that can be used for human welfare in a variety of ways. The value-added biocompounds present in microbes can be obtained using a solvent extraction process or as extracellular metabolites released into the culture medium [1]. Bioactive compounds of microbial origin can be sourced directly from primary metabolism (e.g., proteins, fatty acids, vitamins, and pigments) or can be synthesized via secondary metabolism. Such compounds can present antifungal, antiviral, antialgal, antienzymatic, or antibiotic actions [2, 3], giving them high commercial value [3]. With the modernization of technology and development of hi-tech fermenters and computerized downstream process technology, microbes have become an environmentally friendly and economically viable source for compounds of interest, especially as production can be optimized in a controlled culture under standard regulatory norms.

In most biotechnology-based industries, shorter product life cycles, complex process development technology, and increased competition have raised the level of concern over minimizing process management of new products, processes, and service development. Industries are keener to reduce the development time taken from concept to customer and, at the same time, improve quality and significantly reduce cost by effective and efficient production. One approach many companies are taking is to improve the efficacy of the product supply chain.

B. Kumara Behera, A. Varma, *Microbial Biomass Process Technologies and Management*, DOI 10.1007/978-3-319-53913-3_1

Therefore, microbiological process technology (MPT) is at the heart of all industries that involve the production of products such as food supplements, biopharmaceuticals, food additives, nutraceuticals, biofuels, and plastics such as biopolymers that are being used in daily life. Through MPT we can manufacture a target product at the required quality and quantity with a controlled expense budget and in the minimum possible time. MPT is designed to obtain target products of both intracellular and extracellular origin of maximum quality without losing the physicochemical nature of the products. Special attention is given to ensure pathogen-free product in all stages of processing, up to an end product in a state of perfect purity. Microbiological tests are conducted on all facets of the manufacturing process, including raw materials, mixing, milling, tableting, capsuling, and packaging (Fig. 1.1). Raw materials, in-process products, and final products are only released to the next stage of production by the quality control unit upon verification that the material or product has passed all test requirements.

Standard MPT under national and international norms can help to maximize interoperability, safety, repeatability, and quality. It can also facilitate commoditization of formerly custom processes. Like any other products in daily use, products of microbial origin are also oriented toward customers' requirements. Standard MPT includes some structured processes and policies for product quality maintenance. The documented collection of policies, methods, and procedures followed by a development team or organization during microbial biomass processing is called the microbial product development methodology. The latest of such methodologies are organized structures of sequential and parallel activities imposed on the development of microbial products. A process, in fact, is a series of definable, repeatable, and measurable tasks leading to a useful result. The benefits of a well-defined process are numerous.

1.1 History of Microbial Processes

Microbes are the earliest form of life on earth, first evolving more than three billion years ago [4–6]. Nonpathogenic microbes of commercial value have been processed and packaged since the earliest days of man's history on earth. Our ancestors discovered how to harness the power of microorganisms to derive different value-added products [7–11], even if they did not know the science behind what they were doing. Use of bacteria, yeasts, molds, and algae in food and as food has also been well practiced since Neolithic times (around 10,000 BC) [10]. Fermentation helps to preserve foods and upgrade their nutritional value and organoleptic qualities [12, 13]. The earliest evidence of MPT is in production of wine from natural resources rich in carbohydrates, using fermentative bacteria as biocatalysts [4]. Since its origin, wine making has been associated with sacred purposes. Georgia (ca 6000 BC) and Iran (ca 5000 BC) were the most ancient sites of wine production [7–9]. Since Biblical times, consumption of ritual wine has been part of Jewish practice and, as part of the Eucharist commemorating the Last Supper of Christ, is also essential to the Christian Church. From the religious point of view,

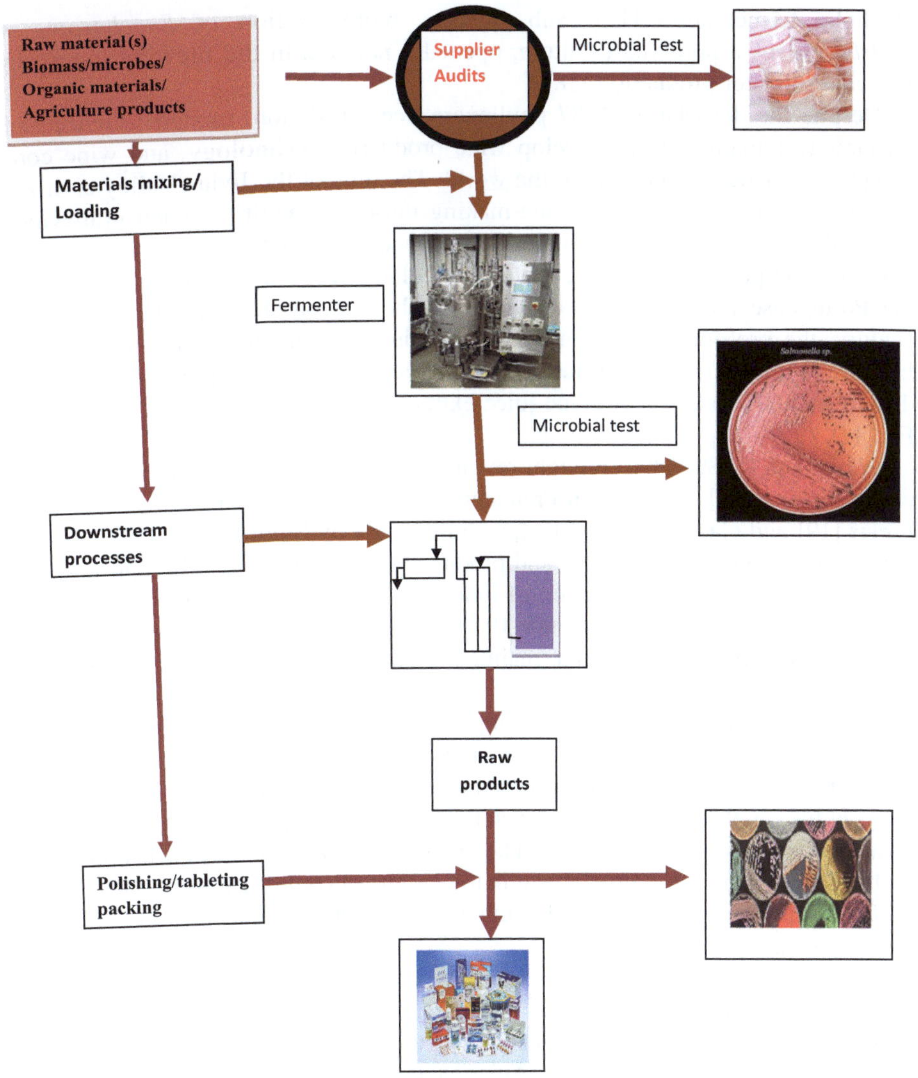

Fig. 1.1 Quality checks from raw ingredient to final product

Islam avoids consumption of wine but, during the Golden Age, alchemists such as Geber pioneered the distillation of wine for medicinal and industrial purposes such as the production of perfume [14, 15]. The Turkish Uyghurs were even responsible for reintroducing viticulture to China, from the Tang dynasty onward. Wine production and consumption increased with time, burgeoning from the fifteenth century onward as part of the European expansion. Ever since the discovery of microorganisms by Robert Hooke and Antoni Leeuwenhoek (1665), researchers have been thinking about the use of bacteria in improving the quality of MPT [16]. In 1857, Louis Pasteur first proved the function of microorganisms in lactic alcoholic fermentation [17]. Subsequently, reorganization of *Saccharomyces carlsbergensis*, a pure yeast culture, opened a new era in the brewing industries for wine and beer production [18].

Despite the devastating 1887 phylloxera lice infestation, modern science and technology remain keen to develop wine production technology, and wine consumption now occurs throughout the world. The time of the Indus Valley civilization witnessed the practice of wine-making through viticulture when grapevines were introduced from Persia. Historical evidence has long been recorded on the wine-making process; however, it was particularly encouraged during the time of the Portuguese and British colonization of the subcontinent. In the nineteenth century, the proliferation of wine-making industries decreased as a result of an increase in religious consciousness in the public. An illustrated presentation of the developmental processes of grape juice extraction for wine-making is depicted in Fig. 1.2.

Weigmann (1889–1896) brought revolutionary changes in milk processing and cream making [11]. Eduard von Freudenreich's (1897) discovery of *Lactobacillus brevis* [19] and classification of lactic acid bacteria by Sigurd Orla (1919) helped control the fermentation process and produce better quality food through fermentation technology [20]. Production of industrial concentrated cultures, frozen cultures, and freeze-dried cultures since 1970 has not only helped milk processing but also consistently improved and maintained the quality of fermentation-based products.

Like wine production, cheese making technology is an interesting process that involves microbes. In fact, the making of cheese dates back more than 4000 years. According to an ancient legend, it was made accidentally by an Arabian merchant who put his supply of milk into a pouch made from a sheep's stomach as he set out on a day's journey across the desert. The rennet in the lining of the pouch, combined with the heat of the sun, caused the milk to separate into curds and whey. That night he found that the whey satisfied his thirst, and the cheese (curd) had a delightful flavor that satisfied his hunger.

Travelers from Asia are believed to have brought the art of cheese making to Europe. Cheese making continued to flourish in Europe and cheese became an established food. In fact, the Pilgrims included cheese in the Mayflower's supplies when they made their voyage to America in 1620. In due course, the cheese production process switched from being a home industry to a sophisticated automatic manufacturing industry (Fig. 1.3). The making of cheese quickly spread in the New World, but until the nineteenth century it remained a local farm industry. It was not until 1851 that the first cheese factory in the USA was built by Jesse Williams in Oneida County, New York. During cheese processing, the role of molds is significant in introducing desired characteristics or in changing the physical properties for cosmetic reasons.

Penicillin, perhaps the most famous of all antibiotic drugs, is derived from a common fungus called *Penicillium* (from *Penicillium chrysogenum*). In the Second World War (1940–1949) this antibiotic could have saved millions of lives. At first, this antibiotic was produced by the bottle culture technique. Extraction of the antibiotic (penicillin G) was carried out manually using a solvent extraction

Fig. 1.2 The history of the wine press: (**a**) The very first wine press was probably the human foot and the use of manual treading of grapes is a tradition that has lasted for thousands of years and is still used in some wine regions today. (**b**) Sixteenth-century basket press. (**c**) Basket press from the Provence region of southeast France. (**d**) Modern pneumatic batch press. Source: en.wikipedia.org

process. With time, antibiotic process technology has completely changed. Highly sophisticated and computerized fermentation technology is now used for manufacture of penicillin G (Fig. 1.4).

The idea of using algae as a source of food, feed, and energy goes back more than half a century. Production of methane gas from algae was proposed in the early 1950s and received a big impetus during the energy crisis of the 1970s, when projects were initiated to produce gaseous fuels (hydrogen and methane). From 1980 to 1996 the US Department of Energy supported the Aquatic Species Program (ASP), a relatively small effort (about $25 million over almost 20 years) with the specific goal of producing oil from microalgae.

Fig. 1.3 Cheese-making (**a**) *Tacuinum sanitatis Casanatensis* (fourteenth century). (**b**) Eighteenth century (Photograph by Granger; https://www.pinterest.com/pin/47147127323257626/). (**c**) Cheese production process as in home industry showing curdling of milk by hand boiling. (**d**) Showing manual separation of cheese (source: www.britannica.com, search by image). (**e**) Modern cheese production process (from Encyclopedia Britannica)

ASP researchers worked primarily on growing algae in open ponds and made significant contributions to our understanding of how to grow algae for fuel. Advances in biotechnology, such as the ability to genetically engineer algae to produce more oils and convert solar energy more efficiently, have unleashed new possibilities not feasible during the ASP years.

Most of the activity in algae research and commercial production has been in the USA. With more than 100 start-up companies, large corporations, and the US Government investing billions in this new industry, the USA is the leader in advancing algae-based fuels. However, algae biofuels are also being researched around the world in both developed and developing nations. The NASA Ames Research Center is carrying out pioneering work in a project studying the possibility of obtaining algal biodiesel, food, fertilizer, and other value-added biochemicals from marine algae as an open system (offshore membrane enclosure for growing algae; OMEGA). OMEGA is a coastal marine system to produce biofuels, clean

Fig. 1.4 Pictorial presentation on development of the pencillin G manufacturing process. Source: penicillin.wikispaces.com. (**a**) Penicillin culture vessels. Flasks being incubated at 24°C in the animal operating theatre of the Sir William Dunn School of Pathology. (Florey et al., Antibiotics, 1949). www.nms.ac.uk700 × 473Search by image. (**b**) Penicillin—Museum of the History of Science: Museum of the .... www.mhs.ox.ac.uk417 × 400Search by image. Early penicillin experiments: (**c**) Penicillin production through deep-tank fermentation—National ... www.acs.org512 × 338Search by image. Citric acid fermentation tanks in Pfizer's Brooklyn facility, circa 1920s. (**d**) Fermenta Biotech Ltd (FBL) set up a biotech plant in 1987 at Kullu, in the foothills of the Himalayas, to manufacture penicillin G amidase enzyme and emerged as the leader in enzyme technologies for β-lactams in India. (**e**) Penicillin—How is it made into a drug?. penicillin.wikispaces.com668 × 404Search by imagerecovery.jpg. (**f**) Penicillin—general bioprocess flow penicillin.wikispaces.com864 × 647Search by image Process_flow_for_penicillin.jpg

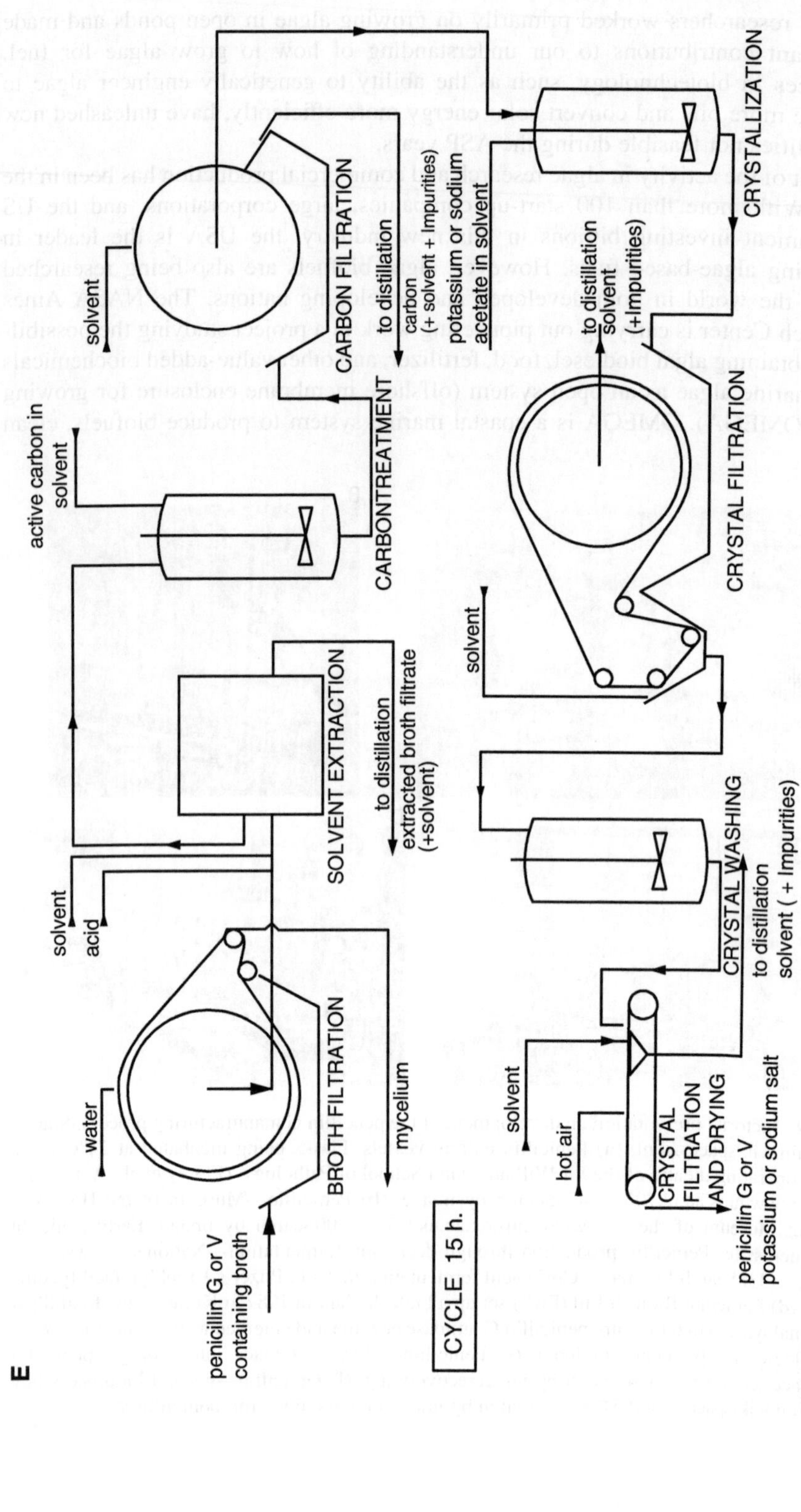

Fig. 1.4 (continued)

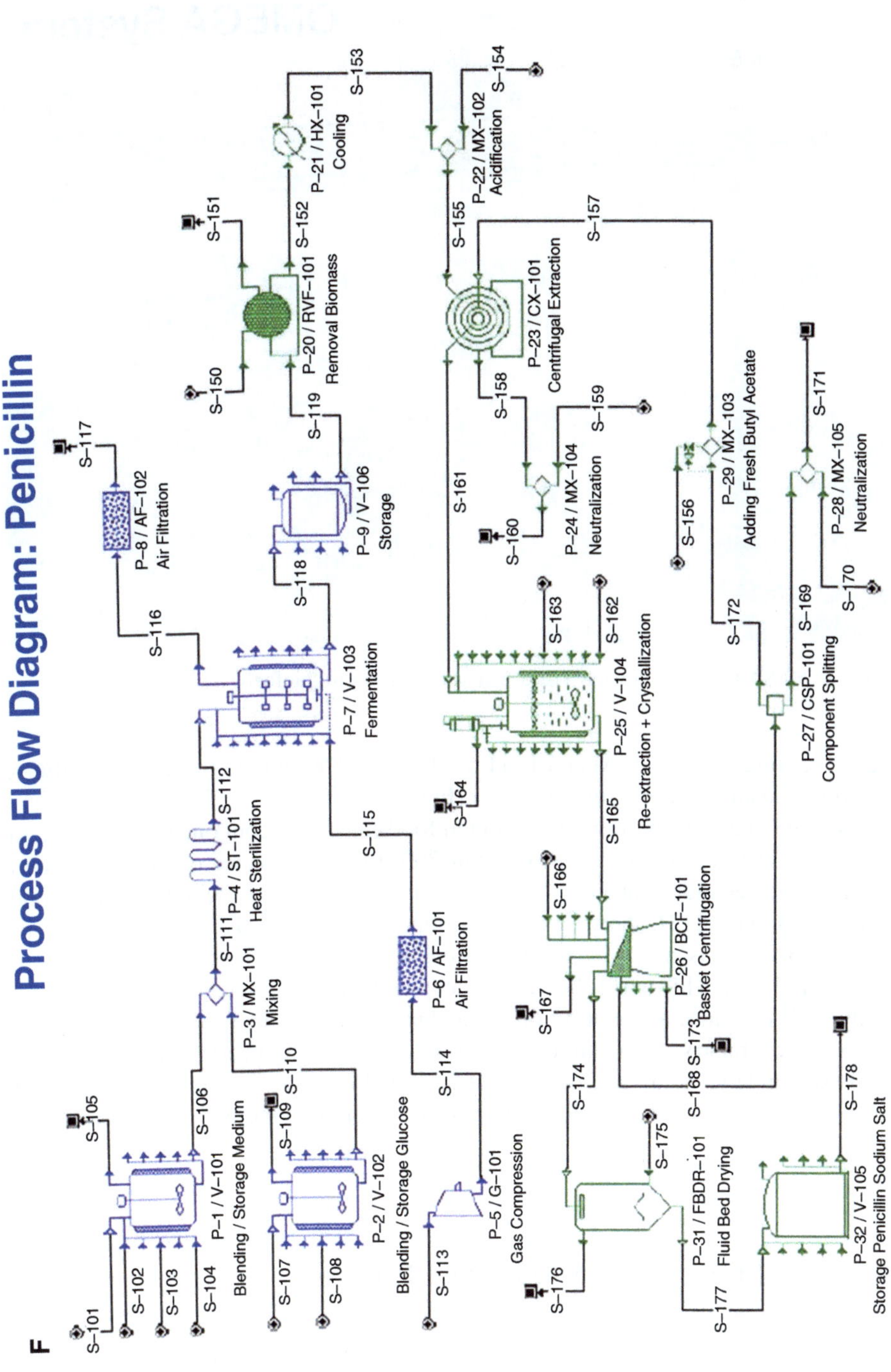

Fig. 1.4 (continued)

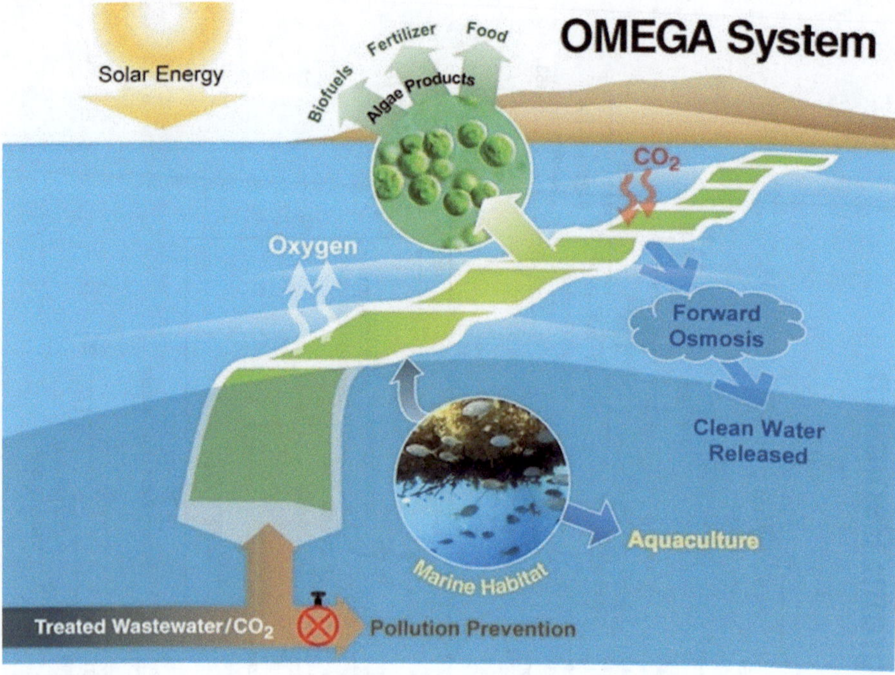

Fig. 1.5 NASA OMEGA Project: The ocean as a platform for biofuels

water, capture CO_2, and expand food production. The goal is to deploy OMEGA worldwide using an open source model (Fig. 1.5).

By 1995, the consumption of fermented foods reached about one third of all food consumed in Central Europe. More than 260 different species of microbial food culture have been identified and described for their beneficial use in fermented food products globally [13], showing the importance of their use. The scientific conceptual discovery was made by Louis Pasteur in the second half of the nineteenth century [13, 17]. This was followed by deep and extensive scientific and technical study with the close association of researchers from various fields, such as physiology, biochemistry, biotechnology, and biochemical engineering. This catalyzed increased understanding and improved traditional food processing with the help of superb microbial strains resulting from genetic engineering and enriched culture techniques.

1.2 Need for Standard Microbial Processes

By definition, MPT is a set of decisions that need to be made in any project operation so that quality products can be consistently manufactured. There is a need to make the correct decisions early in a project as it can be hard to make changes during later phases, because the microbial products are susceptible to any mechanical changes if a new alternation is introduced at any stage of processing.

The critical issue in microbial process development is that the theoretical concept does not ensure reliability in manufacturing products of target quality. However, it is mandatory to control process development through a well-defined and systematic process. The old fashioned "code and test" approach is now impractical because of the increased complexity and huge size of projects. Ad hoc development approaches in a megaproject development can cause disastrous outputs. They can lead to ambiguous communications, imprecise observations, brittle designs, inaccurate risk assessment, insufficient testing, uncontrolled change propagation, and subjective progress assessment.

To have perfect control over a turnkey project development, formally defined processes are needed. Standard process development protocols are required to provide visibility into the projects. Visibility, in turn, aids timely management control and mid-course corrections against expected errors and crisis. It helps developers to weed out faults early, before they cause failure. This also avoids cascading of faults into later phases, where accumulation of an increased number of faults leads to failure. A formal development methodology also helps to organize workflow and outputs to maximize resource utilization and to define everybody's roles and responsibilities clearly. Individual productivity increases as a result of specialization and, at the same time, the team's productivity increases because activities are coordinated. The adoption of a formal development process with well-defined, unambiguous, and complete policies leads to the benefits of having the correct requirements to generate the correct product to meet the customer's needs. The inclusion of necessary features, in turn, reduces the post-development cost.

A variety of microbial bioproducts are increasingly being used in our daily life. We have to make sure that process operation works well to keep up with the demand for microbial products in an effective and efficient manner. Measurements in an MPT project ensure that the project stays in control. High quality products are the result of high quality development processes. The absence of process and product quality increases problems during microbial product development. Process and product quality assurance (PPQA) are process-driven approaches with specific steps for achieving development goals. PPQA considers design, development, and production. Quality assurance prevents defects from occurring by providing methods and rules. It starts in the microbial product development process in the early stages of the system development life cycle (SDLC) and grants confidence to customers regarding the microbial product development process and the product. PPQA is a systematic and planned pattern of all actions necessary to provide adequate confidence that a product or an item conforms to established technical requirements.

The purpose of PPQA is to provide management and staff with an objective insight into processes and related work products. Process and product quality have become the key to success in the biotechnology and biopharmaceutical industries.

We know that microbial products help us in shaping our life for better sustainability, even under stress conditions. So, it is necessary to better understand, evaluate, and control the processes, products, and microbial biomass-based projects from the perspective of controlling, evaluating, tracking, forecasting, and understanding. A valid measurement process can help organizations to make better and timely decisions in order to achieve success in producing quality products of high purity.

In microbial process development, measurement is a helpful support tool for microbial product management. Measurement is an important factor for effective management of processes. Measurement is a mechanism for evaluation, identification, and prediction for various microbial biomass-based products and processes. The best way to enhance microbial biomass processes is to measure specific attributes of the process and then define a set of related metrics based on the attributes of the process. Those metrics can then be used to define indicators that can lead to improvement planning. Measurement of microbial products has an important function in understanding and controlling bioproduct development and processes. In microbial process measurement, numbers or symbols are assigned to attributes of the selected entities in real life so that some defined rules can be used to characterize the attributes. Microbial process measurement is important for understanding, controlling, and improving processes. Measurements are used to assess achievement of goals, determine status according to plans, establish principles for comparison with future evaluations, monitor improvement efforts, and gain an understanding of resources, environmental processes, and products. The main measurement objective is to control and monitor the performance of software processes.

At present, microbial process measurement is at a state in which terminologies, methods, and principles are still being defined and combined. We should not expect to find quantitative laws that are mostly valid and usable, and have the same accuracy and precision as the laws of physical science. The definition of usable and internationally valid measures is a long-term and ideal research goal, which will not be realized in the near future. MPT is not governed by the quantitative laws of physical science; sometimes microbial processes even disobey the laws of thermodynamics. Because microbial process measurements are often indirect, they are open to controversy. The goal of microbial process measurement is to maintain quality biological products in the manufacturing process.

1.3 Microbes and Microbial Products

Microbes are versatile organisms, found in diversified habitats, and have been serving humanity in almost all aspects of requirements helpful for sustaining life. Food supplements, biopharmaceuticals, food additives, nutraceuticals, various types of biofuels, and plastics such as biopolymers (Table 1.1) are some important products regularly used in daily life that are obtained from microbes [21–94]. Microbial cells are the most important key players in the wellness industries. The global market for microbes and microbial products was valued at $143.5 billion

Table 1.1 Value-added microbial products from different types of microbes

Type of product	Microbes involve	References
Polysaccharides		
Alginate	*Azotobacter vinelandii*	[21, 22]
Gellan	*Sphingomonas paucimobilis*	[23, 24]
Pullulan	*Aureobasidium pullulans*	[25, 26]
Kefiran	*Lactobacillus kefiranofaciens*	[27, 28]
Lentinan	*Lentinula edodes*	[29, 30]
Pleuran	*Pleurotus ostreatus*	[31, 32]
Xanthan	*Xanthomonas capastris*	[33, 34]
Enzymes		
Glucose isomerase	Different microorganisms	[35, 36]
Glucose oxidase	*A. niger; G. oxidans*	[37, 38]
Cholesterol oxidase	*Rhodococcus* sp.	[39, 40]
Amylases	Different microorganisms	[41, 42]
Xylanases	Different microorganisms	[43, 44]
Phytases	Different microorganisms	[45, 46]
Cellulases	Different microorganisms	[47, 48]
Vitamins and related products		
Vitamin B2 (Riboflavin)	*Eremothecium ashbyii, Ashbya gossypii*	[49, 50]
Vitamin B7 (Biotin)	r*E. coli; Rhizopus nigricans*	[51, 52]
Vitamin B9 (Folic acid)	*Bifidobacterium* (different species)	[53, 54]
Vitamin B12	Different microorganisms	[55, 56]
Vitamin C (Ascorbic acid)	*G. oxidans*	[57]
Vitamin K2	*Geotrichum candidum, Flavobacterium* sp.	[57]
Probiotics	*Lactobacillus* (different species)	[58, 59]
	Saccharomyces boulardii	[60, 61]
	Kluyveromyces lactis	[62]
	Bifidobacterium (different species)	[63, 64]
	Lactococcus (different species)	[65]
Starter culture	*S. cerevisiae*	[66]
	Penicillium roqueforti	[67]
	Different lactic acid bacteria (LAB)	[68, 69]
Bioplastics		
Polyhydroxyalkanoate (PHA)	Different microorganisms	[70, 71]
Polylactate (PLA)	Different microorganisms	[72, 73]
Organic acids		
Citric acid	*Aspergillus niger*	[74, 75]
Gluconic acid	*A. niger. Gluconobacter oxidans*	[76, 77]
Oxalic acid	*A. niger*	[78, 79]
Amino acids		
L-Arginine	*Brevibacterium glavum, Bacillus subtilis*	[80–83]
L-Aspartic acid	*Alcaligenes metacaligenes, E. coli*	[84–86]
L-Glutamic acid	*Corynebacterium glutamicum*	[87, 88]

(continued)

Table 1.1 (continued)

Type of product	Microbes involve	References
L-Lysine	*C. glutamicum*	[89]
DL-Methionine	*C. glutamicum, E. coli*	[90, 91]
L-Phenylalanine	*C. glutamimcum, rE. coli*	[92, 93]
L-Threonine	*E. coli*	[94]
L-Tryptophan	*C. glutamicum*	
Antibiotics		
Penicillins	*Penicillium chrysogenum*	[95, 96]
Cepaholsporins	*Penicillium chrysogenum*	[97, 98]
Cyclosporins	*Tolypocladium inflatum*	[99, 100]
Erythromycin	*Saccharopolyspora erythreae*	[101, 102]
Rifamycins	*Amycolatopsis mediterranei*	[103, 104]
Natamycin	*Streptomyces natalensis*	[105, 106]
Recombinant proteins	**Recombinant strains**	
Insulin	r*E. coli,* r*P. pastoris,* r*S. cerevisiae*	[107, 108]
Human growth hormones	r*E. coli*	[109, 110]
Viral surface antigen	r*P. pastoris*	[111, 112]
	i	[113, 114]

in 2014. The 2015 market is projected to approach $154.7 billion, and the 2020 market nearly $306 billion, with a compound annual growth rate (CAGR) of 14.6% from 2015 to 2020 [95].

The high market demand means that good yield and productivity are crucial for the economics and viability of MPT. These could be achieved by means of metabolic engineering through genetically modified microbes. Metabolic engineering is the practice of optimizing genetic and regulatory processes within microbial cells to increase the production of a certain substance. The development of recombinant DNA technology and other related technologies has provided new tools for improving yields by means of genetic manipulation of biosynthetic pathways. Industrial microorganisms such as *Escherichia coli* and the actinomycetes have been developed as biocatalysts to provide new processes or optimize existing processes for the biotechnological production of chemicals from renewable plant biomass. Factors such as oxygenation, temperature, and pH have been traditionally controlled and optimized in industrial fermentation to enhance metabolite production. Metabolic engineering of bacteria shows great scope in industrial applications because such techniques can help to increase product formation, speed up the process, save energy, stop formation of byproducts, and develop strains with resistance to environmental stress.

1.3.1 Fungi

Fungi make up one of the most important groups of organisms on the planet. Societies have utilized fungi for centuries in a wide variety of ways by capitalizing on their metabolism and metabolites (chemicals made from metabolism) produced. Fungi are very useful organisms in biotechnology. They are important experimental organisms because they are easily cultured, occupy little space, multiply rapidly, and have a short life cycle. Many fungi are used as model organisms for genetics, cell biology, and molecular biology. The now famous "one gene, one enzyme" hypothesis for the ascomycete fungus *Neurospora* won Beadle and Tatum the Nobel Prize in 1958. Currently there are about 1600 antibiotics commercially produced and a number of medical drugs manufactured using various fungi. These microorganisms are well known for the production of bioactive metabolites such as antibiotics, enzymes, alkaloids, and many products of biotransformation (where less or non-bioactive compounds are transformed into a more active form).

Multibillion dollar industries are involved in high value-added biochemicals such as anticholesterol statins, antibiotics such as penicillin, the immunosuppressant cyclosporins, and steroids. Statins are used to reduce cholesterol levels and prevent cardiovascular disease. The group of statins derived from fermentation include lovastatin (first isolated from *Aspergillus terreus* and the first statin approved by the FDA in 1987), pravastatin (isolated from *Nocardia autotrophica*), and mevastatin (from the fungi *Hypomyces*, *Paecilomyces*, and *Trichoderma*, and a fermentation product of *Penicillium citrinum*). Since its discovery in 1941, the antibiotic penicillin from the fungus *Penicillium notatum* (often called *P. chrysogenum*) has revolutionized human health and treatment of disease. Cephalosporins are another group of antibiotics originally produced by the fungus *Cephalosporium* (synonym of *Acremonium*). First discovered as powerful immunosuppressants in the 1970s, cyclosporins are a primary metabolite of several fungi, including *Trichoderma*, *Tolypocladium*, and *Cylindrocarpon*. Cyclosporins are widely used during and after bone marrow and organ transplants in humans. The steroid progesterone is produced industrially by the fungus *Rhizopus nigricans*. Steroids such as cortisone (used in arthritis treatment) and prednisone are manufactured with the help of molds.

Yeasts and other fungi play an important role in drug production, food processing, biocontrol, enzyme biotechnology, and research and development. The use of yeasts (e.g., *Saccharomyces cerevisae*) to make alcohol and carbon dioxide uses the fermentation process to break down sugars. Up to 50% of the sugar can be converted to alcohol, but rarely surpasses 15% because the fungi are sensitive to high concentrations of alcohol. In the beer making industry, cereal grains are fermented to make the final product. Wine is composed of fermented grapes and hard cider is essentially fermented apples. Sake is produced by rice fermentation using *Aspergillus oryzae*, followed by an additional fermentation step utilizing bacteria and yeasts. With bread making, fermentation utilizes sugar to produce carbon dioxide and alcohol. The carbon dioxide produces the bubbles and causes bread to rise, while the alcohol produced evaporates off during baking. A variety of foods and drinks are fermented before consumption. Some of the largest industries in the West are based around production of alcohol and bread. In the East, the production of soy sauce and other fermented soybean products are probably the largest industries. The peoples of Asia have developed a wide variety of interesting fermented foods, sauces, and drinks using fungi. Other examples and the applicable fungi include koji (*Aspergillus*); miso and soy bean paste (*Aspergillus*); sufu and Chinese cheese (*Rhizopus*); nyufu or fuyu, bean cake or bean cheese (*Rhizopus*); shoyu or soy sauce (*Aspergillus*, *Saccharomyces*); and tempeh (*Rhizopus*).

Many other fungi also produce antibiotic substances, which are now widely used to control diseases in human and animal populations [96]. The discovery of antibiotics revolutionized health care worldwide.

People have been using yeast in bread baking for centuries. The biochemical byproducts of yeast sugar metabolism (carbon dioxide and alcohol) are essential in baking and brewing. Baker's yeast (*Saccharomyces cerevisiae*), when added to baker's dough, produces carbon dioxide pockets that make the bread rise. Brewer's yeast, another strain of *Saccharomyces*, takes advantage of the yeast's characteristic of switching to anaerobic fermentation when deprived of oxygen to produce alcohol as a byproduct of incomplete sugar breakdown. The use of yeast in biotechnology industries has been increasing as yeasts such as *S. cerevisiae* and *Pichia pastoris* are becoming important biofactories for the production of different types of recombinant protein (e.g., insulin, albumin, and hepatitis B surface antigen) and different types of therapeutic protein [97, 98]. Actinomycetes are a very important source for the production of antibiotics. More than 70% of known antibiotics are produced by this type of microorganism. Moreover, actinomycetes are also an important source for the production of different types of enzyme inhibitors used in the treatment of many diseases.

Edible mushrooms are common vegetables that provide an important source of dietary fiber and complete protein. Fungal proteins provide all of the essential amino acids, a consequence of their close relation to animals. Studies have shown that the protein content of the edible penny bun mushroom (*Boletus edulis*, also known as porcino, cep, king bolete) rivals and even exceeds that of some meats. This fact is especially important for people who subsist on wild-collected foods and have limited access to other sources of protein. Moreover, some mushrooms used as food may have medicinal properties, providing a smattering of health benefits.

Microbial metabolites are also important components in many cosmetic products, either as major active ingredients or as filling, stabilizing, flavoring, or coloring agents. These metabolites include peroxide-inducible protective factors (from *S. cerevisiae*), chitosan, octadecandionic acid, rhizobium gum, exopolysaccharides, exfoliation-promoting enzymes, flavors, fragrances, and natural pigments [99, 100]. In addition, some new low molecular weight microbial metabolites such as ectoines are currently used as components of lotions and creams based on their ability to prevent skin dehydration, protect against UV solar radiation, and act as anti-aging agents [101, 102].

The mycorrhizal relationship between fungi and plant roots is essential for the productivity of farm land. Without the fungal partner in root systems, 80–90% of trees and grasses would not survive. Mycorrhizal fungal inoculants are available as soil additives from gardening supply stores and are promoted by supporters of organic agriculture.

As animal pathogens, fungi help to control the population of damaging pests. These fungi are very specific for the insects they attack and do not infect animals or plants. Fungi are currently under investigation as potential microbial insecticides, with several already on the market. For example, the fungus *Beauveria bassiana* is a pesticide being tested as a possible biological control agent for the recent spread of emerald ash borer.

In nature, the breakdown of plant materials is primarily by fungi, by the means of secreted fungal enzymes. Driven by the need for non-food-based bioenergy, the industrial-scale conversion of biowaste has been researched and developed over the last 10 years. After all this research, the conclusion is that the most efficient and gentle way of converting recalcitrant lignocellulosic materials for industrial purposes is through the use of fungal enzymes [103]. Through such conversions, the building blocks of organic materials are kept intact, ready to be used in the value cascade. The enzymatic conversion of biowaste and sidestreams provides the basis for an entirely new and more efficient use of natural resources, paving the way for a larger bioeconomy sector in a more biobased society.

The microbial fertilizer market has grown extensively during the last few decades. This type of fertilizer is composed of different microbial consortia, which are able to survive in the rhizosphere and support plant growth by providing macro- and microelements in utilizable forms. The most famous are the plant growth-promoting rhizobacteria, which includes various species such as *Azotobacter*, *Alcaligenes*, *Arthrobacter*, *Acinetobacter*, *Bacillus*, *Pseudomonas*, *Enterobacter*, *Rhizobium*, *Bradyrhizobium*, and *Serratia* [104–106]. In addition, other microorganisms such as mycorrhiza help maintain soil fertility and absorption of essential elements for better and healthier plant growth [107]. These types of beneficial soil microbes support plant growth through the conversion of natural soil elements from non-utilizable form to utilizable form. This is carried out through different microbial processes such as nitrogen fixation (converting atmospheric nitrogen to ammonia) and phosphate solubilization (converting insoluble organic or rock phosphorus to a soluble form such as orthophosphate).

At present, three types of cheese, blue cheese, soft ripened cheese (such as Camembert and Brie), and rind-washed cheese, depend on molds for quality improvement during manufacture. *Penicillium roqueforti*, is used for blue cheese preparation. The mold is added while the cheese is still in the loosely pressed curd form. As the cheese matures, the mold grows by spreading its blue-colored filaments, leaching flavor into the cheese; examples include Stilton, Roquefort, and Gorgonzola [108]. Soft ripened cheeses such as Brie and Camembert are manufactured by introducing *P. camemberti* to grow on the outside of the cheese, which causes aging from the outside in. The mold forms a soft white crust, and the interior becomes runny with a strong flavor [109]. Rind-washed cheeses such as

Limburger also ripen inward, but here, as the name suggests, they are washed with brine and other ingredients such as beer and wine, which contain mold. This also makes them attractive to bacteria, which adds to the flavor [110].

The wholesale cheese industry was born and showed phenomenal growth during the latter half of the 1800s. As cheese demand continued to grow and spread rapidly, manufactured and processed cheese production increased dramatically. Total natural cheese production grew from 418 million pounds in 1920 to 2.2 billion pounds by 1970. Rising demand for cheese throughout the 1970s and 1980s brought total natural cheese production to more than 6 billion pounds by the beginning of the 1990s. Processed cheese also experienced a surge in consumer demand, with annual production exceeding 2 billion pounds a year by the beginning of the 1990s. Currently, more than one-third of all milk produced each year in the USA is used to manufacture cheese. Recent increases in the overall demand for farm milk have, in large part, been a result of continued growth of the cheese industry. As consumer appetites for all types of cheese continue to expand, so will the industry.

To improve the surface structure and flavor of sausages, *P. chrysogenum* and *P. nalgiovense* are used. They also contribute to shortening of the ripening period and preservation of the natural quality. This expands the shelf life of the meat product [111–113]. In the past, soy sauce was made by mixing soybeans and other grains with a mold (*A. oryzae* or *A. sojae*) and yeast and leaving the mixture to ferment in the sun [114]. Today, soy sauce is made under controlled conditions. The key flavor ingredients formed in this process are salts of the amino acid glutamic acid, notably monosodium glutamate [115].

1.3.2 Bacteria

Bacteria are suitable for industrial use because they grow rapidly, usually have simple nutrient requirements, and can produce a wide range of products. Bacteria are also an important resource for food supplements, antibiotics, probiotics, drugs, vaccines, starter cultures, insecticides, enzymes, fuels, and solvents.

Lactic acid bacteria such as *Lactobacillus*, *Lactococcus*, and *Streptococcus* are used in the manufacture of dairy products such as cheeses (including cottage cheese and cream cheese), cultured butter, sour cream, buttermilk, yogurt, and kefir. Lactic acid bacteria and acetic acid bacteria are used in pickling processes for foods such as olives, cucumber pickles, and sauerkraut. Lactic acid bacteria convert the unstable malic acid [18] that is naturally present in wine into stable lactic acid. This malolactic fermentation gives the stability that is characteristic of high-quality wines that improve on storage [116].

Bacterial fermentations are also used in processing of teas, coffee, cocoa, soy sauce, sausages, and an amazing variety of foods in our everyday lives. Bacterial food cultures are responsible for the aroma, taste, and texture of cheese and fermented milk products such as yogurt, ayran, doogh, skyr, and ymer. They contribute to development of the flavor and color of such fermented products as salami, pepperoni, and dried ham.

Like fungi, bacteria are used to manufacture antibiotics, vaccines, and medically useful enzymes. Most antibiotics are made by bacteria that live in soil. Actinomycetes such as *Streptomyces* produce tetracyclines, erythromycin, streptomycin, rifamycin, and ivermectin. *Bacillus* and *Paenibacillus* species produce bacitracin and polymyxin. Bacterial products are used in the manufacture of vaccines for immunization against infectious diseases. Vaccines against diphtheria, whooping cough, tetanus, typhoid fever, and cholera are made from components of the bacteria that cause the respective diseases. It is significant to note here that the use of antibiotics against infectious disease and the widespread practice of vaccination (immunization) against infectious disease are two twentieth-century developments that have drastically improved the quality of life and the average life expectancy of individuals in developed countries.

Bacteria play significance roles as microfood supplements in human diet for flavor and value addition. Such types of bacterial food cultures can be of two types, starter and probiotic. Generally, starters are added either at the beginning of fermentation or between the fermentation and ripening process to develop the desired metabolic activity. By this technique, flavor, color, texture, safety, preservation, nutritional value, wholesomeness, and/or health benefits are added to the processed food [117–119]. However, probiotics also have a functional role, which refers to the ability of certain microbes to confer health benefits on the consumer [120, 121]. The bacterial strains used as starter are exclusive in nature, but sometimes a bacterium can be used both as starter and probiotic [122, 123]. Food microbiologists are introducing better microbial technology to economize food processing [124].

By means of genetic engineering, superior strains of bacteria have been developed to produce human hormones such as insulin, enzymes such as streptokinase, and human proteins such as interferon and tumor necrosis factor. These products are used for the treatment of a variety of medical conditions and diseases, including diabetes, heart disease, tuberculosis, AIDS, and SLE. Botulinum toxin and *Bacillus thuringiensis* insecticide are bacterial products used in medicine and pest control, respectively. Using *Agrobacterium tumefaciens*, plants have been genetically engineered so that they are resistant to certain pests, herbicides, and diseases.

1.3.3 Algae

Algae are the most promising long-term, sustainable sources of oils for biofuel, food, feed, and other co-products. The multiplication rate of algae means that they can double their numbers every few hours, be harvested daily, and have the potential to produce volumes of biomass and biofuel many times greater than those of our most productive crops [125]. Algae store energy in the form of oils and carbohydrates, which, combined with their high productivity, means that they

Table 1.2 Comparison of liquid fuel yields in different crops (gallons of oil per acre per year)

Crop	Yield
Corn	18
Soybeans	48
Safflower	83
Sunflower	102
Rapeseed	127
Oil palm	635
Microalgae	5000–15,000

Source: http://oakhavenpc.org/cultivating_algae.htm

can produce 2000–5000 gallons of biofuels per acre per year [126]. For high productivity, algae require more CO_2, which can be supplied by emissions sources such as power plants and ethanol facilities. Algae thrive on organic carbon or CO_2 and nutrients such as nitrogen and phosphorus. Growth conditions and the availability of sunlight, carbon, and nutrients affect the metabolism of algae and whether they produce lipids or carbohydrates. Algae cultivation can use land that in many cases is unsuitable for traditional agriculture, as well as water sources that are not useable for other crops (seawater, brackish water, and wastewater). As such, algae-based fuels complement biofuels made from traditional agricultural processes. Algae have higher potential to convert solar energy [127] for production of biodiesel than other crops because algae need less area for cultivation (indoor or outdoor systems) than crops (Table 1.2). To produce a certain amount of biodiesel in an indoor system, algae need 1000 times less water than any petro crop.

Macroalgae (seaweeds) are grown in the sea, or even on land with seawater, and their sugars can be converted into biofuels and chemicals [128]. However, manipulation of nutrients has not proved successful in increasing algal productivity. Researchers have found that when algae naturally produce hydrocarbons (molecules that can most readily substitute for today's petroleum uses), growth and reproduction are limited.

Biotechnology research goals therefore include finding ways to increase the reproductive rate, improve metabolism of inputs, and enhance the production of desired oils, fuel-grade alcohols, or proteins in useful species. Researchers have found that many algae species are adaptable to genetic engineering, expressing complex proteins and accumulating recombinant proteins to very high levels. Biotechnology is already employed in sequencing and annotating the genomes of algal species. Genomic data aids researchers to understand the metabolic processes through which algae convert carbon and nutrients into lipids or carbohydrates. Greater understanding of algal metabolism and reactions to growth conditions will inform further research. Genetic engineering techniques currently used in plant and microbial biotechnology, including synthetic biology and metabolic engineering, are employed to enable algae to more predictably produce the desired lipids for biofuels, alcohols, proteins, enzymes, and other molecules, or carbohydrate-rich biomass for bioprocessing.

Biotechnology research is important, not only in the initial stages of developing algal biofuels, but also in optimizing algal strains for the mechanical engineering and processing needs of biofuel production. Algal strains used in industrial processes must appropriate harvest and molecule recovery requirements, such as resistance to the high heat and pressure used in mechanical separation, which are probably not naturally occurring traits.

Ubiquitous and easy to grow, algae have long been a promising biomass-to-fuel candidate in the eyes of researchers. Now algae is a burgeoning sector in biofuels, with several high-profile start-ups, including Craig Venter's Synthetic Genomics, and the interest of big-time investors like Bill Gates and ExxonMobil. Of course, hurdles still exist that must be overcome to make a competitive fuel. Algal biofuels still cost too much to produce, over $8 a gallon according to the DOE [129]. Furthermore, most existing strains do not yield oil in the quantities needed to quickly scale up to commercial production of biofuels. Companies also need to worry about contaminating local ecosystems and the amount of water needed to grow cultures in large batches. Despite these challenges, inroads, and actual fuel, are being made in the nascent field. Five projects that lead the pack today are Algenol Biofuels [129], Solix Biofuels [130], Sapphire Energy [131], Solazyme [132], and Seambiotic [133]. These companies are seeking to scale-up commercial production of algae and are pursuing several engineering approaches (using closed systems and open pond systems) for the design of an economic system for growing algae.

At present, it is estimated that the manufacturing cost of microalgal biomass is $2.95/kg for photobioreactors and $3.80/kg for open ponds. These costs exclude the cost of CO_2. However, assuming that the biomass contains 30% oil by weight, with an increase in algal biomass productivity to more than 10,000 tonnes per year, the cost of production per kilogram could reduce to about $0.47 and $0.60 from photobioreactors and open ponds, respectively. Numerous funding programs have been created with the aim of promoting the use of renewable energy. In Canada, the eco-Agriculture Biofuels Capital Initiative (ecoABC) provides $25 million per project to assist farmers in constructing and expanding renewable fuel production facilities. In Europe, the Seventh Framework Programme (FP7) is the main instrument for funding research. Similarly, NER 300 is an unofficial, independent portal dedicated to renewable energy and grid integration projects [134].

1.4 Microbial Process Economics

1.4.1 What Is Microbial Process Economics

Microbiological processes have been harnessed to obtain value-added microbial products. Production of microbe-derived metabolites requires processes for culturing the microbes and deriving the target product(s) from microbial biomass or from culture broth. After recovering the crude biomaterials, further downstream processing to purify the metabolite(s) using sophisticated equipment can be substantially more expensive than culturing the microbes. So, before manufacturing any microbe-derived product, it is crucial to find the most effective and commercially relevant option. Commercial production of intracellular or extracellular microbial metabolites requires the following: (1) large-scale monoseptic production of the appropriate microbe, (2) recovery of the biomass from a relatively dilute broth, (3) extraction of the metabolite from the biomass, and (4) purification of the crude extract.

Success in commercialization of any microbe-derived product depends on the involvement, actions, and operations that are associated with the production and sale of microbes and microbial biomass-derived biomaterials. This framework serves as a broad placeholder for all economic processes. To increase the demand for microbial biomass-based products, the microbiological processes must be sound and economically profitable. This means that the cost of production must be lower than the income from marketing. It is therefore necessary to assess the economic benefit of investment in a new microbial process and to ensure that the venture is economically viable and sustainable. Maintaining the sustainability of a biotechnology industry should be an ongoing process throughout the entire life span of the industry, accomplished by modifying and adapting the manufacturing process and quality of the demanded products with time. This is only possible by replacing equipment as and when necessary and following safety legislation. Suitable microbial process engineering is responsible for maximization of profit in correlation with market demand and quality control.

In MPT, expenditure means money paid by an individual, organization, corporation, or government in exchange for microbiological products (intracellular or extracellular metabolites) and services. Income is received thereafter by an individual, organization, corporation, or government in exchanges for microbial products and services. An individual receives salaries, tips, or wages; a corporation receives earnings and sales; and a government receives taxes.

Man-made capital exchanges are economic processes that act upon man-made capital in the form of depreciation, destruction, or investment. Raw material production means the creation, collection, or extraction of raw material for the production of biomaterials and services, especially directly from the natural environment. This includes all activities such as harvesting, mining, lumbering, and fishing. It also includes appreciation, extraction of nonrenewable resources, harvesting of biota, introduction of biota, landscape and habitat alteration, resource development, and resource discovery. Waste production refers to byproducts not used for any other means and released into soil or discharged into water or air.

1.4.2 The Economic Nature of Microbial Processes

Microbiological processes are involved in mass culture of industrially important microbes and extraction of intracellular or extracellular metabolites. In its simplest

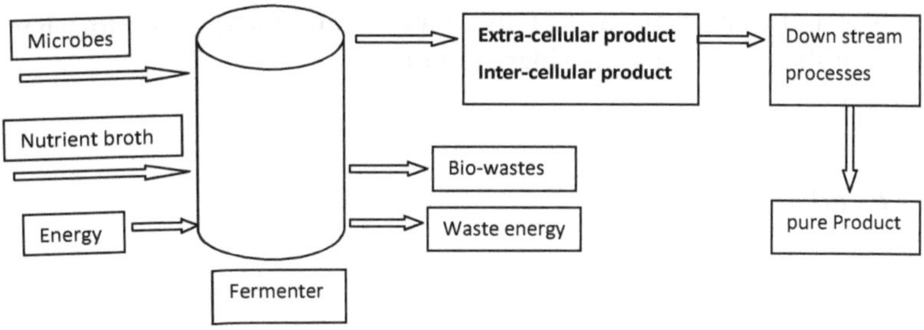

Fig. 1.6 The simplest form of a microbiological process consists of a series of material and heat flows

form, a microbiological process consists of a series of material and heat flows that can be represented by a simple model, as depicted in the Fig. 1.6.

The energy flows or processes for transfer of biological materials have a cost value, either as investment or as income in the form of biological products. For example, the energy utilized in a biochemical process has a cost that reflects its generation and transmission. It is also possible that a biochemical process generates a source of energy that can be utilized or sold to a third party, thereby generating income.

In a microbiological process, the most important factors are the type of microbe used and the supply of specific nutrients for the maximum production of target metabolite. The stoichiometry of the chosen transformation process determines the material and heat flows for the process, and ultimately the economics for that process. This can be expressed as:

$$\text{Type of microbe used A} \\ + \text{nature of nutrient in broth B} \underset{\text{Biocatalyst}}{\overset{\textit{Heat of reaction}}{\rightleftarrows}} \text{Target metabolite C} + \text{Wastes D}$$

The stoichiometry of a microbial processing event determine the degree to which the feed materials (mainly nutrient broth) are helpful for forming target metabolites and the thermodynamics determines the overall heat consumed. The effluent generated during the fermentation process contains unconverted feed, unused nutrient residue, and biowaste. The target metabolite is separated from the broth left over after fermentation, and the biological wastes are subjected to recycling for energy generation or disposed of in the environment after suitable treatment.

All of this processing requires additional unit operations, which have energy requirements in the form of electricity, steam, cooling water, refrigeration, etc. These demands have to be met by generating energy on site or importing from third parties. Therefore, the choice of reaction route defines the process flow sheet containing all the necessary unit operations, with associated costs and revenue streams. The influence that each of the material and energy streams has on the economics of the process are considered in turn.

1.4.3 *Feedstock*

In microbial process engineering, second-generation feedstocks used for the fermentation process are challenging in several ways. Using second-generation feedstock, microbial processes have been harnessed to generate useful bioproducts (metabolites) for health care, food supplements, natural antioxidants, biopharmaceuticals, etc. The means of establishing which products to make and how to optimize the processes are mainly based on the nature of the microbes used. Generally, selection of industrial production is primarily based on the potential of a microorganism to produce product, but not on its substrate utilization capacity. Recently, scientific opinion has strongly favored industrial production hosts that are either known to produce large amounts of a certain product or can easily be engineered to achieve this goal. Moreover, there is a preference for microorganisms that are well characterized physiologically as well as genetically, and therefore have the potential to become microbial production platforms. The most interesting and impressive example of this development is the genetic modification of yeast *S. cerevisiae* to utilize xylose, arabinose, and cellulose [135–140].

Selection of a product-generating microbe based on substrate utilization potential has therefore become increasingly important. All industrial production hosts are able to grow on hydrolysates of agricultural waste or waste glycerol, with a few exceptions. In general, the method of feedstock pretreatment has little effect on the performance of the microbial production host. At present, the use of second-generation feedstocks for the production of bulk microbe-based products has been gaining popularity in industrial biotechnology.

1.4.3.1 Biocatalysts

Biocatalysis can be defined as the use of natural substances to speed up (or catalyze) chemical reactions. The natural substances can be one or more enzymes or cells. The employment of enzymes and whole cells has been important for many industries for centuries. The most obvious uses have been in the food and drink businesses, where the production of wine, beer, cheese, etc., is dependent on the effects of microorganisms. In microbial processes, both the microorganism and microbe-derived enzymes can act as biocatalysts. The selection of biocatalyst in a target-oriented fermentation process is usually made on the basis of product yield and selectivity. Optimization of biocatalyst can reduce investment and operating costs. The important advantage of biocatalysts are that they are environmentally acceptable, being completely degraded in the environment. Furthermore, the enzymes act under mild conditions, which minimize problems of undesired side reactions such as decomposition, isomerization, racemization, and rearrangement, which often plague traditional methodologies. In addition, immobilized enzymes demonstrate very high stability and re-usability and can be used to conduct reactions in continuous mode in microreactors [141].

Microorganisms such as yeast (after appropriate genetic modification) are used as whole-cell biocatalysts for production of target product, both in good quality and quantity. For instance, yeast whole-cell biocatalysts for lipase-catalyzed reactions were constructed by intracellularly overproducing *Rhizopus oryzae* lipase in *S. cerevisiae* MT8-1(a).

A bacterial–yeast consortium (*Brevibacillus laterosporus* and *Galactomyces geotrichum*) was used as a proficient biocatalyst. It decolorized 92% of sulfonated azo dye Reactive Red 198 (RR 198) within 18 h at a dye concentration of 50 mg/L compared with 58 and 42% decolorization using *Brevibacillus laterosporus* and *Galactomyces geotrichum* alone, respectively [142]. Both yeasts and bacteria as whole-cell systems have been widely used as biocatalysts [143–146] at the electrode of microbial fuel cells (MFCs). These microorganisms can transfer electrons to the electrode directly or through natural nanowires (pili), mediate electron transfer by a natural or artificial redox mediator, and finally direct redox transformation of excreted metabolites at the electrodes (Fig. 1.7)

Use of yeast as a whole-cell biocatalyst is a classic example of MPT. Metabolically engineered yeast cells have long been used for ethanol production from complex lignocellulose biomass sources. The experimental strategies for degradation of the cellulose and xylose components of lignocellulose are well reviewed [147].

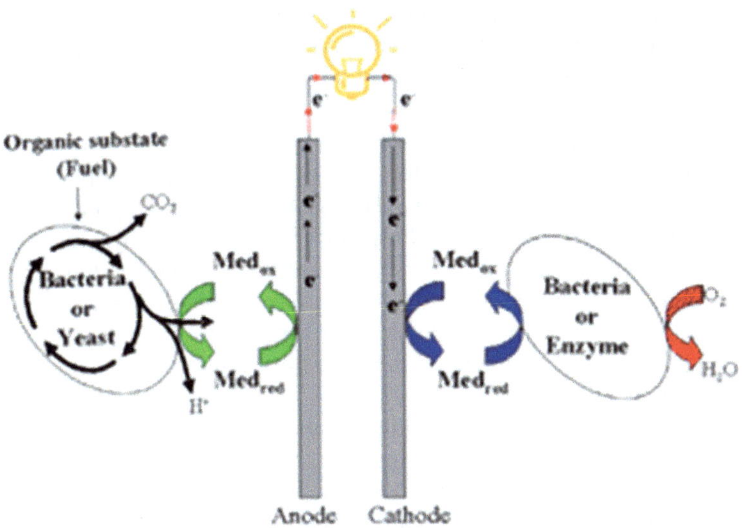

Fig. 1.7 Whole-cell system of bacteria or yeast used as biocatalyst to produce electricity

Baker's yeast *S. cerevisiae* is used as a whole-cell biocatalyst in many different industrial settings, such as production of ethanol, butanol, isobutanol, succhinic acid, artemisinin, and recombinant protein [148]. Additionally, research has been carried out on the production of phenolic compounds such as flavonoids and stilbenoids [148]. It is also a suitable biocatalyst for the enantioselective conversion of (*S*)-1-PEA to (*R*)-1-PE because there is no competing endogenous transaminase activity [147] and very little ACP reductase activity.

1.4.3.2 Bioenergy

In chemical processes, energy streams are necessary in a variety of forms, for example, steam for heating and electricity for bringing the equipment to a functional state and maintaining the water temperature needed at different stages of the process. However, in microbial processes, there is a need for biochemical energy to regulate the biological processes (metabolism), in addition to forms of energy as required in a chemical process. However, there is limited variation in microbial processes and there are several challenges, especially in the operation of energy-requiring microbial processes.

In microbial processes, operating enzymes are used to optimize biochemical reactions and are helpful in developing quality products in quantity. There is no doubt that many industries have attempted to use bioprocesses involving enzymatic transformation to produce chemicals; however, most of the enzymes that have been successfully developed are simple enzymes such as hydrolases and hydratases. In order to give more support to enzyme-catalyzed processes, a combination of enzyme-catalyzed reactions and their accessory components, such as a cofactor regeneration system, was recently developed. The best example of such construction of a microbial platform for an energy-requiring bioprocess is enzymatic carbonyl reduction for chiral alcohol synthesis with a cofactor [NAD(P)H] regeneration system [149–151].

This basic principle was applied to energy-requiring bioprocesses using ATP or phosphorylated high-energy substrates. In this process, two metabolic characteristics of baker's yeast help as driving forces to develop a microbial platform for an energy-requiring bioprocess: efficient ATP regeneration and temporary extracellular accumulation of fructose 1,6-diphosphate (FDP) induced by organic solvent (toluene or acetone) in the presence of glucose and phosphate (Fig. 1.8) [152–154]. Through this unique enzymatic process, 2′-deoxyribonucleosides (dNSs) are synthesized from glucose, acetaldehyde, and a nucleobase via a C–C coupling reaction on the basis of a high-energy substrate provided by the glycolytic pathway as an energy-generating system.

There are opportunities for generation of useful forms of energy that can be recycled in a microbial process. For instance, the heat generated by exothermic reactions can be used to generate steam for use in the process, thereby minimizing heat import requirements, with the balance exported to be used by a third party and thus attracting value.

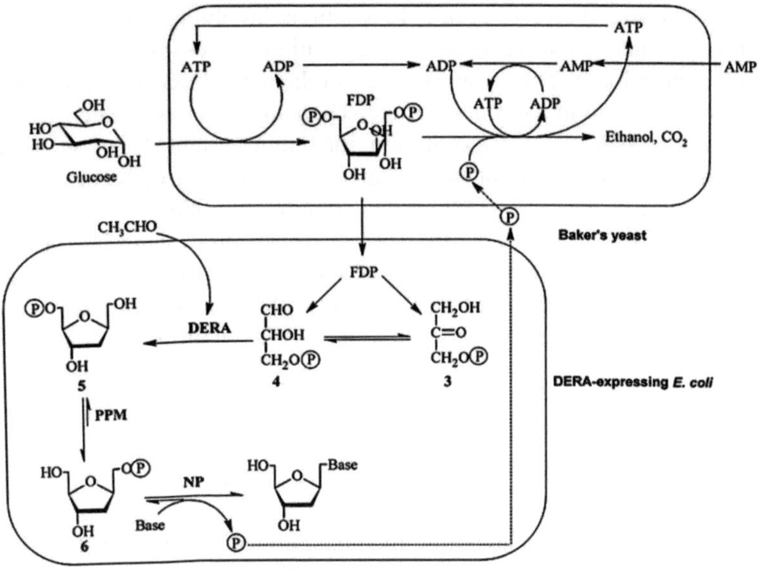

Fig. 1.8 2-Deoxyribose 5-phosphate (DR5P) synthesis from glucose and acetaldehyde through the fermentation energy of baker's yeast and deoxyriboaldolase (DERA)-expressing *E. coli*

1.4.3.3 Target-Oriented Product

Microbial product development is a creative and interdisciplinary activity that transforms market opportunity and technological innovation into successful products. In the biotechnology industry, to produce quality products of international standard it is essential to achieve economic success. For this a biotechnology industry must do two things: (1) creates an entirely new product that either adds to an existing product line or occupies an entirely new niche; and (2) modifies or updates an existing product. Successful product development in a sustainable manner is essential for a microbial processing industry if it hopes to exist for any length of time. New biological products, whether they take the form of new applications, new innovations, or entirely new goods, are an essential component of business success.

The value of any biopharmaceutical is evaluated in in terms of its desired function. Better quality of product not only upgrades its value but also determines the investment and operating costs by dictating the downstream separation step to attain purity of an international standard.

1.4.3.4 Microbiological Waste

In the past, fermentation industries were among the worst offenders in adding heavy pollution loads to sewage treatment plants or streams. Waste products are produced in different forms as the product of side reactions involving feedstock impurities, combustion products, or solvents used in separation processes (e.g., water, purges, and fugitive emissions). The waste streams are hence defined to a large degree by the choice of treatment in the process route. There are costs associated with making waste streams suitable for disposal into the environment (e.g., water effluent treatment and disposal tariffs or taxes imposed by national or local government). Today, the concept of waste handling has completely changed. With the help of genetically modified microbes, waste from microbial processing industries can be detoxified and used for generation of biogas (methane). In this process, the left-over residual mass can be used as biofertilizer. Thus, the cost of manufacturing can be regulated by integrating waste recycling processes with microbial processing activity.

1.4.3.5 Interactions

To develop any microbial metabolite, the standardization of overall heat transfer and material balance is essential. It is clear from the discussion above that process optimization from an economic prospective involves highly interlinked factors and needs trade-offs to be made between competing factors. To manufacture any metabolite at a competitive rate it is necessary to monitor feed stock value, monitor various physical factors like temperature and pressure, and ensure a timely supply of gases such as oxygen, nitrogen, and carbon dioxide.

It has been noticed that a large number of technically feasible options require materializing a specific process objective for a new project. Generally, a standard format is absolutely required to assess the economics of a project starting from its lay-out and ending with products manufactured at a competition market price.

1.5 Economic Evaluation of Microbial Processes

The term “techno-economic” is a combination of two different words, technology and economic. The term refers to the impact of technology on economics or, in another words, how the economy is being affected by technology. Techno-economic analysis reveals both the positive and negative effects of technology on economics. MPT comprises the microbial products and processes created by biotechnologists.

The economic aspects of any microbial process industry is concerned with the production of quality products in sufficient quantity and their passing to end users through an efficient and effective supply chain. The net profit from the operation of a microbial processing unit equals the total income minus all the costs associated with its operation, including the administration function. For a new microbial processing plant or modification to an existing process plant, the cash flow across the entire life cycle of the project must be considered in order to determine the economic performance.

Private sector looks primarily at the cash flow profile of the project in assessing the revenue earning potential. The cash flow profile of infrastructure projects exhibits the timing of the cash flows over the life cycle of the project. The life cycle of an infrastructure project can be divided into the three major phases of appraisal, construction, and operation and maintenance. The typical cash flow profile of an infrastructure project has cash outflows during the development and construction phases of the project, followed by cash inflows during the operation phase (Fig. 1.9). During the appraisal and construction phases of the project life cycle, cash flow is negative in nature because of expenditures incurred toward the execution of appraisal-related studies such as feasibility studies, engineering design, and construction of the infrastructure facilities. Commissioning of the completed infrastructure facilities marks the commencement of project operation phase. The provision of infrastructure services generates revenues for the project, along with the expenditures for operation and maintenance of the facilities. The cash flows in this phase are the net project revenues after deducting the operation and maintenance expenditures.

Before starting any project, there should be sufficient confidence in the project outcome prior to investment of large sums of capital. Actual return from a project could take time to materialize; therefore, cash flow must be estimated ahead of time prior to any major expenditure. As the project progresses, the cash flow system must be refined and re-evaluated to ensure that the capital investment remains worthwhile. It is necessary to have updated information on the net present value (NPV) of a running project. This is the difference between the present value of cash inflows and the present value of cash outflows. NPV is used in capital budgeting to analyze the profitability of a projected investment or project.

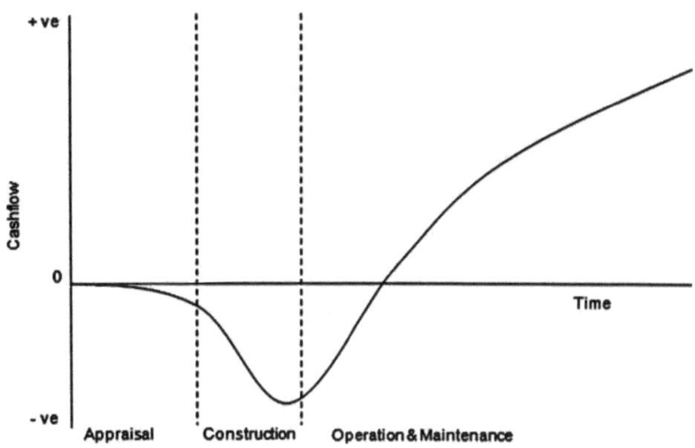

Fig. 1.9 Typical cash flow profile of infrastructure projects, showing cash outflows during the development and construction phases of the project followed by cash inflows during the operation phase

The following is the formula for calculating NPV:

$$\mathrm{NPV} = \sum_{t-1}^{T} \frac{C_t}{(1+r)^t} - C_o$$

where C_t is net cash inflow during the number of time periods t, C_o is total initial investment cost, r is discount rate, and it is number of time periods.

A positive NPV indicates that the projected earnings generated by a project or investment (in present dollars) exceeds the anticipated costs (also in present dollars). Generally, an investment with a positive NPV will be profitable and one with a negative NPV will result in a net loss. This concept is the basis for the NPV rule, which dictates that the only investments that should be made are those with positive NPV values.

1.5.1 Capital Expenditure and Operating Expenses

Capital expenditures (CAPEX) are the funds that a business uses to purchase major physical goods or services to expand the company's ability to generate profits. These purchases can include hardware (such as printers or computers), vehicles to transport goods, or the purchase or construction of a new building. The type of industry a company is involved in largely determines the nature of its capital expenditures. The asset purchased could be a new asset or something that improves the productive life of a previously purchased asset. If the asset's useful life extends more than a year, then the company must capitalize the expense, using depreciation to spread the cost of the asset over its designated useful life, as determined by tax regulations. Capital expenses are most often depreciated over a 5- to 10-year period, but may be depreciated over more than two decades in the case of real estate.

The amount of fixed capital needed to set up a microbial process-based industry is quite variable. Generally, biotechnology industries, biopharmaceutical units, or any microbial metabolite manufacturing units require high fixed-capital investment. Common examples include units producing antibiotics, microbe-derived antioxidants, vaccines, and microbe-based drugs. Fixed-capital investments do not typically depreciate in the even way that is shown on income statements. Some devalue quite quickly, whereas others have nearly infinite "usable" lives. However, the depreciation method allows investors to see a rough estimate of how much value fixed-capital investments are contributing to the current performance of the company.

An operating expense results from the ongoing costs a company pays to run its basic business. In contrast to capital expenditure, operating expenses are fully tax-deductible in the year they are made. As operational expenses make up the bulk of a company's regular costs, management examines ways to lower operating expenses without causing a critical drop in quality or production output. Sometimes an item that would ordinarily be obtained through capital expenditure can have its cost assigned to operating expenses if a company chooses to lease the item rather than purchase it. This can be a financially attractive option if the company has limited cash flow and wants to be able to deduct the total item cost for the year.

1.5.2 *Revenue and Profit*

Revenue can most easily be thought of as the top line of an income statement or profit and loss statement. Revenue is the total amount of income generated by the sale of microbe-derived products or services related to the company's primary operations. The money a company makes from selling a microbe-based product before accounting for any expenses is its revenue. If the company also has income from investments or from a subsidiary company, that income is not considered revenue because it does not come from the sale of microbial product.

Profit, conversely, is the infamous bottom line. This is called net profit, because it is the amount of income that remains after accounting for all expenses, debts, additional income streams, and operating costs. Between the top and bottom line, the term "profit" may emerge in the context of gross profit and operating profit. These are steps on the way to calculating net profit. Gross profit is revenue minus the cost of goods sold. Operating profit is gross profit minus all other fixed and variable expenses associated with operating the business, such as rent, utilities, and payroll. When most people refer to a company's profit, they are referring to the net income remaining after expenses, or the net profit. Although revenue and profit both refer to the money a company makes, it is possible for a company to generate revenue but have a net profit loss. This occurs when debts or expenses outstrip earnings.

1.6 Microbial Product Development

Microbial product development is the process by which a manufacturer does one of two things: (1) creates an entirely new product that either adds to an existing product line or occupies an entirely new niche; or (2) modifies or updates an existing product. Successful product development is essential for any business if it hopes to exist for any length of time.

1.6.1 *Why Develop New Products*

Our lifestyles change with time and nature of habitat. Accordingly, we change our habits regarding food, food supplements, nutraceuticals, and various other products responsible for supporting our health. So, innovation is crucial for the long-term health of a microbial processing business. No longer can a company rest on its laurels and keep doing "business as usual"; one must always be looking at new ways to serve customers. New products are an organization's most important source of revenue. New products bring in higher sales, increased customer loyalty, and, ultimately, higher profits. One only needs to look at Apple back in 2001 when it released the iPod. It was a great product for its time that was extremely successful. Apple was not satisfied, however, and continued to focus on new innovative products. They released new products such as the iPhone and iPad, which directly contributed to Apple's market share increasing at the rate that it has.

It is true that the launch of a new product offers an organization its greatest opportunity to increase revenue and profitability. However, new product development is not a simple, uniform, or sequential process. Its complexity and risk arises from its many phases, many stakeholders, and the intricacy of the product being developed. Increased product complexity, changing consumer demands, market globalization, extended supply chains and design networks, and regulatory compliance are all concerns that affect new product development. Executives that are considering how to improve this complex process must find the right balance between improving execution, efficiency, and timeliness with the genius of innovation and problem solving. Executives that tip the balance too far in either direction end up with a great product that is late or a poor product that is on time; both result in commercial failure.

The global microbial identification market is estimated at $896.5 million at the end of 2014 and is expected to grow at a compound annual growth rate (CAGR) of 5.9% from 2014 to 2019, reaching $1194.1 million by 2019. This growth is mainly attributed to the technological advances in products, high prevalence of infectious disease, growing food safety concerns, government initiatives, and higher healthcare expenditure in the emerging geographies. Furthermore, the efforts of various players to provide innovative products by collaboration and high focus on research and development is likely to drive this market. However, a complex regulatory framework and the high cost of identification instruments are likely to hinder the growth of this market.

North America dominates the microbial identification market. However, the Asia–Pacific region is poised to grow at the highest CAGR during the forecast period as a result of government support and economic development. Apart from this, factors such as governmental focus on antimicrobial resistance, bioterrorism surveillance, and focus on the emerging geographies are expected to provide the required impetus for the growth of this market.

The prominent players in the microbial identification market include Becton, Dickinson and Company, (USA), Biolog Inc. (USA), bioMérieux SA (France), Bruker Corporation (USA), Charles River Laboratories Inc. (USA), MIDI Inc. (USA), QIAGEN N.V. (the Netherlands), Shimadzu Corporation (Japan), Siemens Healthcare (Germany), Sigma-Aldrich Corporation (USA), and Thermo Fisher Scientific Inc. (USA).

Biotech companies have raised more than $120 billion in financing, most of which is used for R&D purposes [155]. This financing and investor climate provides the "fuel in the tank" to drive scientific ideas from discovery research to clinical stage testing and market approval.

Funding is available, but as early-stage biotech companies advance product candidates to clinical stage trials and beyond, the tactics and strategies executed by these companies may well depend on some combination of corporate alliances, financing strategy, and strategic planning. Success at each stage of the life cycle of an early-stage biotech company nonetheless necessitates a thorough and methodical consideration of the checklist of key strategic elements outlined herein.

1.6.2 New Microbial Product Development

The promise of start-up biotechnology companies is enormous, but so are the risks and the uncertainty of product development. A few key strategic elements are crucial to the growth and success of an early-stage biotechnology company. These elements serve as a checklist, and a necessary one because the early-stage company can often be a business case study for high attrition rates among product candidates, huge technical risk, and very long product-development time lines of 8–12 years.

The promise of positive NPV market returns and commercial success are fleeting for most start-up biotechnology industries. The biopharmaceutical development process is complex, calls for significant investment in financial and human capital, and involves risks in project execution, the regulatory process, and scientific technical attrition. Furthermore, the cost of gaining market approval for a single product is estimated at $800 million. Of the companies that set out to develop a drug beyond proof of principle, most fail as a result of lack of product efficacy or safety, or insufficient cash for clinical trials or business operations [156].

Industry data indicate 20% likelihood for a compound to advance from initiation of phase 1 trials to market approval. With these odds, there is a need to mitigate product development risks, from start-up capital through to early- and late-stage clinical trials. Furthermore, leadership and commitment from the company's executives are essential in determining tactics and strategies for developing product candidates, to set the right course for clinical development, and to manage capital. Ask any CEO who has founded a company and then kept it in business: it is the toughest job she or he has ever done. It calls for flawless operational execution, talented management, perseverance, and scientific expertise and vision.

As business experts, analysts, executives, and entrepreneurs all know, there is no one way to organize a company for effective new product development. As Gruenwald [157] noted, the ultimate methodology that is chosen "depends on the nature of the corporation and its goals. It depends on the existing structural order of things. It depends on the corporation's management style. It depends on the caliber, motivations, and growth potential of the staff in place at the time of installing the new products organization. It depends on past performance by organizations charged with the responsibility. It depends on the orientation of the corporation, if this is not to change. (Are the present strengths or weaknesses centered in certain areas?)."

Nonetheless, analysts point to several factors that are fairly universal in determining whether a business will enjoy measurable success in new product development efforts. These include fundamentals such as comprehensive market and cost analysis, support from top management, enthusiasm among workers, clear lines of authority, and past experience. Three essential issues in new product development (focus, resources, and leadership) are discussed in detail below:

1. Focus: First, a small business needs to focus on its goals. Limited time and resources mean that hard decisions must be made and a strategic plan needs to be developed. Companies should "do the right things right" by using the best information available to choose the right technologies and decide on what new products to invest in. Small companies often grow quickly and can pick and choose among many seemingly strong new product avenues, but the key is to decide what the company does very well and then concentrate on that area or areas.
2. Resources: Another key to new product development for small businesses is to secure the resources and skills needed to create and market the new product. Small companies may lack the in-house resources needed to create a new product, making it seem out of reach, but analysts note that small business owners have other avenues that they can often pursue. If the product idea is good enough, the company may decide to look outside its own walls for partnership and outsourcing opportunities. "When the need is not within the capability of your company," states Gruenwald [157], "but beneficial arrangements can be made with other companies to joint-venture, contract-supply, license/acquire, or, in rare instances, to merge. Pools of expertise can also be acquired by recruiting within the subject industry and by the use of technical and marketing consultants."

   One key to resource management is not to undertake too many projects at one time. Every company has a finite amount of resources to allocate to new product development, but small businesses often face especially tight budgets in this regard. Budget in this instance does not just mean money, it also means time. Too many projects means that otherwise talented workers cannot spend enough time on any one project and, as a result, all projects suffer and fall off schedule, leaving gaping openings for competitors or causing market windows to close.
3. Leadership: The third and final step a small company needs to follow is to find the leadership needed to bring a new product from the idea stage to completed product. This leader often takes the form of a "product champion" who can bring both expertise and enthusiasm to the project. (In small business environments, this product champion is often the entrepreneur/owner himself.) A strong product champion is able to balance all the issues associated with a product—economic factors, performance requirements, regulatory issues, management issues, and more—and create a winning new product.

1.6.3 How to Launch a New Microbial Product

Once the product-line architecture has been established and a new product is being developed, it is time for a company to think about how to successfully launch the product in its target market. This is the stage where an advertising or public relations agency can come into play, especially for small businesses without the internal resources to handle such a job themselves. When using an outside agency to launch a product, a company should:

- Have a well-defined concept on the microbial metabolite (which is where product-line architecture comes into play).
- Provide the agency with background information on its products and goals
- Obtain or apply new patents as needed, immediately
- Microbial processes should be strictly under current good manufacturing practice norms of international standard
- Have a formal business plan in place that defines funding of the project
- Determine who should approve the marketing or advertising plan that the agency creates (the fewer people communicating with the agency, the better)
- Determine the proper timing for the launch

1.6.4 Speed-to-Market and Product Development

Time-based strategies, such as first-mover or fast-follower strategies have become the latest key to competitive advantage in the current environment of fast-changing technologies and customer demands. In today's technology-fueled business environment, the always-important speed-to-market factor has become perhaps the most crucial factor in new product development. Improved communication (especially the Internet), increased globalization, and rapid changes in technology have put tremendous pressure on companies to get their product to market first. To improve speed-to-market, a company should first make sure that it is making the best possible use of available technology. If it is, then there are other steps that can be taken to speed product development through efficient, market-oriented product planning that takes the customer into account.

1.6.5 Service Companies and New Products

It is well known that services differ from manufactured goods in a number of significant ways. Services are largely intangible, produced and consumed simultaneously, heterogeneous, and perishable (Fig. 1.10). These differences can lead to changes in service management [156].

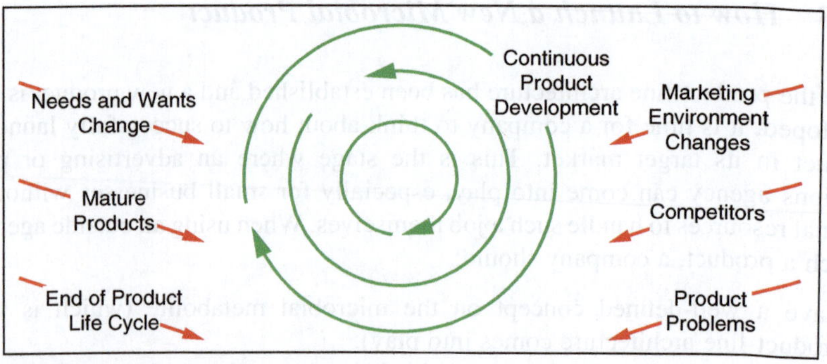

Fig. 1.10 Reasons for new product development

Every business needs to innovate to stay ahead of the competition. No business can continue to offer the same unchanged product, otherwise sales would decrease and profits reduce. Consumer "needs and wants" continuously change. Firms should respond to these changes through their products and services; otherwise, consumers will switch to competitor products that satisfy their needs and wants. For example, consumers are becoming more health conscious, which is forcing companies to introduce products with low sugar, salt, and fat. Coca-Cola Zero, which contains no sugar, is a classic example of new product development even though Coca-Cola's existing product range already contains diet coke. Both diet coke and Coca-Cola Zero contain no sugar, but they taste different.

If a product is at the end of its product life cycle, the company may want to introduce new and improved updated versions. Microsoft has done this by moving from the Xbox to Xbox 360, and now Xbox 360 limited editions allow Microsoft to refresh the product through small changes. The product might be at the maturity stage of its product life cycle, and need modifications to stimulate an increase in sales. Nintendo have replaced its DSi console with the 3DS console, which contains additional features such as an extra camera so that you can film in 3D, a 3D screen that does not require glasses, a joystick, and motion sensors.

There may be environmental changes that the company wants to capitalize on. Music companies are now selling more music via internet downloads than through traditional retail shops. Record companies were pushed into selling music through the internet following the success of the internet site Napster, which offered illegal music downloads. In April 2006, the song "Crazy" by Gnarls Barkley made history by becoming the first song to achieve the number one spot in the UK charts through music download sales only. In March 2011, Mercury Records stopped releasing singles on CDs as by then 99% of single sales were through downloads.

Competitors may force change. This is very apparent in the technology market, where new products are constantly being introduced to a target market that welcomes change and innovation. Technology consumers are not afraid to try new products, in fact they often want the latest gadget to show to friends and colleagues.

If a product is successful, then competitors will attempt to develop similar products. In fact, Google say that they developed the Android operating system to prevent the technology market for products such as mobile phones and tablets being dominated by one supplier. If all of your products are experiencing poor sales or suffering from a negative reputation, then it is time to change your product offering. In 2001, the introduction of the iPod MP3 player reversed the fortunes of Apple Computers. Since then, Apple has introduced the successful iPhone and iPad and increased its share price from $9.07 per share (Oct 2001) to over $400 per share.

References

1. Bhagavathy S et al (2011) Green algae *Chlorococcum humicola*-a new source of bioactive compounds with antimicrobial activity. Asian Pac J Trop Biomed 1:S1–S7
2. Volk RB (2008) A newly developed assay for the quantitative determination of antimicrobial (anticyanobacterial) activity of both hydrophilic and lipophilic test compounds without any restriction. Microbiol Res 163:161–167
3. Priyadarshani I, Rath B (2012) Commercial and industrial applications of micro algae—a review. J Algal Biomass Util 3:89–100
4. Schopf J (2006) Fossil evidence of Archaean life. Philos Trans R Soc Lond B Biol Sci 361:869–885
5. Altermann W, Kazmierczak J (2003) Archean microfossils: a reappraisal of early life on Earth. Res Microbiol 154:611–617
6. Cavalier-Smith T (2006) Cell evolution and Earth history: stasis and revolution. Philos Trans R Soc Lond B Biol Sci 361:969–1006
7. McGovern PE et al (2004) Fermented beverages of pre- and proto-historic *China*. Proc Natl Acad Sci 101:17593–17598
8. Robinson J (ed) (2006) "Teinturier". Oxford companion to wine, 3rd edn. Oxford University Press, Oxford, pp 688–689
9. Cavalieri D et al (2003) Evidence for *S. cerevisiae* fermentation in ancient wine. J Mol Evol 57(Suppl 1):S226–S232
10. Prajapati JB, Nair BM (2003) The history of fermented foods. In: Farnworth ER (ed) Handbook of fermented functional foods. CRC Press, Boca Raton, pp 1–25
11. Jespersen L, Josephsen J (2004) Starter cultures and fermented products. In: Hui YH, Meunier-Goddik L, Josephsen J, Nip W-K, Stanfield PS, Toldra F (eds). Handbook of food and beverage fermentation technology. CRC Press, Boca Raton. isbn:978-0-8247-4780-0
12. Holzapfel WH et al (1995) Biological preservation of foods with reference to protective cultures, bacteriocins and food-grade enzymes. Int J Food Microbiol 2:343–362
13. Burdichon F et al (2012) Food fermentations: microorganisms with technological beneficial use. Int J Food Microbiol 154:87–97
14. Green Prophet (2011) Why Muslims don't drink alcohol. Health. www.greenprophet.com
15. Moon0312 - issuu (2014) The world of wine by. https://issuu.com/moon0312/docs/singe_page_the_world_of_wine1
16. Ges HH (2004) The discovery of microorganisms by Robert Hooke and Antoni van Leeuwenhoek, fellows of the Royal Society. Notes Rec R Soc Lond 58:187–201
17. Farley J et al (1974) Science, politics and spontaneous generation in nineteenth-century France: the Pasteur-Pouchet debate. Bull Hist Med 48:161–198
18. Ribéreau-Gayon P et al (2006) Handbook of enology. Vol 1: The microbiology of wine and vinifications. Wiley, Chichester

19. Web site (2012) La biodiversité des microorganismes des produits laitiers. Les microorganismes bénéfiques de l'industrie de transformation du lait. INRA. https://en.wikipedia.org/wiki/Microbial_food_cultures
20. Hui YH et al (2004) Handbook of food and beverage fermentation technology; technology & engineering. https://books.google.co.in/books?isbn=0203913558
21. Remminghorst U, Rehm BHA (2006) Bacterial alginates: from biosynthesis to applications. Biotechnol Lett 28:1701–1712
22. Sabra W et al (2001) Bacterial alginate: product quality and process aspects. Appl Microbiol Biotechnol 56:315–325
23. Bajaj IB et al (2007) Gellan gum: fermentative production, downstream processing and applications. Food Technol Biotechnol 45:341–354
24. Arockiasamy S, Banik M (2008) Optimization of gellan gum production by *Sphingomonas paucimobilis* ATCC 31461 with non ionic surfactants using central decomposite design. J Biosci Bioeng 105:204–210
25. Leathers TD (2003) Biotechnological production and application of pullulan. Appl Microbiol Biotechnol 62:468–473
26. Cheng KC et al (2011) Pullulan: biosynthesis, production, and applications. Appl Microbiol Biotechnol 92:29–44
27. Cheirsilp B et al (2003) Enhanced kefiran production by mixed culture of *Lactobacillus kefiranofaciens* and *Saccharomyces cerevisiae*. J Biotechnol 100:43–53
28. Zajšek K et al (2011) Characterization of the exopolysaccharide kefiran produced by lactic acid bacteria entrapped within natural kefir grains. Int J Dairy Technol 64:544–548
29. Nikitina VE et al (2007) *Lentinula edodes* biotechnology- from lentinan to lectins. Food Technol Biotechnol 45:230–237
30. Wasser SP (2002) Medicinal mushrooms as a source of antitumor and immunomodulating polysaccharides. Appl Microbiol Biotechnol 60:258–274
31. El Enshasy H et al (2012) Pleuran: immunomodulator polysaccharide from *Pleurotus ostreatus*: structure, production and application. In: Andres S, Baumann N (eds) Mushrooms: type, properties and nutrition. Nova, New York, pp 153–172
32. Gregori A et al (2007) Cultivation techniques and medicinal properties of *Pleurotus* sp. Food Technol Biotechnol 45:238–249
33. Garcia-Ochoa F et al (2000) Production, recovery and properties. Biotechnol Adv 18:549–579
34. El Enshasy H et al (2011) Enhanced xanthan production process in shake flask and pilot scale bioreactors using industrial semi-defined medium. Afr J Biotechnol 10:1029–1038
35. Bhosale SH et al (1996) Molecular and industrial aspects of glucose isomerase. Microbiol Rev 60:280–300
36. Kovalenko GA et al (2011) Immobilization of a recombinant strain producing glucose isomerase inside SiO2-Xerogel and properties of prepared biocatalysts. Appl Biochem Microbiol 47:151–157
37. El-Enshasy H et al (2001) *GpdA*-promoter controlled production of glucose oxidase by recombinant *Aspergillus niger* using non-glucose carbon sources. Appl Biochem Biotechnol 90:1–10
38. El Enshasy H (2006) Agitation effects on morphology and protein productive fractions of filamentous and pelleted growth forms of recombinant *Aspergillus niger*. Process Biochem 41:2103–2112
39. Sojo M (1997) Cell linked and extracellular cholesterol oxidase activities from *Rhodococcus erythropolis* isolation and physiological characterization. J Appl Microbiol Biotechnol 47:583–589
40. Lashkarian H et al (2010) Extracellular cholesterol oxidase from *Rhodococcus sp.*: isolation and molecular characterization. Iran Biomed J 14:49–57
41. El Enshasy H (2007) Bioprocess development for the production of α-amylase by *Bacillus myloliquefaciens* in batch and fedbatch cultures. J Microbiol Res 7:560–568

42. El Enshasy H et al (2012) Bioprocessing technologes in integrated biorefinery from production of biofuels, biochemicals, and biopolymers from biomass. Wiley, New York
43. Beg QK et al (2001) Microbial xylanases and their industrial applications: a review. Appl Microbiol Biotechnol 56:326–338
44. Biely P (1985) Microbial xylanolytic systems. Trends Biotehnol 3:286–290
45. Greiner R, Konielzny U (2006) Phytase for food application. Food Technol Biotechnol 44:125–140
46. Bajaj BK, Wani MA (2011) Enhanced phytase production from *Nocardia* sp. MB 36 using agro-residues as substrates: potential application for animal feed production. Eng Life Sci 11:620–628
47. Maki ML et al (2011) Characterization of some efficient cellulose producing bacteria from paper mill sludges and organic fertilizers. Int J Biochem Mol Biol 2:146–154
48. Sukumaran RK et al (2005) Microbial cellulases-production, application and challenges. J Sci Ind Res 64:832–844
49. Kalingan AE, Liao CM (2002) Influence of type and concentration of flavinogenic factors on production riboflavin by *Eremothecium ashbyii* NRRL 1363. Bioresour Technol 82:219–224
50. Jiménez A et al (2005) Metabolic engineering of purine pathway for riboflavin production by *Ashbya gossypii*. Appl Environ Microbiol 71:5743–5751
51. El-Refai HA et al (2010) Application of multi-factorial experimental designs for optimization of biotin production by a *Rhizopus nigricans* strain. J Am Sci 6:179–187
52. Brown SW et al (1991) The production of biotin by recombinant strains of *Escherichia coli*. J Chem Tech Biotechnol 50:115–121
53. D'Aimmo MR et al (2012) The potential of bifidobacteria as a source of natural folate. J Appl Microbiol 112:975–984
54. Pompei A et al (2007) Folate production by *bifidobacteria* as potential probiotic property. Appl Environ Microbiol 73:179–185
55. Martens JH et al (2002) Microbial production of vitamin B12. Appl Microbiol Biotechnol 58:275–285
56. El Enshasy H et al (2008) Production of vitamin B12 by methanol utilizing bacteria *Rubrobacter motiliticus* nov. var. *gelatinoliquificicans* in shake flask and bioreactor cultures under different cultivation conditions. Deutsche Lebensmittel Rundschau 104:384–388
57. Vandamme EJ (1992) Production of vitamins, Coenzymes and related biochemicals by biotechnological process. J Chem Tech Biotechnol 53:313–327
58. Pan DD et al (2010) Characterisation of *Lactobacillus fermentum* SM-7 isolated from koumiss, a potential probiotic bacterium with cholesterol-lowering effects. J Sci Food Agric 91:512–518
59. Ding WK, Shah NP (2007) Acid, bile and heat tolerance of free and microencapsulated probiotic bacteria. J Food Sci 72:446–450
60. Czerucka D, Rampal P (2002) Experimental effects of *Saccharomyces boulardii* on diarrheal pathogens. Microbes Infect 4:733–739
61. El Enshasy H, Shereef A (2008) Probiotic/biotherapeutic yeast *Saccharomyces boulardii* adapted to dryness stress: optimization of high cell density cultivation of yeast. Deut Lebensmittel Rundschau 104:389–394
62. Tiago FCP et al (2009) Physiological characterization of non-*Saccharomyces* yeasts from agro-industrial and environmental origins with possible probiotic function. World J Microbiol Biotechnol 25:657–666
63. Sanders KME (2006) Summary of probiotic activities of *Bifidobacterium lactis* HN019. J Clin Gastroenterol 40:776–783
64. Yoshida Y et al (2010) Clinical effects of probiotic *Bifidobacterium breve* supplementation in adult patients with atopic dermatitis. Yonago Acta Med 53:37–45
65. Ljungh Å, Wadström T (2006) Lactic acid bacteria as probiotics. Curr Issues Intest Microbiol 7:73–90

66. Moslehi-Jenabian S et al (2010) Beneficial effects of probiotic and food borne yeasts on human health. Nutrients 2:449–473
67. Fairclough A et al (2011) Factors affecting *Penicillium roquefortii* (*Penicillium glaucum*) in internally mould ripened cheeses: implications for pre-packed blue cheeses. Int J Food Sci Technol 46:1486–1490
68. Liu DM et al (2006) Survivability of *Lactobacillus rhamnosus* during the preparation of soy cheese. Food Technol Biotechnol 44:417–422
69. Rivera-Espinza Y, Gallardo-Navaro Y (2010) Non-dairy probiotic products. Food Microbiol 27:1–11
70. Poirier Y et al (1995) Production of polyhydroxyalkanoates, a family of biodegradable plastics and elastomers in bacteria and plants. Nat Biotechnol 13:142–150
71. Steinbüchel A (2001) Perspectives for biotechnological production and utilization of biopolymers: metabolic engineering of polyhydroxyalkanoate biosynthesis pathways as a successful example. Mol Biosci 1:1–24
72. Weber CJ et al (2002) Production and application of biobased packaging materials for the food industry. Food Addit Contam 19:172–177
73. Yang J et al (2004) Improvement of recombinant hirudin production by controlling NH_{4+} concentration in *Pichia pastoris* fermentation. Biotechnol Lett 26:1013–1017
74. Papagianni M (2007) Advances in citric acid fermentation by *Aspergillus niger*: biochemical aspects, membrane transport and modeling. Biotechnol Adv 25:244–263
75. Jernejc K, Cimerman A (1992) Composition of *Aspergillus niger* mycelium during growth on productive and unproductive substrates. J Biotechnol 25:341–348
76. El Enshasy H (2006) Optimization of gluconic acid production by glucose oxidase gene in batch and fed-batch cultures. Deutsche Lebensmittel Rundschau 102:1–6
77. Dronawat SN et al (1995) The effects of agitation and aeration on the production of gluconic acid by *Aspergillus niger*. Appl Biochem Biotechnol 51/152:347–354
78. Bohmann JT et al (1998) Oxalic acid production by *Aspergillus niger*. Part II: Optimization of fermentation with milk whey as carbon source. Bioproc Biosys Eng 19:337–342
79. Strasser H, Burgstaller W, Schinner F (1994) High-yield production of oxalic acid from metal leaching process by *Aspergillus niger*. FEMS Microbiol Lett 119:365–370
80. Utagawa T (2004) Production of arginine by fermentation. J Nutr 134(10 Suppl):2854S–2857S
81. Momose H, Ishida M, Terabe M et al (1982) Production of L-arginine by fermentation. Japanese Patent 57-5693
82. Sato T et al (1975) Engineering analysis of continuous production of L-aspartic acid by immobilized *Escherichia coli* cells in fixed beds. Biotechnol Bioeng 17:1797–1804
83. Plachy J, Sikyta B (1977) Production of L-aspartic acid from fumaric acid by *Alcaligenes metalcaligenes* CCEB 312. Folia Microbiol 22:410–414
84. Nakamura J et al (2007) Mutations of *Corynebacterium glutamicum* NCg11221 gene, encoding a mechanosensitive channel homolg, induce L-glutamic acid production. Appl Environ Microbiol 73:4491–4498
85. Kurihara K (2009) Glutamate: from discovery as a food flavor to role as basic taste (umami). Am J Clin Nutr 90:719S–722S
86. Sano C (2009) History of glutamate production. Am J Clin Nutr 90:728–732
87. Leuchtenberger W, Huthmacher K (2005) Biotechnological production of amino acids and derivatives: current status and prospects. Appl Microbiol Biotechnol 69:1–8
88. Anastassiadis S (2007) L-Lysine fermentation. Rec Pat Biotechnol 1:11–24
89. Kumar D, Gomes J (2005) Methionine production by fermentation. Biotechnol Adv 23:41–61
90. Shu CH, Lia CC (2002) Optimization of L-phenylalanine production of *Corynebacterium glutamicum* under feedback inhibition by elevated oxygen transfer rate. Biotechnol Bioeng 77:131–141
91. Förberg C, Häggström L (1988) Phenylalanine production from a rec *Escherichia coli*-strain in fed-batch culture. J Biotechnol 8:291–300

92. Chen N et al (2009) Optimization of fermentation conditions for the biosynthesis of L-threonine by *Escherichia coli*. Appl Biochem Biotechnol 158:595–604
93. Lee MH et al (2006) Improved L-threonine production of *Escherichia coli* mutant by optimization of culture conditions. J Biosci Bioeng 101:127–130
94. Ikeda M, Katsumata R (1999) Hyperproduction of tryptophan by *Corynebacterium glutamicum* with the modified-pentose phosphate pathway. Appl Environ Microbiol 65:2497–2502
95. http://www.prnewswire.com/news-releases/8 Oct (2015) New York, Oct 8, 2015/ PRNewswire/Microbial products: technologies, applications and global markets—Reportlinker Review. *Microbial-products-technologies*
96. Marquina D et al (2002) Biology of killer yeasts. Int Microbiol 5:65–71
97. Kjeldsen T (2000) Yeast secretory expression of insulin precursors. Appl Microbiol Biotechnol 54:277–287
98. Idiris A et al (2010) Engineering of protein secretion in yeast: strategies and impact on protein production. Appl Microbiol Biotechnol 86:403–417
99. Laughlin TJ, Ferrell TM (1987) Biotechnology in the cosmetics industry. Nat Biotechnol 5:1035–1037
100. Grabenhofer R (2007) Biotechnology in cosmetics: concepts, tools and techniques. Allured Publishing, Carol Stream, IL
101. Anzali S et al (2010) Elucidation of antiaging effects of ectoine using cDNA microarray analysis and signaling pathway evaluation. Int J Cosmet Sci 32:319
102. Buenger J, Ectoin DH (2004) An effective natural substance to prevent UVA-induced premature photoaging. Skin Pharmacol Physiol 17:232–237
103. Lange L (2010) The importance of fungi for a more sustainable future on our planet. Fungal Biol Rev 24:90–92
104. Kloepper JW, Lifshitz R, Zablotowicz RM (1989) Free-living bacterial inocula for enhancing crop productivity. Trends Biotechnol 7:39–44
105. Rodriguez H, Fraga R (1999) Phosphate solubilizing bacteria and their role in plant growth promotion. Biotechnol Adv 17:319–339
106. Saharan BS, Nehra V (2011) Plant growth promoting rhizobacteria: a critical review. Life Sci Med Res 21:1–30
107. Johansson JF, Paul LR, Finlay RD (2004) Microbial interactions in the mycorrhizosphere and their significance for sustainable agriculture. FEMS Microb Ecol 48:1–13
108. Nelson JH (1970) Production of Blue cheese flavor via submerged fermentation by *Penicillium roqueforti*. J Agric Food Chem 18:567–569
109. Leclercq-Perlat MN et al (2004) Controlled production of Camembert-type cheeses. J Dairy Res 71:346–354
110. Fergal PR, Patrickf FF (1999) Aspects of enzymology and biochemical properties of Brevibacterium linens relevant to cheese ripening: a review. J Dairy Sci 82:891–909
111. Ludemann V (2009) Toxicological assessment of *Penicillium nalgiovense* strains for use as starter cultures in the manufacture of dry fermented sausages. J Food Prot 72:1666–1670
112. Marianski S, Mariański A (2009) The art of making fermented sausages. Bookmagic LLC
113. Lund B et al (2000) Microbiological safety and quality of food. Aspen Publishers, Gaithersburg
114. Machida M et al (2008) Genomics of *Aspergillus oryzae*: learning from the history of koji mold and exploration of its future. DNA Res 15:173–183
115. Wikipedia, the free encyclopedia (2015) Monosodium glutamate—https://en.wikipedia.org/wiki/Monosodium_glutamate
116. Lonvaud-Funel A (1999) Lactic acid bacteria in the quality improvement and depreciation of wine. Antonie von Leeuwenhoek 76:317–331
117. Raw-fermented sausages (2012) Food and Agriculture Organization of the United Nations (FAO)

118. Wu Z-Y et al (2009) Developing new sacchariferous starters for liquor production based on functional strains isolated from the pits of several famous Luzhou-flavor liquor brewers. J Inst Brew 115:111–115
119. Mogra R, Choudhry M (2008) Effect of starter culture on the development of curd. J Dairy Foods Home Sci 27:130–133
120. WHO. Health and Nutritional properties of probiotics in food including powder milk with live lactic acid bacteria (PDF). Report of a Joint FAO/WHO Expert Consultation on Evaluation of Health and Nutritional Properties of Probiotics in Food Including Powder Milk with Live Lactic Acid Bacteria. World Health Organisation of the United Nations (WHO). Retrieved 8 Aug 2012
121. Rijkers GT (2011) Health benefits and health claims of probiotics: bridging science and marketing. Br J Nutr 106:1291–1296
122. Heller KJ (2012) Probiotic bacteria in fermented foods: product characteristics and starter organisms. Am Soc Clin Nutr 73:374S–379S
123. Vinderola CG, Reinheimer JA (2003) Acetic acid starter and probiotic bacteria: a comparative "in vitro" study of probiotic characteristics and biological barrier resistance. Food Res Int 36:895–904
124. Battcock M, Azam S (2012) Fermented frutis and vegetables. A global perspective. Chap. 5—Bacterial fermentations. Food and Agriculture Organization of the United Nations (FAO)
125. Al-Qasmi M et al (2012) A review of effect of light on microalgae growth. In: Proceedings of the World Congress on Engineering, vol 1. WCE, London
126. Watry G (2015) Funds bolster algae biofuel development. Digital Reporter. www.rdmag.com/article
127. Schenk P (2008) Second generation biofuels: high-efficiency microalgae for biodiesel production. Springer, Berlin
128. Paul D (2016) Algae for biofuels. www.dorianpaul.com/sciencecornerzero/algaeforbiofuels.html
129. Jacquot J (2009) Companies making fuel from algae now—popular mechanics. www.popularmechanics.com/science/energy/a4677/4333722/
130. Lane J (2012) Solix BioSystems to expand commercial algae production with $31M Series C venture round; Solix: Biofuels Digest. www.biofuelsdigest.com/bdigest/tag/solix
131. Wikipedia, the free encyclopedia (2016) https://en.wikipedia.org/wiki/Algeno
132. Wikipedia (2016) Solazyme—the free encyclopedia. https://en.wikipedia.org/wiki/Solazyme
133. Kloosterman K (2011) Food and fuel from greenhouse gas? With the help of algae, Israel's Seambiotic is turning carbon dioxide emitted by power plants into fuel and nutraceuticals. www.israel21c.org/the-top-12-ways-israel-feeds-the-world
134. Darzins A et al (2010) Current status and potential for algal biofuels production. A report to IEA Bioenergy Task; Report T39
135. Tantirungkij M et al (1993) Construction of xylose-assimilating *Saccharomyces cerevisiae*. J Ferment Bioeng 75:83–88
136. Kotter P, Ciriacy M (1993) Xylose fermentation by *Saccharomyces cerevisiae*. Appl Microbiol Biotechnol 38:776–783
137. Toivari MH et al (2001) Conversion of xylose to ethanol by recombinant *Saccharomyces cerevisiae*: importance of xylulokinase (*XKS1*) and oxygen availability. Metab Eng 3:236–249
138. Ho NWY et al (1998) Genetically engineered *Saccharomyces* yeast capable of effective cofermentation of glucose and xylose. Appl Environ Microbiol 64:1852–1859
139. Jin YS et al (2003) Optimal growth and ethanol production from xylose by recombinant *Saccharomyces cerevisiae* require moderate D-xylulokinase activity. Appl Environ Microbiol 69:495–503

140. Nakamura N et al (2008) Effective xylose/cellobiose co-fermentation and ethanol production by xylose-assimilating *S. cerevisiae* via expression of β-glucosidase on its cell surface. Enzyme Microb Technol 43:233–236
141. Atul B (2012) Enzyme-catalyzed polymerization of end-functionalized polymers in a microreactor. Macromolecules 45:7000–7008
142. Kurade MB et al (2015) Bacterial–yeast consortium as an effective biocatalyst for biodegradation of sulphonated azo dye Reactive Red 198. RSC Adv 5:23046–23056
143. Schaetzle O et al (2008) Bacteria and yeasts as catalysts in microbial fuel cells: electron transfer from micro-organisms to electrodes for green electricity. Energy Environ Sci 1:607–620
144. Jang JK (2013) Electricity generation by microbial fuel cell using microorganisms as catalyst in cathode. J Microbiol Biotechnol 23:1765–1773
145. He Z (2009) Electricity production coupled to ammonium in a microbial fuel cell. Environ Sci Technol 43:3391–3397
146. Huang Y (2012) Electricity generation from a floating microbial fuel cell. Bioresour Technol 114:308–313
147. Weber N (2014) Exploiting cell metabolism for biocatalytic whole-cell transamination by recombinant *Saccharomyces cerevisiae*. Appl Microbiol Biotechnol 98:4615–4624
148. Mattanovich D et al (2014) Yeast biotechnology: teaching the old dog new tricks. Microb Cell Fact 13:34–38
149. Kataoka M (2003) Novel bioreduction system for the production of chiral alcohols. Appl Microbiol Biotechnol 62:437–445
150. Kataoka M (1998) *Escherichia coli* transformant expressing the glucose dehydrogenase gene from *Bacillus megaterium* as a cofactor regenerator in a chiral alcohol producing system. Biosci Biotechnol Biochem 62:167–169
151. Kataoka M (1999) Stereoselective reduction of ethyl 4-chloro-3-oxobutanoate by *Escherichia coli* transformant cells co-expressing the aldehyde reductase and glucose dehydrogenase genes. Appl Microbiol Biotechnol 51:486–490
152. Horinouchi N (2006) Efficient production of 2-deoxyribose 5-phosphate from glucose and acetaldehyde by coupling reaction of the alcoholic fermentation system of baker's yeast and deoxyriboaldolase-expressing *Escherichia coli*. Biosci Biotechnol Biochem 70:1371–1378
153. Tochikura T (1978) Production of nucleotide-related substance by yeast and utilization of fermentation energy to biosynthetic process. Hakko Kogaku Kaishi (in Japanese) 56:508–526
154. Tochikura T et al (1967) Fermentation and metabolism of nucleic acid-related compounds in yeast. J Ferment Technol 45:511–529
155. Coghill K. Biotechnology industry ends 2005 with breakthrough product approvals and steady financing. Bionews. 2006 Feb–Mar 8
156. DiMasi JA (2001) Risks in new drug development: approval success rates for investigational new drugs. Clin Pharmacol Ther 69:297–307
157. George G (1995) New product development: responding to market demand. NTC Publishing. www.sciencedirect.com/science/journal/07376782/10/3

Chapter 2
Upstream Processes

Microbial biomass processing is a discipline in industrial biotechnology concerned with establishing consistency and reproducibility in microbial culture techniques, starting from slant/plate culture to inoculum preparation for pilot or turnkey-level fermentation processes, to the manufacture of value-added products for the biopharmaceutical, food processing, biopolymer manufacturing and other industries [1, 2]. Industrial microbial processes result in the control of safety, mixing, and environmental conditions and in the use of selected microorganisms to obtain higher productivity with lower cost. Microbial process technology is design according to the norms of international and national standards to produce quality products, keeping control over the cost of goods.

Microbial biomass processing is directed toward the application of chemical engineering principles for the development and optimization of biological processes (Fig. 2.1). A wide range of biological processes are studied, including those using bacterial, fungal, and enzymatic biocatalysts. A major focus is the elucidation of process strategies for enhancement of the organism's performance, together with optimization of product accumulation and isolation to maximum purity. Any industrial fermentation operation can be divided into the following three steps (Fig. 2.2):

1. Upstream processes, which mainly consists of (a) medium preparation for growth and production, (b) medium sterilization and maintenance of an aseptic environment, and (c) inoculum preparation
2. Optimization of operating conditions
3. Downstream processes to (a) separate and purify the fermentation product and (b) treat and dispose of effluents produced by the process

Depending on the circumstances, the term "fermentation" can have two meanings. In a physiological sense, fermentation refers to a process that produces energy by breaking down energy-rich compounds under anaerobic conditions. This is the basis of food fermentation, which is commonly used in the food industry to produce

B. Kumara Behera, A. Varma, *Microbial Biomass Process Technologies and Management*, DOI 10.1007/978-3-319-53913-3_2

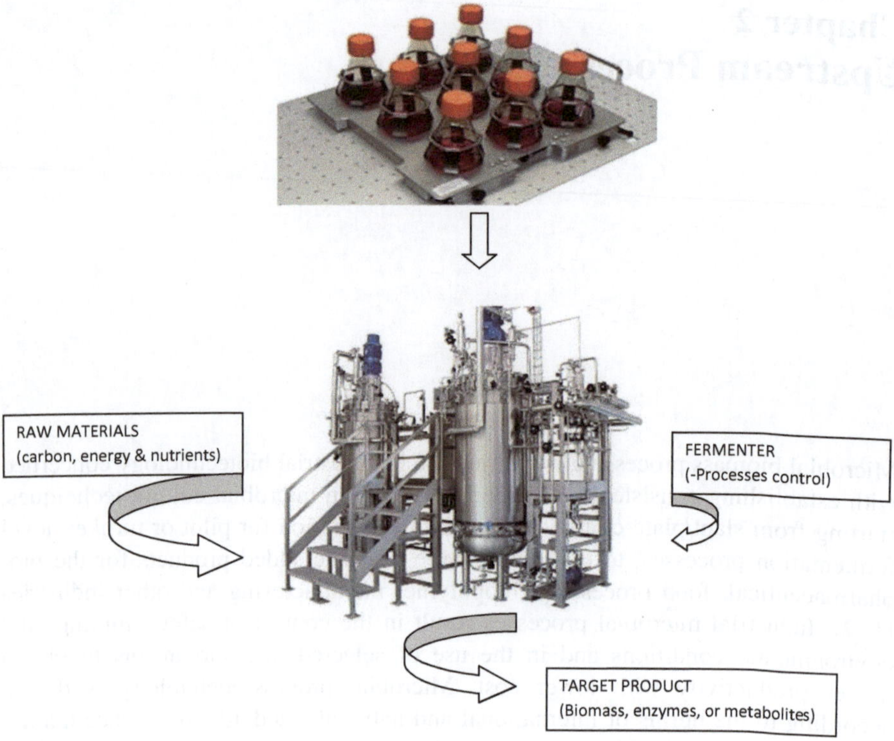

Fig. 2.1 Schematic presentation of a biotechnological processes

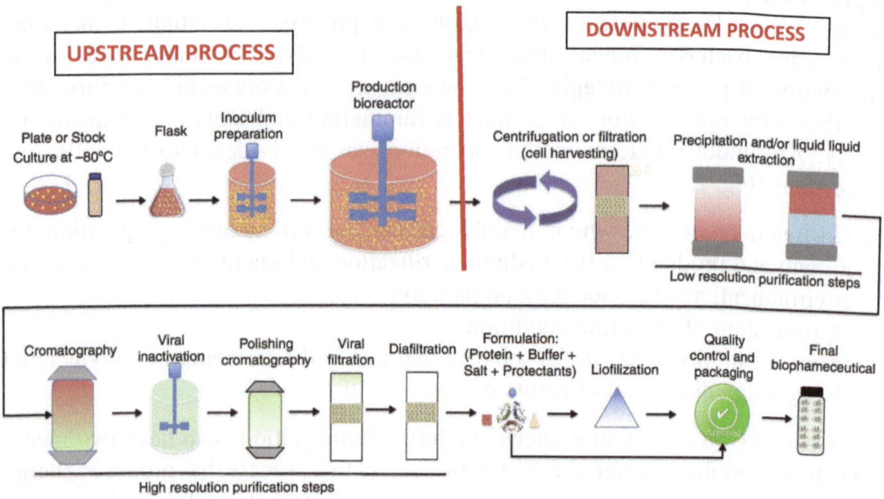

Fig. 2.2 Scheme of entire fermentation process, showing microbial culture preparation, upstream processes, and downstream processes

diverse food products and also as a food preservation method. In a biotechnological context, this term is used in a much broader sense. There, fermentation means a process in which microorganisms, cultured on a large scale under aerobic or anaerobic conditions, convert a substrate into a product that is useful to man [3].

Certain organisms perform fermentation to obtain the energy they need to carry on their life processes (most organisms obtain the energy for these processes through aerobic respiration, in the presence of free oxygen). Various microorganisms, including yeasts and certain molds and bacteria, obtain their energy through fermentation. Many of the fermentation processes result in products that are important in medicine, food preparation, and other fields.

The specific product resulting from fermentation is determined by the type of microorganism carrying on the process and the substance in which the fermentation occurs. For example, wine is the product of yeast fermentation in fruit juice, whereas beer is the product of yeast fermentation in grain. Antibiotics (drugs used to fight infectious diseases) are obtained from both bacterial and mold fermentations. Fermentation by various microorganisms is used to produce substances called enzymes, which are used in many medical and industrial processes to speed up chemical reactions. Vinegar and cheese are products of bacterial fermentation. Yeast fermentation is used to make leavened bread [4].

In the first stage of the upstream process, laboratory work is carried out. Microorganisms that are most suitable for production of a particular compound are selected. Strains of the same organism are made to improve the yield and quality. Media are prepared. Depending on the biomass chosen, the medium is formulated such that it is best suited for optimal growth of the microorganisms used.

Then comes the sterilization procedures, wherein the instruments that are to be used for fermentation and other processes are rid of contaminants such as fungi, virus, bacteria, etc. This ensures that the nutrients supplied in the medium are available only to the desired microbes and not to other unnecessary microbes present, if any. A particular amount of inoculum is prepared, which is the amount of microbes to be used to start the fermentation in the fermenter. The amount should be sufficient enough to begin the fermentation.

2.1 Upstream Processes at Different Stages

Upstream processing includes preparation and formulation of the fermentation medium; sterilization of air, fermentation medium, and fermenter; inoculum preparation; and inoculation of the medium (Fig. 2.2).

2.1.1 *Medium Preparation*

The upstream process is defined as the entire process from early cell isolation and cultivation, to cell banking and culture expansion of the cells until final harvest (termination of the culture and collection of the live cell batch). Aside from technological challenges concerning the scalability of the culture apparatus, a number of raw material supply risks have emerged in recent years, including the availability of current good manufacturing practice grade microbial products derived from inside the cells or as exogenous products isolated from the fermentation medium. The upstream part of a bioprocess refers to the first step in which microbes are grown in bioreactors.

2.1.1.1 Types of Culture Medium

Culture media are of fundamental importance for obtaining pure cultures, growing and counting microbial cells, and cultivating the select microorganisms. Without high-quality media, the possibility of achieving accurate, reproducible, and repeatable microbiological test results is reduced [5]. A microbiological culture medium is a substance that encourages the growth, support, and survival of microorganisms. Culture media contain nutrients, growth promoting factors, energy sources, buffer salts, minerals, metals, and gelling agents (for solid media) [6]. Culture media have been used by microbiologists since the nineteenth century. Even with the increased use of rapid methods, the majority of techniques found in pharmaceutical quality control laboratories required growth media [7]. For the assessment of culture media, no one definitive standard exists.

Organisms that cannot grow in artificial culture medium are known as obligate parasites. *Mycobacterium leprae*, *Rickettsias*, *Chlamydias*, and *Treponema pallidum* are obligate parasites.

Culture media can be classified into several categories depending on their composition or use. According to composition, there are two broadly classified groups: solid and semisolid culture media, commonly called "agar," and liquid culture media, commonly called "broth":

1. **Agar**

   Solid medium contains agar at a concentration of 1.5–2.0% or some other, mostly inert, solidifying agent. Solid medium has physical structure and allows bacteria to grow in physically informative or useful ways (e.g., as colonies or in streaks). Solid medium is useful for isolating bacteria or for determining the colony characteristics of the isolate. Semisolid media are prepared with agar at concentrations of 0.5% or less. They have a soft custard-like consistency and are useful for the cultivation of microaerophilic bacteria or for determination of bacterial motility.

2. **Liquid (broth) medium**

   Liquid media contain specific amounts of nutrients but do not contain gelling agents such as gelatin or agar. They serve various purposes, such as propagation of a large number of organisms, fermentation studies, and various other tests (e.g., sugar fermentation tests). Such media can be further divided into such categories as growth media (designed to grow most heterotrophic microorganisms), transport media (for preserving microorganisms), enrichment media (designed to increase the numbers of desired microorganisms), and selective growth media [8–10].

The manufacture of media is done either in-house (whereby a dehydrated formulation is used) or, more commonly, is purchased ready-made [11]. Where media are purchased ready-to-use, the microbiologist has a responsibility to audit the manufacturer of the media [12]. For certain plate media, such as those used in clean rooms, the media should be sterilized by irradiation [13].

On the basis of use, media can be classified into seven types as follows:

1. **Defined media**

   A defined medium (also known as chemically defined medium or synthetic medium) is a medium in which all the chemicals used are known; no yeast, animal, or plant tissue is present. Chemically defined media are of value in studying the minimal nutritional requirements of microorganisms, for enrichment cultures, and for a wide variety of physiological studies. Some examples of nutrient media include plate count agar, nutrient agar, and soy agar (Table 2.1a–c). Defined media are usually composed of pure biochemicals off the shelf; complex media usually contain complex materials of biological origin such as blood, milk, yeast extract, or beef extract, the exact chemical composition of which is obviously undetermined. A defined medium is also referred to as an enrichment medium (Table 2.2).
2. **Complex media**

   An undefined medium (also known as a basal or complex medium) is a medium that contains a carbon source such as glucose for bacterial growth, water, various salts needed for bacterial growth, and a source of amino acids and nitrogen (e.g., beef or yeast extract).This is an undefined medium because the amino acid source contains a variety of compounds, with the exact composition being unknown (Table 2.3).
3. **Minimal media**

   A defined medium is a minimal medium if it provides only the exact nutrients (including any growth factors) needed by the organism for growth (Table 2.4). The use of defined minimal media requires the investigator to know the exact nutritional requirements of the organisms in question. Minimal media are those that contain the minimum nutrients possible for colony growth, generally without the presence of amino acids, and are often used by microbiologists and geneticists to grow "wild-type" microorganisms. Minimal media can also be used to select for or against recombinants or exoconjugants. Minimal media typically contain water, a carbon source for bacterial growth (a sugar such as

Table 2.1a Minimal medium for the growth of *Bacillus megaterium*

Component	Amount	Function of component
Sucrose	10.0 g	C and energy source
K_2HPO_4	2.5 g	pH buffer; P and K source
KH_2PO_4	2.5 g	pH buffer; P and K source
$(NH_4)2HPO_4$	1.0 g	pH buffer; N and P source
$MgSO_4 \cdot 7H_2O$	0.20 g	S and Mg^{++} source
$FeSO_4 \cdot 7H_2O$	0.01 g	Fe^{++} source
$MnSO_4 \cdot 7H_2O$	0.007 g	Mn^{++} source
Water	985 mL	
pH 7.0		

An example of a chemically defined medium for growth of a heterotrophic bacterium

Table 2.1b Defined medium (also an enrichment medium) for the growth of *Thiobacillus thiooxidans*, a lithoautotrophic bacterium

Component	Amount	Function of component
NH_4Cl	0.52 g	N source
KH_2PO_4	0.28 g	P and K source
$MgSO_4 \cdot 7H_2O$	0.25 g	S and Mg^{++} source
$CaCl_2 \cdot 2H_2O$	0.07 g	Ca^{++} source
Elemental sulfur	1.56 g	Energy source
CO_2	5%[a]	C source
Water	1000 mL	
pH 3.0		

[a]Aerate medium intermittently with air containing 5% CO_2

glucose or a less energy-rich source like succinate), and various salts, which may vary among bacteria species and growing conditions, but generally provide essential elements such as magnesium, nitrogen, phosphorus, and sulfur to allow the bacteria to synthesize proteins and nucleic acids.

4. **Supplementary minimal media**

   Supplementary minimal media are minimal media that also contain a single selected agent, usually an amino acid or sugar. This supplementation allows the culturing of specific lines of auxotrophic recombinants.

5. **Selective and differential media**

   Other concepts employed in the construction of culture media are the principles of selection and enrichment (Table 2.5). A selective medium is one that has a component (or components) added to it that inhibits or prevents the growth of certain types or species of bacteria and/or promotes the growth of desired species. Bismuth sulfite agar inhibits Gram-positive and most Gram-negative bacteria and is used to isolate *Salmonella typhi*. Brilliant green agar inhibits Gram-positive and most Gram-negative bacteria and is used to isolate *Salmonella* species. Sabouraud glucose agar has a pH of 5.6, inhibits most bacteria, and is used to isolate fungi. One can also adjust the physical conditions of a culture

Table 2.1c Chemically defined medium for growing a fastidius *Chemoheterotrophic* bacteria such as *Neisseria gonorrhea*

Constituent	Amount	Constituent	Amount
Carbon and energy sources		Aminoacids	
Glucose	9.1 g	Cysteine	1.5 g
Starch	9.1 g	Arginine, proline (each)	0.3 g
Sodium acetate	1.8 g	Glutamic acid, methionine (each)	0.2 g
Sodium citrate	1.4 g	Asparagine, isoleucine, serine (each)	0.2 g
Oxaloacetate	0.3 g	Cystine	0.06 g
Salt		Organic growth factors	
Potassium phosphate, dibasic (K_2HPO_4)	12.7 g	Calcium pantothenate	0.02g
Sodium chloride (NaCl)	6.4 g	Thiamine	0.02g
Potassium phosphate, monobasic (K_2PO_4)	5.5 g	Nicotinamide-adenine dinucleotide	0.01 g
Sodium bicarbonate ($NaHCO_3$)	1.2 g	Uracil	0.006 g
Potassium sulphate (K_2SO_4)	1.1 g	Biotin	0.005 g
Sodium sulphate (Na_2SO_4)	0.9 g	Hypoxanthine	0.003 g
Magnesium chloride ($MgCl_2$)	0.5 g	Reducing agent	
Ammonium chloride (NH_4Cl)	0.4 g	Sodium thioglycolate	0.00003 g
Potassium chloride (KCl)	0.4 g	Water	1 L
Calcium chloride ($CaCl_2$)	0.006 g		
Ferric nitrate [$Fe(NO_3)_3$]	0.006 g		

Source: R.M Atlas, Handbook of Microbiological, Ann Arbor. MI CRC Press, 1993

Table 2.2 Defined medium also known as enrichment medium is a highly nutritive medium which can nourish and support the growth of gram-positive as well as gram-negative bacteria. It can also be supplemented with blood for enriched growth or study the haemolysis

Composition	
Ingredients	g/L
Peptic digest of animal tissue	40.000
Yeast extract	6.000
Dipotassium phosphate	3.000
Agar	15.000
Final pH (at 25 °C) 7.0 ± 0.2	

medium, such as pH and temperature, to render it selective for organisms that are able to grow under these particular conditions. If a microorganism is resistant to a certain antibiotic, such as ampicillin or tetracycline, then that antibiotic can be added to the medium to prevent cells that do not possess the resistance from growing. Media lacking an amino acid such as proline in conjunction with

Table 2.3 Complex medium for the growth of fastidious bacteria

Component	Amount	Function of component
Beef extract	1.5 g	Source of vitamins and other growth factors
Yeast extract	3.0 g	Source of vitamins and other growth factors
Peptone	6.0 g	Source of amino acids, N, S, and P
Glucose	1.0 g	C and energy source
Agar	15.0 g	Inert solidifying agent
Water	1000 mL	
pH 6.6		

Table 2.4 Minimal medium. It is recommended for the isolation and characterization of nutritional mutants of *Escherichia coli*

Composition	
Ingredients	g/L
Dextrose	1.000
Dipotassium phosphate	7.000
Monopotassium phosphate	2.000
Sodium citrate	0.500
Magnesium sulphate	0.100
Ammonium sulphate	1.000
Agar	15.000
Final pH (at 25 °C) 7.0 ± 0.2	

Table 2.5 Selective enrichment medium of *Listeria* species in accordance with FDA/IDF-FIL

Composition	
Ingredients	g/L
Peptone from casein	17.000
Peptone from soyameal	3.000
Glucose	2.500
Sodium chloride	5.000
Dipotassium hydrogen phosphate	2.500
Yeast extract	6.000
Acriflavine	0.010
Cycloheximide	0.050
Nalidixic acid	0.040
Final pH (at 25 °C) 7.3 ± 0.2	

Escherichia coli, which is unable to synthesize it, were commonly used by geneticists before the emergence of genomics to map bacterial chromosomes.

An example of the use of selective media is restriction of Gram-positive and Gram-negative bacterial growth by eosin methylene blue (EMB) and bile salts, respectively. In general, yeast and mold (YM) deter bacterial growth, whereas

Table 2.6 Mannitol salt agar (MSA)

Ingredients	Content (g/L deionized water)
Beef extract	1.0
Peptone	10.0
Sodium chloride	75.0
D-mannitol	10
Agar	15.0
Phenol red	0.025

MacConkey agar, buffered charcoal yeast extract agar, and xylose lysine desoxycholate (XLD) inhibit certain Gram-negative bacteria. Hektoen enteric agar (HE), mannitol salt agar (MSA), and Terrific Broth (TB) restrict the growth of Gram-positive bacteria. Baird–Parker agar acts as a selective inhibitor for staphylococci.

Differential media are used to distinguish between different organisms. Blood agar is used to identify organisms that lyse red blood cells (*Streptococcus pyogenes*). A culture medium may also be a differential medium if allows the investigator to distinguish between different types of bacteria on the basis of some observable trait in their pattern of growth on the medium.

6. **Enriched media**

Enriched media are media that have been supplemented with highly nutritious materials such as blood, serum, or yeast extract for the purpose of cultivating fastidious organisms. Some special-purpose media are as follows:

a. **Mannitol salt agar**

Mannitol salt agar (MSA) is both a selective and differential media used for the isolation of pathogenic staphylococci from mixed cultures (Table 2.6).

Composition of MSA:

- Enzymatic digest of casein (source of nitrogen, vitamin and carbon)
- Enzymatic digest of animal tissue (source of nitrogen, vitamin and carbon)
- Beef extract (Source of nitrogen, vitamin and carbon)
- D-Mannitol, which is the only carbohydrate source present in the medium
- Sodium chloride
- Phenol red (indicator)
- Agar

Final pH: 7.4 ± 0.2 at 25 °C

On MSA, only pathogenic *Staphylococcus aureus* produces small colonies surrounded by yellow zones. The reason for this color change is that *S. aureus* has the ability to ferment mannitol, producing an acid, which, in turn, changes the indicator color from red to yellow. The growth of other types of bacteria is usually inhibited. This growth differentiates *S. aureus* from *S. epidermidis*, which forms colonies with red zones or both zones (Fig. 2.3).

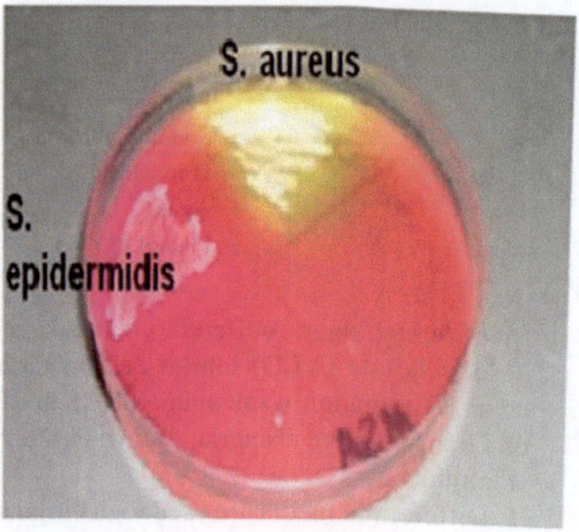

Fig. 2.3 Plating of MSA medium showing growth of Gram-positive bacteria that ferment mannitol. *S. Aureus* grows as white colonies and causes the surrounding medium to appear *yellow*. Gram-positive, non-mannitol fermenter *S. epidermis* appears as a *white zone*. With courtesy of Natural Capsule Limited, cGMP Laboratory

Table 2.7 MacConkey's agar (MAC)

Ingredients	Content (g/L deionized water)
Bacto peptone	17.0
Protease peptone	3.0
Lactose	10.0
Bile salts mixture	1.5
Sodium chloride	5.0
Agar	13.5
Neutral red	0.03
Crystal violet	0.001

b. **MacConkey's agar**

MacConkey's agar (MAC) was the first solid differential media to be formulated; it was developed in the twentieth century by Alfred Theodore MacConkey (Table 2.7). MacConkey agar is a selective and differential media used for the isolation and differentiation of non-fastidious Gram-negative rods, particularly members of the family Enterobacteriaceae and the genus *Pseudomonas* (Fig. 2.4).

Composition of MacConeky's agar:

- Enzymatic digest of gelatin, casein, and animal tissue, which provides nitrogen, vitamins, minerals, and amino acids essential for growth.
- **Lactose:** fermentable carbohydrate providing carbon and energy.
- **Bile salts:** selective agents that inhibit Gram-positive organisms.
- **Crystal violet:** Gram-positive bacteria are generally inhibited by crystal violet.

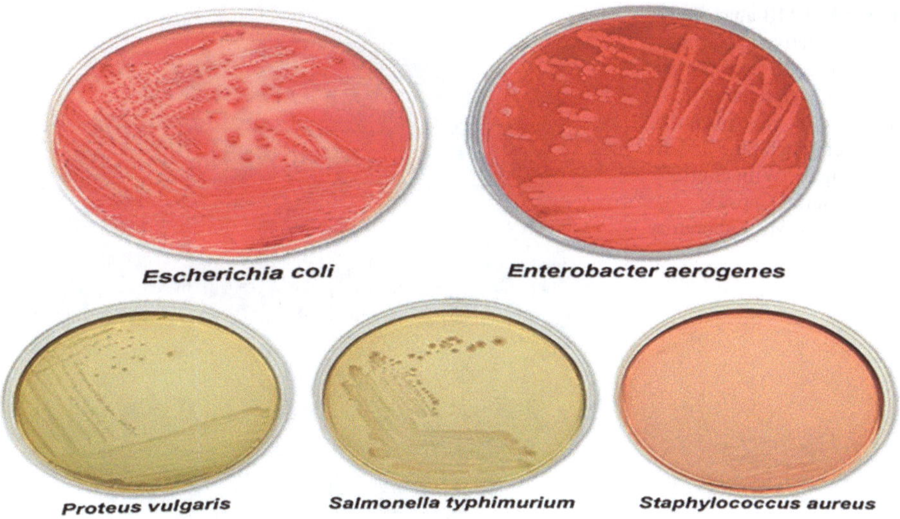

Fig. 2.4 MacConkey's agar (MAC). With courtesy of Natural Capsule Limited, cGMP Laboratory

Table 2.8 Eosin methylene blue (EMB) agar (Levine)

Ingredients	Content (g/L deionized water)
Peptone	10.0
Lactose	10.0
Dipotassium phosphate	2.0
Eosin Y	0.4
Methylene blue	0.065

- Sodium chloride: supplies essential electrolytes for transport and osmotic balance.
- **Neutral red**: pH indicator that is red in color at pH below 6.8. When lactose is fermented, the pH of the medium decreases, changing the color of neutral red to pink.
- Agar: solidifying agent.

Note: Remember the ingredients written in **bold** letters.

c. **Eosin methylene blue agar (Levine)**

Eosin methylene blue agar (EMB) is both a selective and differential medium used for the detection and isolation of Gram-negative intestinal pathogens (Table 2.8).

Composition of EMB agar:

- **Lactose**: a disaccharide that can be fermented by some bacterial enzymes to produce acidic end products.

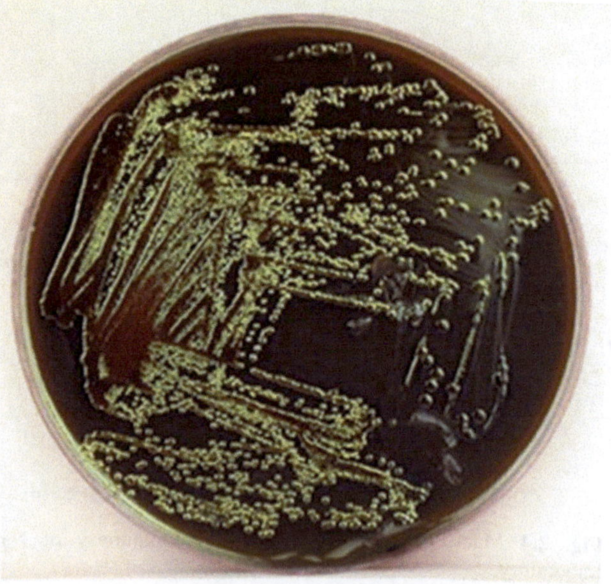

Fig. 2.5 EMB agar. Only Gram-negative bacteria grow on EMB agar (Gram-positive bacteria are inhibited by the dyes eosin and methylene blue added to the agar). With courtesy from Natural Capsule Limited, cGMP Laboratory

Table 2.9 Phenylethyl alcohol agar: (Ingredients per liter of deionized water)

Ingredients	Content (g/L deionized water)
Trypticase	15.0
Phytan	5.0
Sodium chloride	5.0
β-phenylehyl alcohol	2.0
Agar	15.0

- **Eosin and methylene blue:** dyes that inhibit the growth of most Gram-positive bacteria. They also react with any acidic products resulting from lactose fermentation to color the colonies. Acid production from lactose fermentation causes precipitation of the dyes on the surface of the colony, resulting in different colors:

  Large amounts of acid → green metallic sheen
  Small amounts of acid → pink
  No fermentation→ colorless

Enterobacter aerogenes produces large colonies that are pink-to-buff around dark centers. *E. coli* produces small, dark colonies with a green metallic sheen. *Pseudomonas*, *Proteus*, *Salmonella*, and *Shigella* sp. produce colorless colonies because they do not ferment lactose (Fig. 2.5).

d. **Phenylethyl alcohol agar**

Phenylethyl alcohol agar (PEA) is a selective medium (Table 2.9) used to cultivate Gram-positive organisms. The active ingredient, phenylethyl

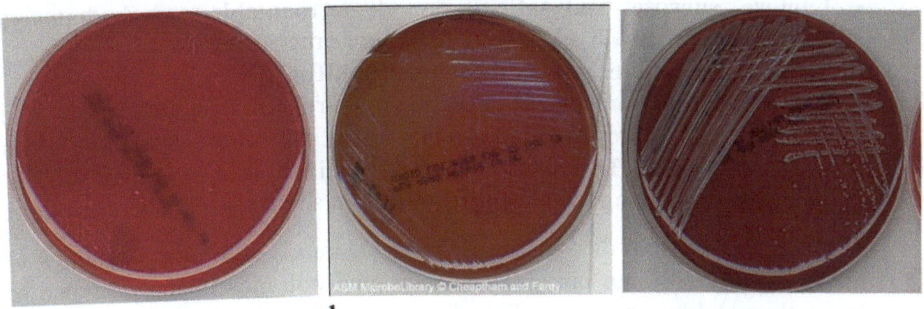

a b c

Fig. 2.6 PEA agar plate showing Gram-positive organisms: (**a**) uninoculated PEA agar plate with 5% sheep blood, (**b**) PEA agar plate with 5% sheep blood inoculated with *Escherichia coli*, and (**c**) PEA agar plate with 5% sheep blood inoculated with *Staphylococcus aureus*. With courtesy from Natural Capsule Limited, cGMP Laboratory

Table 2.10 Hektoen enteric (HE) agar

Ingredients	Content (g/L deionized water)
Proteose	12
Yeast extract	3.0
Bile salt	9.0
Lactose	12.0
Saccharose	12
Silicin	2.0
Sodium chloride	5.0
Sodium thiosulfate	5.0
Ferric ammonum citrate	1.5
Agar	14.0
Bromthymol blue	0.065
Acis fucssin	0.1

alcohol, inhibits or markedly reduces growth of Gram-negative organisms by interfering with DNA synthesis (Fig. 2.6).

e. **Hektoen enteric agar**

Hektoen enteric (HE) agar is a moderately selective medium (Table 2.10) used in qualitative procedures for the isolation and cultivation of Gram-negative enteric microorganisms, especially *Shigella* and *Salmonella* from a variety of clinical and nonclinical specimens

Composition of HE agar:

- **Bile salts:** inhibit the growth of most Gram-positive organisms.
- **Bromothymol blue and acid fuchsin dyes:** lower toxicity than many other enteric media, resulting in improved recovery.

- **Lactose, sucrose, and salicin:** provide fermentable carbohydrates to encourage the growth and differentiation of enterics.
- **Sodium thiosulfate:** provides a source of sulfur.
- **Ferric ammonium citrate:** provides a source of iron that allows the production of hydrogen sulfide from sodium thiosulfate, which provides a source of sulfur. This also allows the visualization of hydrogen sulfide.
- Produced by reacting with hydrogen sulfide gas to form a black precipitate.

Coliforms capable of overcoming the moderately inhibitory qualities of the media develop into orange or salmon-pink colonies in the presence of bromothymol blue indicator. *Shigella* species develop into green-colored colonies with darker blue-green centers. *Salmonella* species appear as blue–green colonies with or without black centers. Producers of H_2S form black-centered colonies in the presence of the ferric ammonium citrate indicator (Fig. 2.7).

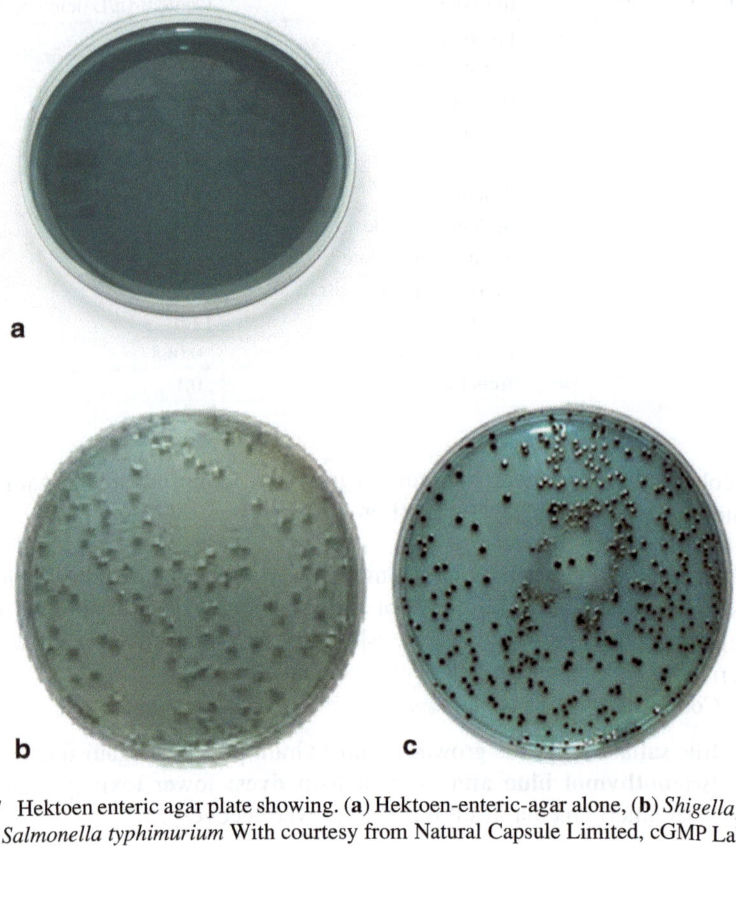

Fig. 2.7 Hektoen enteric agar plate showing. (**a**) Hektoen-enteric-agar alone, (**b**) *Shigella flexneri*, and (**c**) *Salmonella typhimurium* With courtesy from Natural Capsule Limited, cGMP Laboratory

Table 2.11 Blood agar

Ingredients	Content (g/L deionized water)
Infusion from beef heart	500.0
Tryptose	10.0
Sodium chloride	5.00
Agar	15

f. **Blood agar**

Blood agar is both a differential and enriched medium. The blood that is incorporated into this medium is an enrichment ingredient for the cultivation of fastidious organisms such as S*treptococcus* species (Table 2.11). A number of streptococcal species produce substances that destroy red blood cells; that is, they cause lysis of the red cell wall with subsequent release of hemoglobin. Such substances are referred to as hemolysins. The activity of streptococcal hemolysins (also known as streptolysins) can be readily observed when the organisms are growing on a blood agar plate.

Different streptococci produce different effects on red blood cells in blood agar. Those that produce incomplete hemolysis and only partial destruction of the cells around colonies are called alpha-hemolytic streptococci. Characteristically, this type of hemolysis is seen as a distinct greening of the agar in the hemolytic zone, and thus this group of streptococci has also been referred to as the viridans group.

Species whose hemolysins cause complete destruction of red cells in the agar zones surrounding their colonies are said to be beta-hemolytic. When growing on blood agar, beta-hemolytic streptococci are small opaque or semi-translucent colonies surrounded by clear zones in a red opaque medium. Two types of beta-lysins are produced, streptolysin O and streptolysin S. Streptolysin O, an antigenic, oxygen-labile enzyme, whereas streptolysin S is a nonantigenic, oxygen-stable lysin. The hemolytic reaction is enhanced when blood agar plates are streaked and simultaneously stabbed to show subsurface hemolysis by streptolysin O in an environment with reduced oxygen tension. Some strains of staphylococci, *E. coli*, and other bacteria may also show beta-hemolysis. Some species of streptococci do not produce hemolysins. Therefore, when their colonies grow on blood agar, no change is seen in the red blood cells around them. These species are referred to as nonhemolytic or gamma-hemolytic streptococci. On blood agar, *S. aureus* usually displays a light to golden yellow pigment, whereas *S. epidermidis* has a white pigment and *S. saprophyticus* either a bright yellow or white pigment. However, pigmentation is not always a reliable characteristic. On blood agar, *S. aureus* is usually, but not always, beta-hemolytic; *S. epidermidis* and *S. saprophyticus* are almost always nonhemolytic (Fig. 2.8).

Note: For preparation of bloods agar, dissolve the above ingredients and autoclave. Cool the sterile blood agar base to 45–50 °C and aseptically add 50 mL of sterile defibrinated blood. Mix thoroughly and dispense into plates

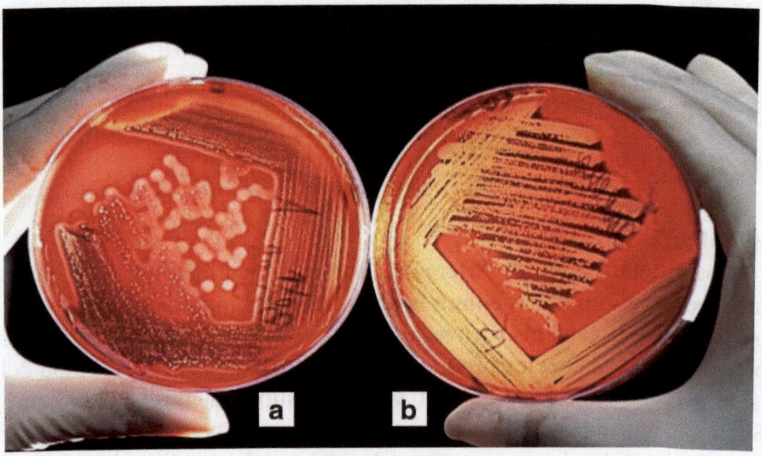

Fig. 2.8 Blood agar plates with cultures of (**a**) *Staphylococcus* (**b**) *Streptococcus*

Table 2.12 Chocolate agar

Ingredients	Content (g/L deionized water)
Proteose peptone	200
Dextrose	0.5
Sodium chloride	5.0
Disodium phosphate	5.0
Agar	15.0

while still liquid. Blood agar base for use in making blood agar also can be purchased. A combination of hemoglobin and a commercial nutrient supplement can be used in place of defibrinated blood.

g. **Chocolate agar**

Fastidious organisms such as *Haemophilus* and *Neisseria* require specially enriched culture media and microaerophilic incubation conditions. "Chocolate" agar is commonly used for primary isolation of *Haemophilus* from clinical specimens. This medium contains hemoglobin derived from bovine red blood cells as well as other enrichment growth factors. Chocolate agar may be made selective for *Haemophilus* species by the addition of bacitracin (Table 2.12).

Two special growth factors, called X and V, are required by some *Haemophilus* species. The X factor is hemin, a heat-stable derivative of hemoglobin. The red blood cells in chocolate agar have been heated until they are lysed, producing the characteristic brown color of this medium. Lysing the blood with heat releases the X factor that otherwise is not available in regular blood agar plates. This is why chocolate agar is the media of choice for culturing *Haemophilus influenza* (Fig. 2.9).

The V factor is a heat-labile coenzyme (nicotinamide adenine dinucleotide; NAD) that is essential in the metabolism of some species that lack

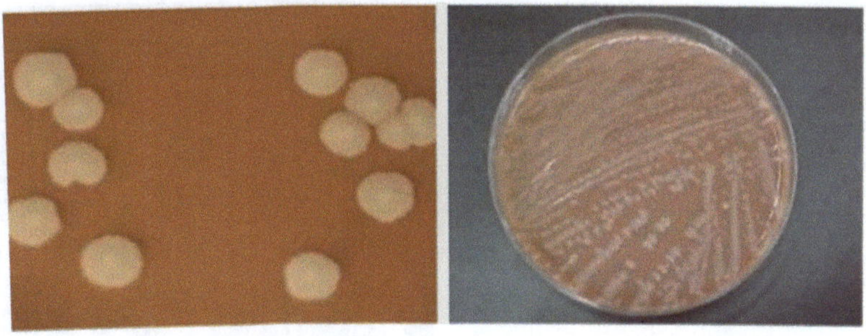

Fig. 2.9 Chocolate agar plates showing *Haemphilus influenza* colonies at different magnifications

it. Yeast extracts contain V factor and are convenient supplements for chocolate agar or other media used for *Haemophilus*. Chocolate agar, however, does not reveal hemolysis data, so species differentiation among the members of *Haemophilus* must be performed in another manner.

Note: For preparation, aseptically add 5% sterile, defibrinated sheep blood to the sterile and molten agar. Heat at 80 °C for 15 min or until a chocolate color develops.

7. **Anaerobic media**

Anaerobic bacteria need special media for growth because they need low oxygen content, reduced oxidation–reduction potential, and extra nutrients. Media for anaerobes may have to be supplemented with nutrients such as hemin and vitamin K. Such media may also have to be reduced by physical or chemical means. Boiling the medium serves to expel any dissolved oxygen. Addition of 1% glucose, 0.1% thioglycollate, 0.1% ascorbic acid, 0.05% cysteine, or red hot iron filings can render a medium reduced. Before use, the medium must be boiled in a water bath to expel any dissolved oxygen and then sealed with sterile liquid paraffin. Petri plates can be incubated in an anaerobic jar or anaerobic chamber. It is advisable to use sodium bicarbonate and sodium borohydride in a small amount of water to produce CO_2 and H^+. A palladium catalyst can be used as it combines with O_2 in the jar and H^+ to remove O_2 (Fig. 2.10a). Oxyrase is a bacterial respiratory enzyme that can be added to media; it combines O_2 with H^+ to remove O_2 by forming H_2O and turns the petri dish into a mini-anaerobic chamber. Anaerobic chambers with air locks are filled with inert gas (Fig. 2.10b).

CO_2 incubators or candle jars are used to grow bacteria requiring an increased CO_2 concentration (capnophiles). Low O_2 and high CO_2 concentrations mimic conditions found in the intestinal tract. CO_2 incubators allow adjustment of CO_2 concentrations to lower or higher than atmospheric when needed by certain aerobic bacteria. Candle jars increase CO_2 concentrations and still leave some O_2 for aerobic capnophiles. Gas packs can also generate CO_2 and are generally used in place of candle jars (Fig. 2.10c).

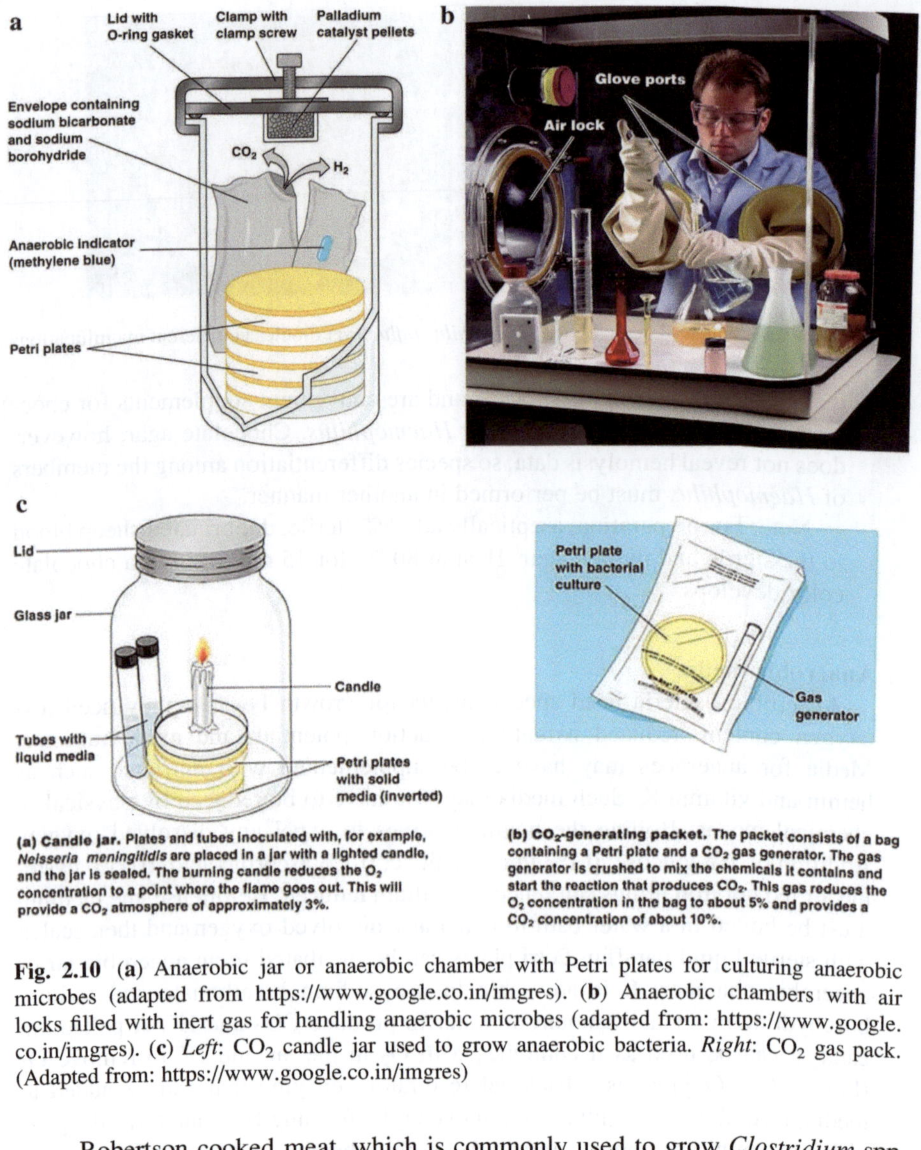

Fig. 2.10 (**a**) Anaerobic jar or anaerobic chamber with Petri plates for culturing anaerobic microbes (adapted from https://www.google.co.in/imgres). (**b**) Anaerobic chambers with air locks filled with inert gas for handling anaerobic microbes (adapted from: https://www.google.co.in/imgres). (**c**) *Left*: CO_2 candle jar used to grow anaerobic bacteria. *Right*: CO_2 gas pack. (Adapted from: https://www.google.co.in/imgres)

Robertson cooked meat, which is commonly used to grow *Clostridium* spp medium, contains a 2.5-cm column of bullock heart meat and 15 mL of nutrient broth. Thioglycollate broth contains sodium thioglycollate, glucose, cystine, yeast extract, and casein hydrolysate. Methylene blue and resazurin are oxidation–reduction potential indicators that can be incorporated into the medium. Under reduced conditions, methylene blue is colorless.

A medium that is used for a large scale fermentation to ensure the sustainability of the operation should have the following characteristics:

- It should be cheap and easily available

- It should maximize the growth of the microorganism, productivity, and the rate of formation of the desired product
- It should minimize the formation of undesired products

Usually, waste products from other industrial processes, such as molasses, lignocellulosic wastes, cheese whey, and corn steep liquor, after modifying by the incorporation of additional nutrients, are used as substrates for many industrial fermentations. It is important that each batch of such media undergoes some form of quality control before it is released for general use, thus providing a measure of confidence that the results issued from microbiology laboratories are accurate. Testing is normally undertaken once all preparatory steps have been completed, including irradiation.

2.1.1.2 Quality Control of Culture Media

Quality control can be divided into two parts:

1. **Physical characteristics**

   The tests undertaken for the physical characteristics of culture media vary depending upon the type of media. Examples of physical tests include:

- Visual test for color: The color of a sterilized medium should be compared to a nonsterilized medium and any differences in color noted.
- Visual test for clarity: The clarity of the media should be examined for optical artefacts, such as crystallization.
- Gel strength: The gel strength should not be over-hard or over-soft, but firm and usable.
- pH of the finished media: This is probably the most important chemical test, because if the pH is outside the recommended range for the media, growth of some of the microorganisms that the media is intended to grow is inhibited [14].
- Checks for damage: Plates and bottles should be examined for damage such as cracks and defects.

2. **Microbiological characteristics**

- The test for media sterility is designed to detect microbial contamination during the manufacturing process. A small number, normally 2% of the batch, of uninoculated items are incubated. The temperature and time selected for the sterility test incubation depend upon the type of media. For general-purpose media, a temperature of 30–35 °C for 3 days is typical. To pass the sterility test, the items must demonstrate no growth.
- Arguably the challenge of culture media with microorganisms is the most important test carried out in the microbiology laboratory. That such a key test is undertaken by the media manufacturer is unquestionable. Additionally, it is common for the purchaser to carry out growth promotion, check for batch-to-batch variability, and assess any issues during shipment [15].

For growth promotion, a panel of microorganisms is required to demonstrate the suitability of the media for its intended use. The pharmacopeia recommends certain microorganisms and these must be traceable to a reputable culture collection, such as the American Type Culture Collection (ATCC) (although the pharmacopeia allows alternative culture collections to be used, there is some ambiguity about strain equivalency). Type cultures should be carefully preserved within the culture collection of the laboratory. This includes ensuring that cultures are held at a temperature low enough to avoid phenotypic variations from occurring and restricting the number of passages between subculture steps to less than five [16].

These microorganisms have been serially subcultured in national culture collections over decades and are conditioned for growth on rich laboratory culture media. They are designed to allow the vendor to assess the media as suitable at the point of manufacture and for the user to verify the media upon receipt.

In addition to type cultures, environmental isolates are commonly used in media testing regimes. These organisms are designed to demonstrate that a particular batch of culture media grows microorganisms that are representative of the types found in the manufacturing environment [17]. Thus, media used for the examination of water have a test panel that includes microbial isolates from water (such as *Pseudomonad*-related bacteria) and media used for environmental monitoring include bacteria transient to human skin (such as staphylococci).

Although the use of such isolates is increasingly becoming a regulatory expectation, the adoption of environmental or plant isolates is not supported by all microbiologists. Arguments for the use of such isolates are that the media are challenged with microorganisms actually encountered within the pharmaceutical environment, and that these are often more representative than the standard cultures. Moreover, the isolates can be varied over time, based on reviews of microflora, so that they remain perpetually relevant. Arguments against their use include the fact that interlaboratory assessments are rendered difficult because each laboratory is using a different organism set. A second point is that once organisms are grown on standard media they become indistinguishable from other laboratory strains. It has been counter-argued that minimally subcultured environmental isolates have aspects of their wild-type attributes conserved. The outcome of this debate is ongoing and clearly further study is needed.

2.1.2 Media Formulation

There is restricted availability of research documents on complex substrates that have been developed for the production of various products. Most fermentation processes, which include the production of fermentation media, are closely guarded trade secrets. However, the technical objectives of media formulation has two main purposes: (1) preparation of the best innoculum (starter culture) that can give a proper boost to pilot scale fermentation, and (2) production of target product in quality and quantity.

Most fermentations require liquid media, often referred to as broth, although some solid substrate fermentations are in operation. Microorganisms used for fermentation processes grow on or in growth medium that satisfies their nutritional needs. Complete analysis is needed to establish the most favorable medium for the growth of microbes used for fermentation. The primary objectives for making a standard protocol for formulating broth for microbial culture are:

- Maximum biomass or product yield for each gram of substrate used
- Maximum concentration of biomass/product
- Maximum rate of product formation
- Minimum risk of undesirable products (byproducts)
- Availability of medium constituents in adequate quantities and consistent quality throughout the year
- Minimum problems in medium preparation and sterilization
- Minimum problems in aeration, agitation, downstream processing, and other production processes
- Minimum problems in safe disposal of effluents and waste treatment

2.1.2.1 Seed Culture Medium Formulation

Seed cultures are stocks of microorganisms carrying target characteristics in dormancy condition. They are used in fermentation processes to obtain a product through fermentation under controlled physical conditions (Fig. 2.11). Inoculum aims to:

- Minimize the loss of viable microorganisms during the recovery from dormancy
- Obtain a genotypically identical copy of the population that was stored
- Increase biomass
- Develop the culture to a physiological state suitable for the performance in the final production stage [18]

A standard protocol should be followed to maintain such factors as the minimum time after inoculation. An adequate quantity must be available to provide an optimum volume ranging from 0.1 to 10% of the medium volume. The original appearance of the microorganism should be in intact form. The inoculum must be free from any contamination and the microbes should express their product formation potential while in the optimum growth period or at a specific growth stage [19]. In seed culture, less attention is given to the cost of ingredients used in media preparation, as 5% of the total fermenter volume is used as seed culture. Some specially prepared dairy products have been used quite extensively in primary and secondary seeds as constituents of the medium. In order to maintain the quality of microbial strains, certain internationally prescribed norms should be followed [20, 21]. A variety of standard molecular fingerprinting techniques are practiced to ascertain the status of microbes within shortest possible time. However, the

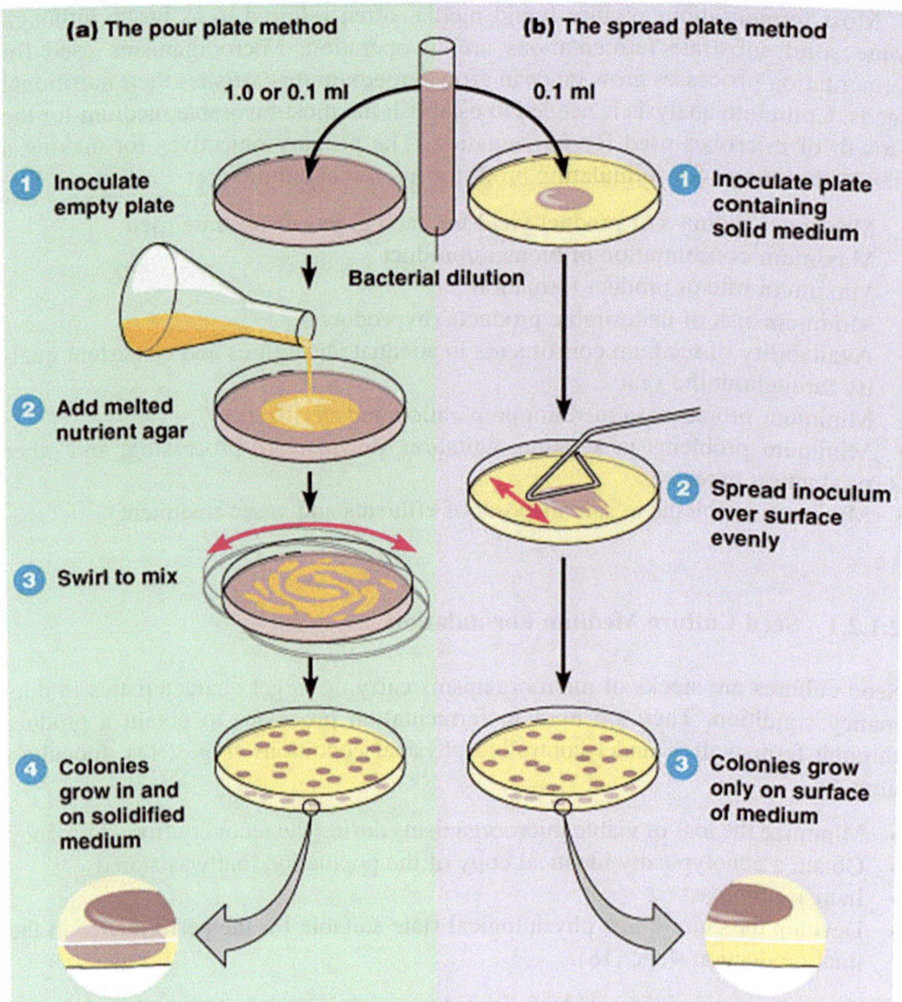

Fig. 2.11 Preparation of microbial seed culture for storage and further use. (Adapted from : https://www.google.co.in/imgres)

classic method recommended by Lincoln [22] is still used, whereby the colony type is tested on agar plates seeded with the master culture before initiating inoculum development. It has been observed that innoculum transfer simply by using the inoculating wire loops can cause wide variations in the number of cells that are seeded to the first shaken flask [23]. It is therefore necessary to prepare the first shake culture in aliquots to be used for preparing several rounds of inocula for regular supply over a long period of time. Proper means of storing these laboratory seeds is taken care of under the prescribed protocol.

It was observed that storage of the stock inoculum of *Streptomyces viridoflavus* in the gas space of a liquid nitrogen refrigerator was the most satisfactory of several methods tested [24]. Heat-shock treatment of stock cultures of *Clostridium butylicum* increased solvent production [25]. However, for *Clostridium acetobutylicum*, neither heat shock nor treatment with aqueous ethanol or butanol improved solvent production during subsequent fermentation, but resulted in significantly improved production when culture transfers during the inoculum development procedure were made at the time of maximum cell motility [26]. So, it is necessary to follow selective inoculum processes to obtain good production, as noticed in the case of production of 1,3-propanediol using *Klebsiella pneumonia* [27]. For filamentous fungi, the morphological and physiological status of the inoculum varies according to the species and the process being conducted [28]. It has been noticed that the pellet morphology can have significant effect on product formation, as observed in the production of penicillin V by *Penicllium chrysogenum*. This could be the result of an improvement in process performance by increasing the stirrer speed in the third preculture. The pellet size in an airlift tower loop reactor was reduced to about half, cell growth was influenced only slightly, but the production phase was extended and the final penicillin concentration was doubled [29].

In solid-state fermentation (SSF) and biofilm fermentation it is recommended to use fungi spore inoculum rather than filamentous samples because cell adhesion is necessary for the development of a biofilm. It has been observed that cell adhesion stimulates physiological responses that are different from those in suspended growth [30, 31]. Sometimes carrier-attached starter cultures shows good results in SSF processes [32].

Selection of a suitable quantity of inoculum of different types of microbes for large-scale fermentation is important in achieving the target quantity of product. The fermenter preculture (seed culture) quantity should be large enough to generate the desired quantity of microbial biomass in the final stage of fermentation within the stipulated time. Generally, the microbial inoculums used in industries contain the following: bacteria 0.1–3%, actinomycetes and fungi 5–10%, and spore suspension 1×10^5 to 5×10^5 spores per liter of culture medium [33]. However, use of 10% seed culture in the final working volume of fermentation is good practice. Adaptation of microbes during lag phase is an important factor in achieving perfect growth during the exponential phase.

When cells are well adapted to the new environment they grow at a constant rate following autocatalytic first-order kinetics, where growth rate is proportional to cell or biomass concentration, as shown in Eq. (2.1) (μ represents specific growth rate in hours^{-1} and x is biomass concentration):

$$\frac{dx}{dt} = \mu x \tag{2.1}$$

This new phase is called the exponential or “log” phase (Fig. 2.12). Exponential growth continues until changes in the environment (i.e., decrease in nutrients and

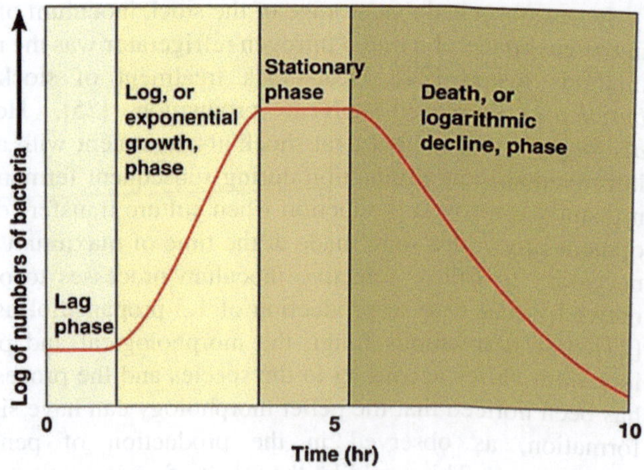

Fig. 2.12 Bacterial cell growth kinetics. During log phase growth, the rate is proportional to cell concentration

increase in waste metabolic products) slow or even stop growth. This is the third phase, known as the "stationary" phase. Continuation of the exponential phase depends on the status of nutrients present in the medium and accumulation of extracellular metabolites. Both log and stationary phases are of interest from the industrial point of view because primary and secondary metabolites are produced in these phases, respectively. The bacterial growth curve can be calculated on the basis of the logarithm value of bacterial population with reference to time (Fig. 2.13)

Genetics also influence further development of the quality of the microorganism being used. The appearance of a variant can affect productivity. Although the mutation rate is very low, the number of variants increases with the number of generations that a given culture has during exponential growth. Given that loss-of-function mutations occur in 1 in 10^5 or 10^6 newly divided cells, and even rare gain-of-function mutations occur in 1 in 10^9 or 10^{10} bacteria, there is always the possibility of beneficial mutations leading to organisms fitter than the inoculated strain [34]. Also, the final proportion of variant in the population increases with fermentation scale [35]. If Eq. (2.1) is expressed in terms of a period of time *t* from a total biomass of inoculum x_0 , and the final biomass concentration is *x* in a given fermentation volume *V*, then Eq. (2.2) is obtained:

$$\frac{Vx}{X_0} = e^{-\mu t} \tag{2.2}$$

Considering that the number of generations *G* is related to the doubling time t_d and the specific growth rate μ by:

$$G = \frac{\mu t}{\ln 2} \tag{2.3}$$

Then, substituting Eq. (2.3) into Eq. (2.2) yields:

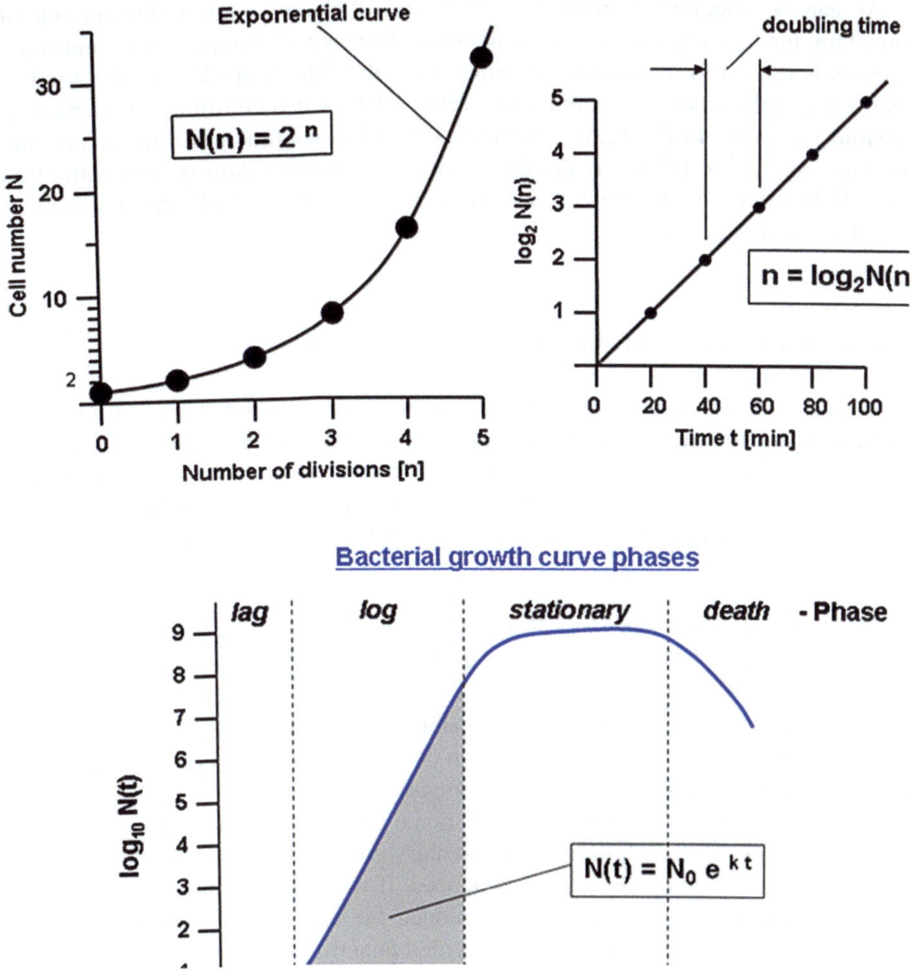

Fig. 2.13 Measurement of bacterial growth, expressed in different forms

$$\frac{Vx}{xa} = e^{Gln2} \tag{2.4}$$

or

$$G = 1.44\left(\ln V + \ln x + \ln\left(x_0\right)\right) \tag{2.5}$$

$$\frac{X_p}{X_{gv+}X_p} = \frac{\alpha + \lambda - 1}{(\propto -1) + \lambda 2^{G(\alpha+\lambda-1)}} \tag{2.6}$$

As can be understood from Eqs. (2.5) and (2.6), the higher the amount of inoculum, the lower the number of generations because the frequency of undesired variants is related to increasing fermentation scale. Equation (2.6) is also used to follow the replacement of recombinant cells (bearing a recombinant plasmid) by plasmid-free cells. In Eq. (2.6), 1 signifies that a loss-of-function mutation occurs once in every 10^5 or 10^6 newly divided cells and p is the probability of plasmid loss per cell division (i.e., per generation). Thus, the probability of plasmid loss can be calculated from eq. (2.6) by changing 1 to p [36].

2.1.2.2 Fermentation Medium (Broth) Formulation

The fermentation medium must provide all the elements needed for cell biomass formation and for metabolite production in the required quantities, and it must provide enough energy for cell metabolism and maintenance. An idea of the medium composition can be developed from the equation based on the stoichiometry for growth and product formation. For an aerobic process, this equation is as follows:

$$\text{Carbon and energy source} + \text{Nitrogen source} + O_2 + \text{Other requirements} \\ \rightarrow \text{Biomass} + \text{Products} + CO_2 + H_2O + \text{Heat}$$

This qualitative equation needs to be quantified in terms of the specific elements involved. Quantification allows calculation of the minimum quantities of nutrients required to produce a given amount of biomass.

There are at least three stages for the rational design of a fermentation medium. First is to design it according to the elementary compositions of a specific microorganism and the main product to be produced. If we know the amount of biomass needed to obtain a given quantity of product, we can estimate the amounts of various nutrients needed to obtain the specified quantity of product [37]. However, some medium components may be needed for product formation but not for biomass production, and often it is not easy to quantify all the nutritional needs precisely. The elemental compositions of various microorganisms have been determined (Table 2.13) but it is desirable that the composition be determined for the specific microorganism to be used for fermentation. In addition, micronutrients such as Fe, Zn, Cu, Mn, Co, Mo, and B may also be needed in small quantities. In most media, some nutrients (e.g., P, K) are added in substantial excess of the requirement, whereas others such as Zn and Cu are used in near-limiting quantities. In many media, the concentration of P is raised deliberately to improve the buffering capacity. Some microorganisms cannot synthesize specific biochemicals (e.g., amino acids, vitamins) and so they must be included in the medium. The production of some biological products requires the addition of inducers or precursors to the medium. Thus, for the production of vitamin B_{12} the addition of cobalt ions, glycine, threonine, α-aminolevulinic acid, or compatible solutes such as betaine (rich in sugar factory molasses) and choline proves to be beneficial [38].

Table 2.13 Major elements, their sources and functions in bacterial cells

Element	% of dry wreight	Source	Function
Carbon	50	Organic compounds or CO_2	Main constituent of cellular material
Oxygen	20	H_2O, organic compounds, CO_2, and O_2	Constituent of cell material and cell water; O_2 is electron acceptor in aerobic respiration
Nitrogen	14	NH_3, NO_3, organic compounds, N_2	Constituent of amino acids, nucleic acids nucleotides, and coenzymes
Hydrogen	8	H_2O, organic compounds, H_2	Main constituent of organic compounds and cell water
Phosphorus	3	Inorganic phosphates (PO_4)	Constituent of nucleic acids, nucleotides, phospholipids, LPS, teichoic acids
Sulfur	1	SO_4, H_2S, S^o, organic sulfur compounds	Constituent of cysteine, methionine, glutathione, several coenzymes
Potassium	1	Potassium salts	Main cellular inorganic cation and cofactor for certain enzymes
Magnesium	0.5	Magnesium salts	Inorganic cellular cation and cofactor for certain enzymatic reactions
Calcium	0.5	Calcium salts	Inorganic cellular cation, cofactor for certain enzymes and a component of endospores
Iron	0.2	Iron salts	Component of cytochromes and certain nonheme iron proteins and a cofactor for some enzymatic reactions

With courtesy from Todar's Online Textbook of Bacteriology

The second stage is to design the medium according to the stoichiometry of growth and product formation. For this, a chemical equation should be established as follows [39]:

$$\begin{aligned} CH_mO_1 + {}_aNH_2 + {}_bO_2 \quad & Y^C_B CH_2O_nN_{q(\text{biomass})} \\ & + Y^c_p CH\ ON_{r(\text{product})} + {}_cH_2O + {}_dCO_2 \end{aligned} \tag{2.7}$$

where, y^c_b is the yield coefficient of biomass based on carbon; Y^c_p is the yield coefficient of product based on carbon; a is the stoichiometric coefficient for nitrogen requirement and b that for oxygen. From Eq. (2.1) it is possible to estimate the true yield coefficient based on substrate for biomass ($Y_{x\backslash s}$) and for product ($Y_{p\backslash s}$) considering the "one carbon formula" (OCF) of biomass and product

$$Y_{x\backslash s} = Y_{\mathrm{b}}^{\mathrm{c}} \frac{\text{OCF (weight) biomass}}{\text{OCF (weight) substrate}} \tag{2.8}$$

$$Y_{p\backslash s} = Y_{\mathrm{p}}^{\mathrm{c}} \frac{\text{OCF (weight) products biomass}}{\text{OCF (weight) substrate}} \tag{2.9}$$

However, there are only a few industrial processes in which there is a straightforward stoichiometric relationship between both sides of Eq. (2.7). Moreover, cell physiology is very complex and metabolic pathways for some metabolites are extremely intricate, which make it impossible to represent growth and product formation in a single chemical equation. Likewise, all cells derive part of the carbon to obtain energy for maintenance from the substrate, which is a cost that reduces the theoretical true yield values.

The third important stage in medium formulation is consideration of yield factors and maintenance of energy. Let us assume that carbon substrate is consumed in least three physiological activities (τ_b) (growth energy, assimilation, and maintenance energy) and that this produces a given amount of biomass (ΔX). Then we can obtain the growth yield factor as follows:

$$Y'_{x\backslash s} = \frac{\Delta X_{\text{total produced}}}{(\Delta S)_{\text{growth energy}+(\Delta S)_{\text{assimilation}}+(\Delta S)_{\text{maintenance energy}}}} = \frac{\Delta X_{\text{produced}}}{(\Delta S)_{\text{and consumed}}} \tag{2.10}$$

$$Y'_{p\backslash s} = \frac{\Delta P}{\Delta S_{\text{total consumed}}} \tag{2.11}$$

Yield factors depend on the type of microbe to be cultured. Fermentation operation conditions also have an important effect on the yield of final product. The energy consumed in the overall process of fermentation also depends on a small fraction of substrate consumed, which can be ignored, or on a significant fraction that can be over 50% of the consumed substrate as in secondary products [19].

The yield factors also depend on operating modes of the fermentation system (Fig. 2.14). The yield and, very frequently, the volumetric productivity of biomass (τ_{b}) or product (τ_{p}), for a given operating time (t_1), a given biomass concentration (x), and under a specific production rate (q_{p}) can be given by:

$$\tau_p = \frac{I}{t_1} \int_0^{t_1} q_p.x\mathrm{d}t \tag{2.12}$$

The carbon requirement for biomass production under aerobic conditions can be estimated from the yield coefficient (Y_p^c), which is estimated as follows [9]:

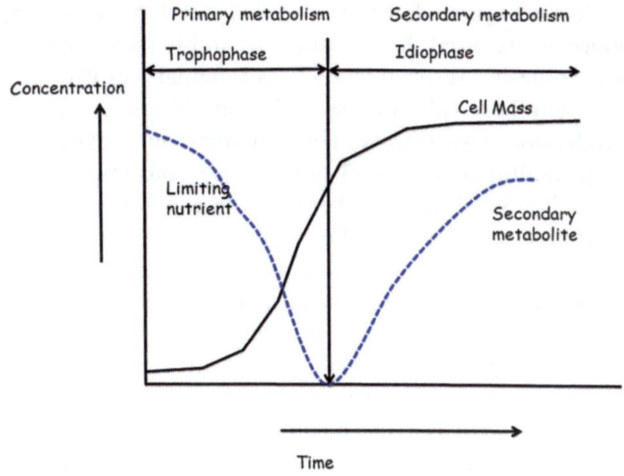

Fig. 2.14 Time course profiles of cell growth and production of primary and secondary metabolites

Cellular yield coefficient $Y_{\mathrm{p}}^{\mathrm{c}}$ = Quantity of cell dry matter produced/quantity of carbon substrate utilized

The value of greatly affected by the carbon substrate. For bacteria, the value of *G* ranges from 0.34 for acetate, through 0.51 for glucose (from molasses), to 1.01 for α-alkanes. Thus 1 g glucose leads to 0.51 g biomass production.

Carbon source is also required for product formation. The theoretical yield of product can be estimated on the basis of material and energy balances for the biosynthetic pathway and its stoichiometry. It should be then compared with the actual yields to critically evaluate the efficiency of conversion of the carbon source into product. However, this often cannot be done because of limited knowledge of the biosynthetic pathways concerned. For example, the stoichiometry equation for penicillin G (PenG) biosynthesis is as follows:

$$\mathrm{A_2C_6H_{12}O_6 + B_2NH_3 + C_2O_2 + d_2H_2SO_4 + e_2PAA} \\ \rightarrow \mathrm{n_2Pen\ G + p_2CO_2 + q_2H_2O}$$

Where, $a_2, b_2, c_2, d_2, e_2, n_2, p_2$, and q_2 are stoichiometric coefficients, and PAA is phenylacetic acid. On solving this equation, we obtain:

$$10/6\ \mathrm{C_6H_{12}O_6 + 2NH_3 + 1/2O_2 + H_2SO_4 + C_8H_8O_2} \\ \rightarrow \mathrm{C_{16}H_{18}O_4N_2S + 2CO_2 + 9H_2O}$$

It can thus be calculated that, theoretically, 1 g of glucose yields 1.1 g penicillin G. It should be pointed out that phenylalanine is not used by the fungus as a carbon source; instead it is used as a precursor for penicillin G biosynthesis. This highlights another principle of medium formulation: Whenever applicable, the precursor, and/or inducer/inhibitor needed for production must be provided.

It was estimated, using a simple batch fermentation model, that 28, 61, and 11% glucose are used for production of biomass, cell maintenance, and penicillin biosynthesis, respectively. Experimental results from fed-batch fermentation revealed these values to be 26, 70, and 6% for growth, maintenance, and penicillin production, respectively. Thus, the maximum experimental conversion of glucose to penicillin is only 0.53 g/g glucose consumed, and the theoretical conversion value is several times higher than the experimental value.

Rational science-based design and formulation of media is still difficult. So, it is necessary to use statistical techniques instead of the traditional costly and time-consuming-approach of changing one component while fixing the other at a certain level. For the development of a new medium, a two-step strategy is recommended. In the first stage, important factors are considered to determine which physical conditions are the most promising for optimization. At this stage, the Planckett–Burman design is most recommended. In the second stage, the most important ingredients are then optimized by the response surface technique [40].

2.1.3 Broth Composition

The choice of medium composition is as crucial to product yield as the choice of high-producing strains of microorganisms. The medium not only provides the nutrients needed for microbial growth, but also for metabolite production. Formulation of medium at the laboratory scale can be done by adding main ingredients like water, carbon source, nitrogen source, minerals, and other supplements in pure form and in the required quantities. This is very easy and supports the growth of microbes, whereas the same medium might not support satisfactory growth of the same organism at the industrial level.

Water is the vehicle of entry for all nutrients into the cell and for the elimination of waste products. It is the basic carrier for nutrients needed to maintain a sustainable form of growth and development. In addition, various elements in the form of carbon, nitrogen, phosphorus, and sulfur are needed to build the basic blocks of microbial life. Vitamins and trace elements are also essential for growth and development, but in microquantities. Like water, oxygen is a primary requirement for microbial life. Even some anaerobic bacteria require aeration at the beginning of culture, as practiced in beer fermentation.

To bring the biochemical metabolic activity into a balanced state, buffering the medium by addition of acids and alkali is essential during fermentation. To have control over foaming, which frequently occurs during fermentation and can cause problems, addition of antifoaming agents is necessary. Other essential additions are growth inducers and growth inhibitors to restrict the growth of unwanted microbes while the fermentation process is in operation.

During formulation of the broth, three main factors should be considered: nutritional requirements, environmental requirements, and techno-economic assessment.

2.1.3.1 Nutritional Requirements of Bacteria

Bacterial nutritional requirements include elements, specific nutrients, and energy requirements. The elemental requirement is essential for maintaining the stoichiometry for growth and product formation, as stated in the following equation:

$$\text{C-source} + \text{N-nitrogen source} + \text{minerals} + \text{specific nutrients} \rightarrow \text{Microbial cell mass} + \text{product(s)} + Co_2 + H_2O + \text{heat}$$

Auxotrophs require specific nutrients in the form of complex medium or specific chemical(s). The specific requirements for vitamins and amino acids are based on the biomass constituents and metabolic activity of microbes used in fermentation. In general, the elemental formulation of microbes is $C_4H_7O_2N$ (dry weight basis 48% C, 7% H, 32% O, and 14% N). For example, baker's yeast is $C_{3.72}H_{6.11}O_{1.95}N_{0.65}S_{0{\cdot}017}P_{0.035}K_{0.6}$. The formulation of broth is decided on the basis of the elements constituting both microbial biomass (see Tables 2.13 and 5.7) and product. The following criteria need to be satisfied for the material to be used at the industrial level:

- It should give maximum yield of product.
- It should give minimum yield of undesired product
- It should be consistently available throughout the year
- It should be cheap

Generally, carbohydrates are used as "carbon sources" for fermentations at the laboratory level. At industrial level, inexpensive carbon sources such as cane molasses, corn steep liquor, and sugar beet juice are utilized. By contrast, some sensitive fermentations make use of glucose, sucrose, and other carbohydrates in their pure form, which ensures the purity and quality of the final product. Sometimes starch is added to the medium for the specific production of amylases. At the laboratory level, peptone, tryptone, or beef extract (a partially digested hydrolysate used in the synthesis of proteins), components of nucleic acids and other essential cellular components are added. However, at the industrial level, the nitrogen source is supplemented with soy meal, ammonia, or nitrate salts. Other elements are added to the fermentation medium, including growth factors, vitamins, antifoaming agents, precursors, inducers, chelating agents, and trace elements such as Fe, Cu, Mn, Mo, and Co. Where growth factors, vitamins, precursors, inducers, and trace elements directly support the growth of microbes and antifoaming agents are added to prevent foam formation, chelating agents may be needed in the case of high concentrations of metal ions.

Microorganisms vary greatly in their nutrient requirements, ranging from autotrophs, which produce all the biochemicals required from simple inorganic nutrients and derive their energy from oxidation of some inorganic component of the medium, to organisms such as lactic acid bacteria, which require many organic compounds for growth. Bacteria can be classified nutritionally on the basis of their energy requirements and their ability to synthesize essential metabolites. Bacteria that derive energy from sunlight are called phototrophs. Those that obtain energy from chemical reactions are called chemotrophs. Bacteria that can synthesize all their organic compounds are called autotrophs. Autotrophs are able to use atmospheric carbon dioxide and nitrogen and are capable of independent existence in water and soil. They are of no medical importance. Some bacteria are unable to synthesize their own metabolites and depend on preformed organic compounds. These bacteria are called heterotrophs and are unable to grow with carbon dioxide as the sole source of carbon. Their nutritional requirements vary widely. Some may require only a single organic substance such as glucose. Others may need a large number of different compounds such as amino acids, nucleotides, lipids, carbohydrates, and coenzymes. Bacteria require a supply of inorganic salts. They require anions such as phosphate and sulfate anions and cations such as sodium, potassium, magnesium, iron, and calcium. Some ions like cobalt may be required in trace amounts. Some bacteria require certain organic compounds in minute quantities. These are called growth factors or bacterial vitamins. Growth factors are called essential when growth does not occur in their absence. Accessory growth factors are those that enhance growth without being absolutely necessary for it. In many cases, bacterial vitamins are the same as the vitamins necessary for nutrition of mammals (e.g., the B group vitamins thiamine, riboflavin, pyridoxine, nicotinic acid, folic acid, and vitamin B12).

Depending on the influence of oxygen on growth and survival, bacteria can be divided into aerobes and anaerobes. Aerobic bacteria require oxygen for growth. They may be obligate aerobes or facultative anaerobes. Obligate aerobes grow only in the presence of oxygen (e.g., *Cholera, Bacillus*). Facultative anaerobes are ordinarily aerobic but can grow in the absence of oxygen, although less abundantly. Most bacteria of medical importance are facultative anaerobes. Anaerobic bacteria, such as Clostridia, grow in the absence of oxygen. Obligate anaerobes may even die on exposure to oxygen. Microaerophilic bacteria are those that grow best in the presence of low oxygen. In the case of aerobes, atmospheric oxygen is the final electron acceptor in the process of respiration (aerobic respiration). In this case, the carbon and energy source may be completely oxidized to carbon dioxide and water. Energy is provided by the production of energy-rich phosphate bonds and the conversion of adenosine diphosphate (ADP) to adenosine triphosphate (ATP). This process is called oxidative phosphorylation.

Anaerobic bacteria use compounds such as nitrates or sulfates, instead of oxygen, as final electron acceptors in the process of respiration (anaerobic respiration). A more common process used by these bacteria in anaerobic metabolism is fermentation. It is defined as the process by which complex organic compounds, such as glucose, are broken down by the action of enzymes into simpler compounds without the use of oxygen. This process leads to the formation of several organic end-products, such as organic acids and alcohols, as well as gas (carbon dioxide and hydrogen). For example, *E. coli* ferments glucose with the production of acid and gas. It also ferments lactose. During the process of fermentation, energy-rich phosphate bonds are produced by the introduction of organic phosphate into intermediate metabolites. This process is known as substrate-level phosphorylation. The energy-rich phosphate groups so formed are used for conversion of ADP to

ATP. All bacteria require some amounts of carbon dioxide for growth. This is obtained from the atmosphere or from bacterial cell metabolism. Some bacteria (e.g., *Brucella abortus*) require much higher levels of carbon dioxide (5–10%) for growth; they are called capnophilic bacteria.

2.1.3.2 Nutritional Requirements of Fungi

All fungi are chemoheterotrophic (chemo-organotrophic) and synthesize the organic compounds they need for growth and energy from pre-existing organic sources in their environment, using the energy from chemical reactions. Because their protoplasm is protected by a rigid wall, fungi must obtain their nutrients by the process of absorption. Small molecules (e.g., simple sugars and amino acids) in solution can be absorbed directly across the fungal wall and plasma membrane. Larger, more complex molecules (e.g., polymers such as polysaccharides and proteins) must first be broken down into smaller molecules, which can then be absorbed. This degradation takes place outside the fungal cell or hypha and is achieved by enzymes that are either released through or are bound to the fungal wall. Because these enzymes act outside the cell they are called extracellular enzymes. Water is essential for the diffusion of extracellular enzymes and nutrients across the fungal wall and plasma membrane, meaning that actively growing fungi are usually restricted to relatively moist (or humid) environments.

Most fungi (seed culture) are grown on media containing natural components (e.g., potato dextrose agar, malt extract agar, or commercial agar). Generally, fungal growth media tend to be rich in carbohydrates and the pH is maintained as slight acidic, in the range 5–6. Fungi can also be grown on minimal media (i.e., salts and glucose), depending up on their specific requirements, which may include one or more vitamins (e.g., thiamine, biotin), organic nitrogen (some fungi cannot use nitrate or ammonium as nitrogen source), and other needs (e.g., iron as heme, sulfur as cysteine).

The above requirements are the minimum, and do not necessarily represent the needs for growth. Some growth requirements change as a result of suboptimal conditions (e.g., aerobic versus anaerobic growth of *Saccharomyces*). The main nutritional requirements are described briefly below:

Carbon and energy sources: Fungi can use a variety of carbon sources to derive cellular energy. Use of these sources approximates a bell curve based upon the structural complexity and degree of utilization of the substrates. Simple molecules (e.g., glucose) diffuse through the membrane wall, and utilization of some carbon sources often depends upon membrane transporters. The complex molecules must first be broken into simpler components by secreted enzymes before transport into the cell. Fungi can also exhibit diauxic growth (i.e., predominant use of one carbon source before switching to a second source).

Nitrogen: Fungi do not fix atmospheric nitrogen. The main nitrogen sources include amino acid transamination and ammonia/ammonium used to produce glutamic acids or aspartic acid. However, ammonium is not a good media component because it lowers the pH, thereby inhibiting growth. Nitrate is converted to ammonia by nitrate reductase or nitrite reductase. Regulatory control of nitrogen uptake ensures that ammonium is used first, before nitrates or amino acids, by preventing the synthesis of nitrogen uptake proteins in the plasma membrane and the synthesis of enzymes used in nitrate utilization.

Phosphorus: Phosphorus is used in significant amounts in the cell, even though it is poorly available in the soil. By increasing the activity of the uptake system, freely available soluble phosphorus generated by phosphatase enzymes is taken up by the fungal hyphae cells. The phosphate is stored in vacuoles as polyphosphates.

Iron: Iron is an essential donor and acceptor of electrons. It normally occurs as insoluble ferric (Fe^{3+}) oxides in the environment and is needed in small amounts.

2.1.3.3 Nutritional Requirements of Algae

Algae media are the solutions or cultures in which algae grow, and there are two major types of algae media: enrichment media and artificial media. An enrichment medium is generally made by adding soil extracts to distilled or natural water, or by simply adding chemical nutrients to sea water or lake/dam water. The artificial medium uses "pure" water and "pure" chemicals and does not include additions of soil extracts or natural lake or sea water. This artificial medium is mostly used under laboratory conditions to exacting standards, although unknown impurities can still be present in even the most carefully prepared artificial medium. The most important aspects of algae growing conditions are nutrient quantity and quality, light, pH, turbulence, salinity, and temperature. Concentrations of algae in media are generally much higher than those found in nature, so the media or culture must be enriched with nutrients to boost the sea water to support these higher cell densities.

A wide range of references are available on algae culture [41–44]. The following media are commonly used in algae culture: BG (blue-green) medium [43], D medium [44], DYIY medium, medium f and fE, GP medium, GSe medium [47], JM (Jaworrski's medium), K medium, MB medium, MJ medium, MLA medium, mineral medium II, and porphyridium medium [44].

2.1.4 Carbon Sources Used in Scale-Up Processes

All fermentation processes need a carbon source but, as mentioned previously, industrial processes require that the source be cheap and easily available all year round. The following list gives brief details of some of the commonly used carbon sources for scale-up processes:

Molasses: Molasses is a byproduct of cane or beet sugar production or the residue remaining after most of the sucrose has been crystallized from the plant extract. It is a dark-colored viscous syrup containing 50–60% (w/v) carbohydrate,

Table 2.14 Generalised set of conditions for culturing algae

Parameters	Range	Optima
Temperature (°C)	16–27	18–24
Salinity (g/L)	12–40	20–24
Light intensity (lux)	1000–10,000 (depends on volume and density)	2500–5000[a]
Photoperiod (light: dark, hours)		16:8 (minimum)
		24.0 (maximum)
pH	7–9	8.2–8.7

[a]Depends on volume and density

primarily sucrose, with 2% (w/v) nitrogenous substances and some vitamins and minerals. Overall composition varies depending on the plant sucrose, location of the crop, climate conditions under which it was grown, and the factory where it was processed. The carbohydrate concentration may be reduced during storage by contaminating microorganisms. Hydrol molasses contain primarily glucose and is a byproduct of maize starch processing.

Malt extract: The concentrated aqueous extract of malted barley forms a syrup that is specifically used for cultivation of filamentous fungi, yeasts, and actinomycetes. The carbohydrate content is about 90% (w/v), with some vitamins and about 5% nitrogenous substances, proteins, peptides, and amino acids. The carbohydrates are present in the form of hexoses (20%), disaccharides (55%), maltotriose (10%), and branched and unbranched dextrins (15–20%). Careful sterilization is needed to prevent overheating, because a brown product can be formed by the Maillard reaction, resulting from the reaction of amino groups and carbonyl groups at low pH. The typical composition of malt extract is given in Table 2.14.

Starch and dextrins: Starch and dextrins are used as cheap carbon sources for filamentous fungal culture, because the amylase produced from these fungi metabolizes starch and dextrin. Maize starch is widely used for this purpose. Generally, the maize starch is converted into sugar syrup, containing mostly glucose. It is first gelatinized and then hydrolyzed by dilute acids or amylolytic enzymes, often microbial glucoamylases that operate at elevated temperatures [45, 46].

Sulfite waste liquor: The sugar-containing wastes derived from the paper pulping industry are primarily used for the cultivation of yeasts. Generally, coniferous and deciduous tree woods are processed to obtaining pulp for paper manufacture. Waste from coniferous trees contains 2–3% (w/v) sugar, 80% hexoses (glucose, mannose, and galactose), and 20% pentoses (mostly xylose and arabinose). Liquor derived from deciduous trees contains mainly pentose. Usually, the liquor requires processing before use because it contains SO_2. The low pH is adjusted with calcium hydroxide or calcium carbohydrate, and the liquors supplemented with sources of nitrogen and phosphorus.

Cellulose: At present, cellulose is mainly used in solid substrate fermentation (mushroom). It is available from agricultural, forestry, industrial, and domestic wastes. The composition consists mainly of cellulose, hemicelluloses, and lignin. Relatively few microorganisms can utilize it directly because the cellulose component is partly crystalline, encrusted with lignin, and provides little surface area for enzyme attachment. Potentially, it is very valuable renewable source of fermentable sugars once hydrolyzed, particularly in the bioconversion to ethanol for fuel use.

Whey: Whey is an aqueous byproduct of the dairy industry. Annual worldwide production is over 80 million tons, containing over 1 million tons of lactose and 0.2 million tons of milk proteins. Lactose is less useful than sucrose although it was formerly used extensively in penicillin fermentation. It has also been used for producing ethanol, single-cell protein, lactic acid, xanthan gum, vitamin B_{12}, and gibberlic acid. *S. cerevisiae* does not ferment lactose.

Alkanes and alcohols: n-Alkanes (C_{10}–C_{20}) are readily metabolized by certain microorganisms. Their industrial use is dependent upon the prevailing price of petroleum.

Methane: Methane is utilized by a few microorganisms but its conversion product, methanol, is often preferred for industrial fermentation. Methanol is completely soluble in wastewater. It has a high percentage carbon content and is relatively cheap. Microorganisms metabolize methanol when it is about 0.1–1% (v/v) in the culture medium; higher concentrations are toxic.

Ethanol: Ethanol is less toxic than methanol but its use is restricted because of its high cost. It is used in biotransformation to acetic acid, which remains a major fermentation process.

Fats and oils: Hard animal fats (composed mainly of glycerides of palmitic and steric acids) are rarely used in fermentation. Plant oils (primarily from cotton seed, linseed, maize, olive, palm, rape seed, and soy) and occasionally fish oil may be used as the primary or supplementary carbon source, especially in antibiotic production. Plant oils are mostly composed of oleic and linoleic acids, but linseed and soy oil also have a substantial amount of linolenic acids. Oils contain more energy per unit weight than carbohydrates and can be particularly useful in fed-batch operations because they occupy less volume than aqueous solutions of carbohydrates (usually less than 50% w/v).

2.1.5 *Medium Sterilization and Maintenance of Aseptic Environments*

2.1.5.1 Maintenance of Aseptic Environments

All culture vessels, media, and instruments used in handling microbes must be sterilized. It is important to keep the air, surface, and floor free of dust. All operations are carried out in laminar air flow, in a sterile cabinet. Infection can come from three main sources:

1. Air contains a large quantity of suspended microorganisms in the form of fungal and bacterial spores
2. Microbes are covered with pathogens on their surface
3. The human body (skin, breath, etc.) carries many microorganisms

In general, the methods of elimination of these sources of infection can be grouped under different categories of sterilization procedures:

1. Preparation of sterile media, culture vessels, and instruments (sterilization is carried out in an autoclave)
2. Preparation of sterile nutrients (by filter sterilization)
3. Aseptic working conditions

2.1.5.2 Sterilization of Media and Air

Water, containers, and the constituents of culture media all contribute to contamination by vegetative cells and spores. The media must be free from contamination before use in fermentation. Sterilization of media is most commonly achieved by applying heat and to a lesser extent by other means (physical methods, chemical treatment, and radiation), as described below.

1. **Sterilization of media**

   Heat is the most widely used sterilization technique. The quality and quantity of contamination (i.e., the type and load of microorganisms), composition of the media and its pH, and size of the suspended particles are the most important factors that influence the success of heat sterilization.

   In general, vegetative cells are destroyed at relatively low temperatures in a short time (around 60 °C for 5–10 min). However, destruction of spores requires a higher temperature and relatively longer time (around 80 °C for 15–20 min). Spores of *Bacillus stearothermophilus* are the most heat resistant. In fact, this organism is exploited for testing the sterility of fermentation equipment.

   Physical methods such as filtration, centrifugation, and adsorption (to - ion-exchangers or activated carbon) are also in use. Of these, filtration is most widely used. Some constituents (vitamins, blood components, antibiotics) of certain culture media are heat labile and destroyed by heat sterilization. In these cases, filter sterilization could be more appropriate. Such components of the medium must be completely dissolved before the medium is subjected to filter sterilization otherwise they would be removed along with the microorganisms. However, there are a two important limitations to use of the filtration technique: (1) Application of high pressure in filtration is unsuitable for industrial use and (2) some of the media components may be lost during filtration.

   Sometimes, a combination of filtration and heat sterilization is applied. For instance, the water used for media preparation is filtered while a concentrated nutrient solution is subjected to heat sterilization. The filtered water is then added for appropriate dilution of the media. Chemical methods (such as disinfectants) and radiation procedures (UV, γ-rays, X-rays) are not commonly used for media sterilization.

Heat sterilization at elevated temperature is mainly carried out by batch or continuous procedures.

Batch sterilization

Culture media are subjected to sterilization at 121 °C in batch volumes in the bioreactor. Batch sterilization can be carried out by injecting the steam into the medium (direct method) or injecting the steam into interior coils (indirect method). For direct batch sterilization, the steam should be pure and free from all chemical additives (which usually come from the steam manufacturing process).

Under thermal sterilization conditions, the death of microorganisms follows first-order kinetics. Considering N as the number of viable microorganisms present in the medium, with a death rate k, for a period of time t:

$$\frac{dN}{dt} = kN \tag{2.13}$$

Upon integration, Eq. (2.7) gives:

$$N = N_{oe^{-kt}} \tag{2.14}$$

The reaction death rate k is a function of the environmental conditions, the organism,and its history [47]. For practical reasons, it is useful to express [Eq. (2.14)] in a logarithmic form:

$$\ln \frac{N}{N_0} = -kN \tag{2.15}$$

where N_0 is the number of viable microorganisms at the beginning of the thermal treatment and N is the final number of viable microorganisms after a period of thermal treatment t. Inverting Eq. (2.15) one obtains an expression that can be used as sterilization criterion, “Del factor,” or design criterion (V):

$$V = \ln \frac{N_0}{N} = \mathrm{kt} \tag{2.16}$$

In batch sterilization, the temperature profile describes three phases, heating up, holding, and cooling down, each having a killing effect on the microorganisms. Thus, the final design criterion results from the combined activity of these phases:

$$V_{\mathrm{Total}} = V_{\mathrm{heating}} + V_{\mathrm{holding}} + V_{\mathrm{cooling}} \tag{2.17}$$

Because many culture media contain heat-labile components, it is important to consider a "quality criterion" (^) [35], assuming that the degradation rate of such a component is first order:

$$\wedge = \frac{C_0}{C} = K2t \qquad (2.18)$$

where C_0 and C are the initial and final concentrations of the heat-labile critical component, respectively. This subject is taken into account in high-temperature, short-term, or continuous sterilization procedures.

There are two disadvantages of batch sterilization:

- Damage to culture media: Alteration in nutrients, change in pH, and discoloration of the culture media are common.
- High energy consumption: It takes a few hours (2–4 h.) for the entire contents of the bioreactor to attain the requisite temperature (i.e., 120 °C). Another 20–60 min are needed for the actual process of sterilization, followed by cooling for 1–2 h. All this involves wastage of energy and, therefore, batch sterilization is quite costly.

Continuous sterilization

Continuous sterilization is carried out at 140 °C for a very short period of time, ranging from 30 to 120 s. This is in contrast to batch fermentation, which is carried out at 121 °C for 20–60 min. The process is based on the principle that the time required for killing microorganisms is much shorter at higher temperature. Continuous sterilization is carried out by directly injecting steam or by means of heat exchangers. In either case, the temperature is very quickly raised to 140 °C, and maintained for 30–120 s. The stages of continuous sterilization and the corresponding temperatures are depicted in Fig. 2.15. The different stages are exchanger, heater, heat maintenance unit, recovery of residual heat, cooling, and fermenter.

In the continuous sterilization process, three types of heat exchangers are used. The first heat exchanger raises the temperature to 90–120 °C within 20–30 s. The second exchanger further raises the temperature to 140 °C and maintains it for 30–120 s. The third heat exchanger brings down the temperature by cooling for the next 20–30 s. The actual time required for sterilization depends on the size of the suspended particles; the bigger the size, the more is the time required.

The main advantage of continuous sterilization is that about 80–90% of the energy is conserved. The disadvantage is that certain compounds in the medium precipitate (e.g., calcium phosphate, calcium oxalate) as a result of the very high temperature differences that occur in the very short time between sterilization and cooling. Starch-containing culture media become viscous in continuous sterilization and are therefore not used.

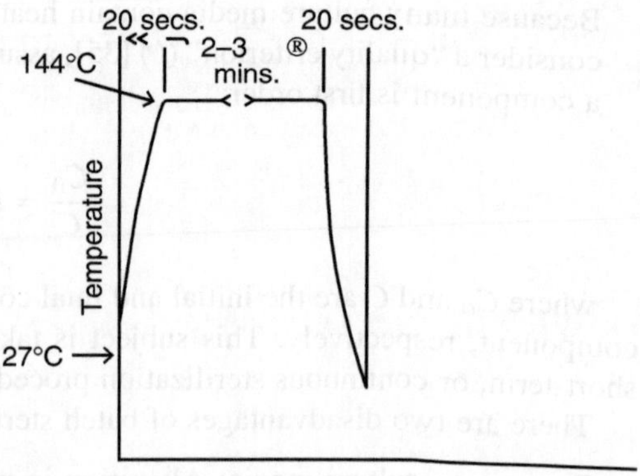

Fig. 2.15 Temperature profile in continuous sterilization

Steam or wet sterilization (autoclaving)

Autoclaving relies on the sterilization effect of superheated steam under pressure, as in a domestic pressure cooker. The size of the equipment used can be as small as a liter or as large as several thousand liters. Most instruments and nutrient media are sterilized with the use of an autoclave. Autoclaves have a temperature range of 115–135 °C. The standard conditions for autoclaving are a temperature of 121 °C and pressure of 15 psi (pounds per square inch) for 15 min to achieve sterility. These figures are based on the conditions necessary to kill thermophilic microorganisms. The time taken for liquids to reach this temperature depends on their volume. It may also depend on the thickness of the vessel. The temperature of 121 °C can only be achieved at 15 psi. The efficiency of the autoclave can be checked in several ways, the most efficient being to use autoclave tape. When the autoclave tape is autoclaved, a reaction causes dark diagonal strips to appear on the tape, indicating that it has been autoclaved successfully.

Precautions:

- Excessive autoclaving should be avoided because it degrades some medium components. Sucrose and agar, in particular, break down under prolonged heating, especially when under pressure and in an acidic environment. A few extremely thermoduraic microorganisms can survive elevated temperatures for some time, but 15–30 min kills even those.
- The level of water at the bottom of the autoclave should be checked.
- The lid of the autoclave should be properly closed.
- The air exhaust should be functioning normally.
- Reduction of pressure after the required time of autoclaving should not be too fast. If the temperature is not reduced slowly, the media begins to boil again. Also, medium in the containers might burst out from their closures because of the fast and forced release of pressure.

- Bottles, when being autoclaved, should not be tightly screwed and their tops should be loose. After autoclaving, these bottles are kept in laminar air flow and the tops tightened after cooling.

Filter sterilization

Some growth regulators (e.g., amino acids and vitamins) are heat labile and are destroyed during autoclaving of the nutrient medium. Therefore, an alternative is sterilized by filtration through a sieve or a filtration assembly using filter membranes of 0.22–0.45 μm pore size.

Irradiation

Irradiation can only be carried out under conditions where UV radiation is available. Consequently, its use is generally restricted to purchased consumables such as petri dishes and pipettes. UV light can be used to kill organisms in rooms or work benches areas where manipulation of cultures is carried out. It is, however, dangerous and should not be turned on while any other work is in progress. UV light of some wavelengths can damage eyes and skin.

2. **Sterilization of air**

In general, industrial fermentations are carried out under vigorous and continuous aeration. For effective fermentation, the air should be completely sterile and free from all microorganisms and suspended particles. There is a wide variation in the quantity of suspended particles and microbes in outdoor air, although fungal spores (50%) and Gram-negative bacteria (40%) dominate. Air and other gases can be sterilized by filtration, heat, UV radiation, and gas scrubbing. Of these, heat and filtration are most commonly used.

Air sterilization by heat

In the past, air was passed over electrically heated elements and sterilized. However, this is quite expensive and, hence, not in use at present.

Air sterilization by filtration

Filtration of air is the most commonly used sterilization procedure in fermentation industries. When air is passed through glass wool containing depth filters, particles are trapped and removed (Fig. 2.16a). This filtration technique primarily involves physical effects such as inertia, blocking, gravity, electrostatic attraction, and diffusion. Glass wool filters can be subjected to steam sterilization and reused. However, there is a limit to their reuse because glass wool shrinks and solidifies during steam sterilization.

The evolving maturity of nanofiber technology has enabled its use in gradient filtration media for air filtration applications, whereby particles are loaded within the depth of the filter structure (Fig. 2.16b). Nanofiber layers enable key performance features such as high porosity, small fiber diameter, small pore size, and high specific surface area. These features enhance the overall performance of gradient filtration composite media, resulting in high mechanical filter efficiency, high dust loading capacity, and low pressure drop. These performance advantages lower the kilowatt-hour usage profile of filters in application, leading to lower system operating costs

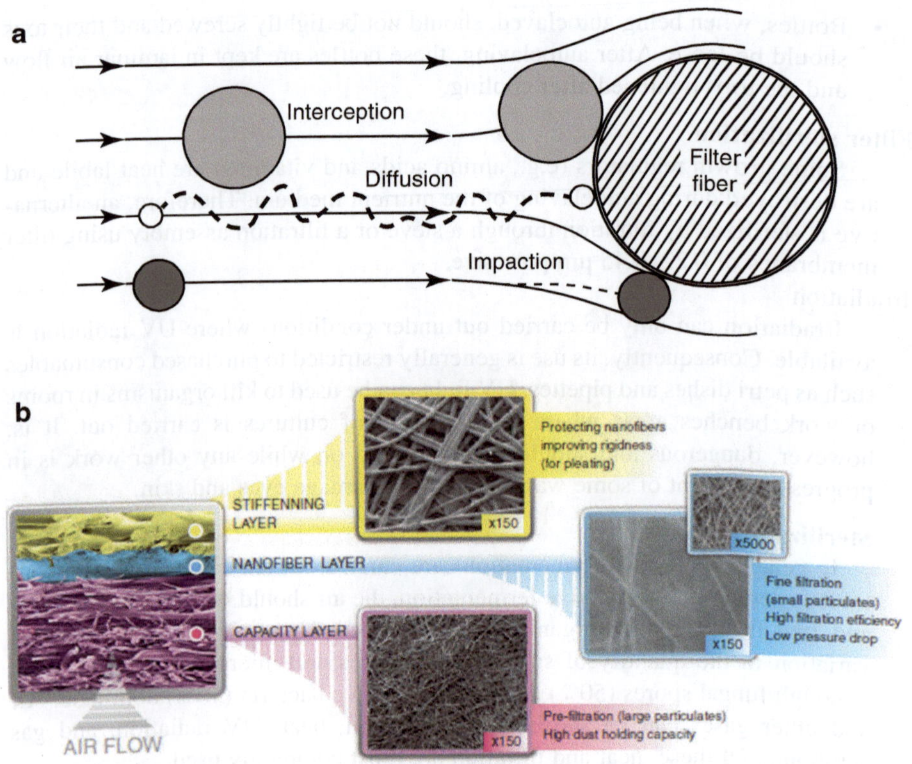

Fig. 2.16 (**a**) Capture mechanisms for a depth filter (adapted from https://www.google.co.in/imgres) (**b**) Nanospider™ electrospun nanofibers in multilayer composite media (source: https://www.google.co.in/imgres)

Membrane cartridge filters are also used to sterilize air before its entry into side culture vessels or fermenters. These filters are removable pleated membrane filters made up of cellulose ester, nylon, or polysulfone. Membrane cartridge filters are smaller in size and simpler for operation and replacement. The most important limitation of air sterilization is that there is no filter that can remove bacteriophages. Bacteriophages are capable of crippling industrial fermentation (e.g., bacteriophages interfere with the production of glutamic acid by *Corynebacterium glutamicum*).

Laminar airflow cabinet

This is the primary equipment used for aseptic manipulation (Fig. 2.17). This cabinet should be used for horizontal airflow from the back to the front, and equipped with gas corks in the presence of gas burners. Air is drawn in by electric fans and passed through a coarse filter and then through a fine bacterial filter (HEPA). The high efficiency particulate air (HEPA) filter is an apparatus designed so that the airflow through the work place flows in

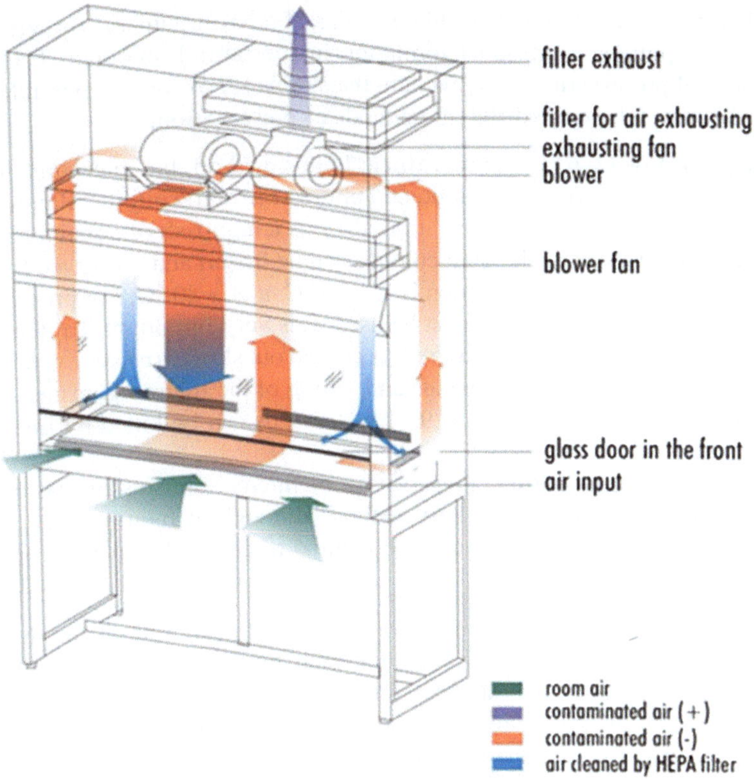

Fig. 2.17 Diagram showing two types of laminar flow showing air circulation

direct lines (i.e., laminar flow). Care is taken not to disturb this flow too much by vigorous movements. Before commencing any experiment, it is desirable to clean the work surface with 70% alcohol. The air filters should be cleaned and changed periodically.

2.1.6 *Stages in the Fermentation Process*

Fermenters or bioreactors are the central part of the biotech process because they provide the link between the raw materials and final products, by conducting the fermentation [48]. The fermentation process basically consists of inoculum preservation, inoculum build-up, pre-fermenter culture, and production fermentation. A brief account of the four stages of fermentation is given below:

1. **Inoculum preservation (culture maintenance)**

   The preservation of high-yielding strains of microorganisms for fermentation is very important for product formation in substantial amounts. The ultimate purpose of preservation is to maintain the strains for as long as possible without cell division. There are different methods of preservation:

   Storage at low (2–6 °C) temperature: In this method, the microorganisms can be stored in a refrigerator in liquid culture or as stab culture. Although this is the easiest method of preservation, there is a high risk of contamination.

   Storage by freezing: Microbial cultures can be frozen and preserved for several years. In freezers, preservation is at −18 °C or at −80 °C. For preservation at −196 °C, liquid nitrogen must be used. It is very important that the freezing (and later thawing when required) is carried out slowly (usually with a change of 1 °C/min) to prevent damage and killing of the microorganisms. If proper care is not taken, as many as 95% of the cells can be killed by freezing and thawing.

   Storage by lyophilization: Preservation of microorganisms by lyophilization (i.e., freeze-drying) is the best method, although, it requires special equipment. In fact, lyophilization is the method of choice for many fermentation biotechnologists.

   The storage of microorganisms can be done by any one of the three techniques described above. However, for each method, optimal conditions for preservation must be worked out for each strain separately. In general, the preserved master strains are cultivated once in two years to check their activity. When needed for use, the working strains can be obtained from the master strains.

2. **Inoculum build-up**

   The preserved cultures have to be revived for industrial use. This can be done by growing the cultures in liquid or on solid media. The actual process and the conditions used for inoculum build-up largely depend on the preservation technique used. There are wide variations in the growth times, which depend on the type of preservation and the organisms used. Some typical times are given below:

   Refrigerated cultures (2–6 °C)

   Bacteria 6–24 h
   Actinomycetes 1–3 days
   Fungi 1–5 days

   Frozen cultures (18 °C, −80 °C, −196 °C)

   Bacteria 6–48 hours
   Actinomycetes 1–5 days
   Fungi 1–7 days

Lyophilized cultures

For all organisms 4–10 days

For proper growth, and to obtain sufficient quantity of inoculum, a series of cultures are prepared. For good fermentation yield, the number of cells and spores, the nutrient medium, temperature, and age of the inoculum are all important. Inoculum build-up is suspended using a surface-active agent such as Tween 80 and then transferred to the bioreactor for fermentation.

3. **Production stage**

The size of the fermenter used mainly depends on the product. For example, a small bioreactor (1–20 L) can be used for producing diagnostic enzymes and substances for molecular biology with recombinant microorganisms, whereas large bioreactors ($\geq$450 L) are employed for producing single-cell protein and amino acids.

For appropriate production by fermentation, several parameters need to be carefully considered and optimized. These include the composition of nutrient medium, carbon and nitrogen sources, batch to batch variations, effect of sterilization on nutrients and on pH, and alterations in temperature and aeration. Temperature, pressure, aeration, and stirring are briefly discussed below:

Temperature: The temperature must be maintained such that there occurs maximal growth of microorganisms with optimal product formation, although this is not always possible. In general, there are two temperature ranges for the fermentations: a mesophile range (20–45 °C) and a thermophile range (>45 °C). Sometimes, two different temperatures are used for the same fermentation process; a higher temperature is employed for good growth (in trophophase) and then the temperature is decreased for optimizing product formation (in idiophase).

Pressure: Appropriate maintenance of hydrostatic pressure, particularly in large bioreactors is very important. This is because pressure influences the solubility of O_2 and CO_2 in the culture medium. An overpressure in the range of 0.2–0.5 bar is generally used.

Aeration: A bioreactor is aerated by the supply of O_2 and, therefore, adjustments must be made to supply the required amount of O_2 to the microorganisms. Usually, the aeration rate is in the range of 0.25–1.25 vvm (volume of air/volume of liquid/minute).

Stirring: The type and the speed of impellers determine the stirring rate in a fermenter. In general, the impeller speed decreases as the size of the fermenter increases. Thus, for a small bioreactor (1–20 L) the impeller speed is in the range of 250–350 rpm, whereas for a large bioreactor (around 450 L) the impeller speed is 60–120 rpm.

2.1.7 The Fermentation Process

Fermentation technology is a field that involves the use of microorganisms and enzymes for production of compounds that have application in the energy, material, pharmaceutical, chemical, and food industries (Fig. 2.18). It also involves propagation of the microorganism and production of the desired product. The fermentation process can be categorized according to various parameters. It can be either aerobic fermentation, carried out in the presence of oxygen, or anaerobic fermentation, carried out in the absence of oxygen. Many industrial fermentations are carried out under aerobic conditions although a few processes, such as ethanol production by yeast, required strictly anaerobic environments.

The fermentation process can also be divided into three basic systems, namely batch, continuous, or fed-batch, depending on the feeding strategy of the culture and medium into the fermenter (Fig. 2.19). Each of these processes has its own advantages and disadvantages. In batch cultivation, the bacteria are inoculated into the bioreactor (always a stirred tank bioreactor). Then, under certain conditions (temperature, pH, aeration, etc.) the bacteria go through all the growth phases (lag, exponential, stationary). Finally, the fermentation is stopped and the product is collected. Then, after cleaning and sterilizing, the fermenter is ready for another batch. In a continuous process, fresh medium is continuously added and the products, along with the culture, are removed at the same rate, thus maintaining a constant concentration of nutrients and cells throughout the process. A fed-batch system is a combination of these two systems, whereby additional nutrients are

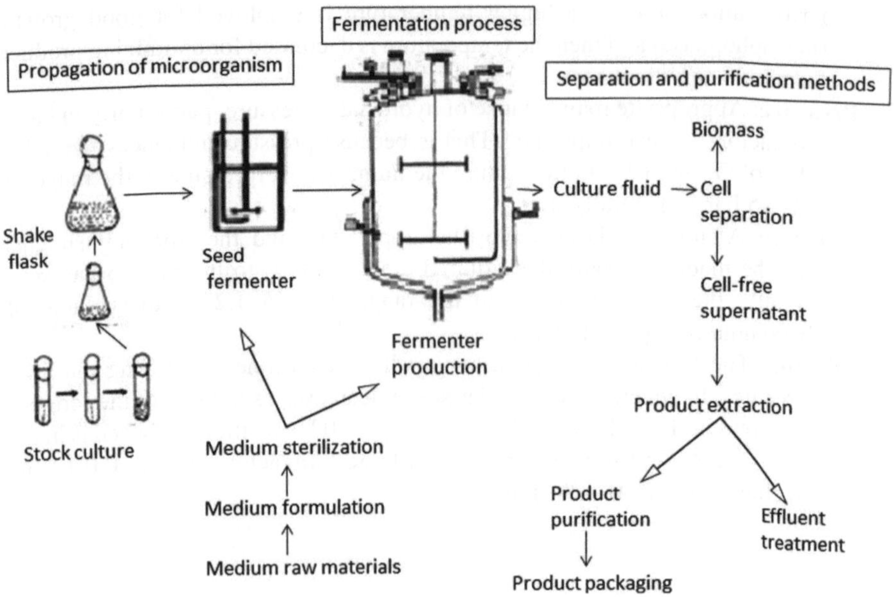

Fig. 2.18 Scheme of the fermentaion process

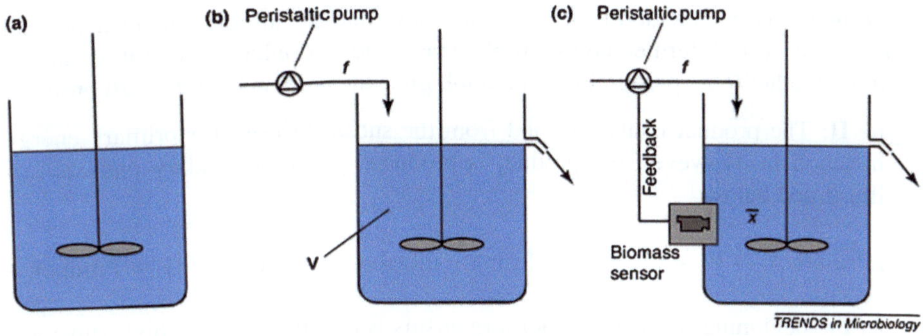

Fig. 2.19 (**a**) Batch culture: a non-steady-state culture with cells at different stages of their growth cycle. Growth rate changes with time in accordance with the classic growth curve, and available nutrient concentrations change with time. Metabolism operates in a non-steady state. (**b**) Chemostat culture: a growth rate-limiting substrate is added at a constant flow rate (f) in the inflowing medium. The volume (V) of the culture is kept constant by overflow of effluent through a side port. Metabolism operates in a steady state. (**c**) Turbidostat culture: a continuous flow culture in which biomass density is measured and controlled at a fixed value ($\bar{x}$) by automatic adjustment of the inflow rate (f) of the complete medium. Metabolism operates in a pseudo-steady state (adapted from [52])

added to the fermenter during fermentation [49, 50]. This extends the time of operation, but the products are harvested at the end of the production cycle as for batch fermentation.

The process can also be categorized as solid-state fermentation (SSF) or submerged fermentation (SmF), depending on the amount of free water in the medium. In SSF, the medium contains no free-flowing water. The organisms are grown in a solid substrate, which is moistened. This is used in certain industrial process such as "koji" fermentation of soybeans, production of amylase and protease by *Aspergillus oryzae* on roasted soybeans and wheat, bioremediation, and detoxification of agro-industrial wastes. In SmF, microorganisms grow submerged in a liquid medium where free water is abundant. This is the method of choice for many industrial operations although SSF is also rapidly gaining interest.

Control of product quality throughout repeated fermentations depends on maintenance of genetic uniformity from the time of strain selection until the product is harvested. Genetic control, therefore, must be maintained for stock cultures and during inoculum preparation and growth.

There are three categories of fermentation, known as types I, II, and III.

Type I: When the product is formed directly from the primary metabolism used for energy production, it is referred to as type I and can be represented as:

$$\text{Substrate A} \rightarrow \text{Product}$$
$$\text{Substrate A} \rightarrow \text{B} \rightarrow \text{C} \rightarrow \text{D} \rightarrow \text{Product}$$

Growth, energy metabolism, and product formation almost run in a parallel manner. In type I fermentation, trophophase and iodophase are not separated from each other (e.g., production of ethanol, gluconic acid, and single-cell protein).

Type II: The product is also formed from the substrate used for primary energy metabolism. However, the product is produced in the secondary pathway, as illustrated below:

Substrate A → B → C → D. . . .Primary metabolism → E → F G → Product

At the beginning, growth of microorganisms is accompanied by high substrate utilization with little or no product formation. Then, growth slows down but the substrate consumption is high, and this is coupled with product formation. In this process, the trophophase and idiophase are separate. Production of some amino acids, citric acid, and itaconic acid are good examples of type II fermentation.

Type III: There is a clear distinction between the primary metabolism and product formation in type III fermentation because they occur at separate times. Substrate consumption and rapid growth occur in the first phase and product formation occurs in the second phase. The product is formed from amphibolic metabolic pathways and not from primary metabolism (e.g., production of vitamins and antibiotics).

Overlap of Different Types of Fermentations

The categorization of fermentations into types I, II and III, as originally outlined [51], is not very rigid. There are intermediate forms based on the composition of the nutrient culture medium, strain of the microorganism used, and product formation. For instance, industrial production of lactic acid falls between types I and II, whereas production of the antibiotic amyloglycoside is intermediate between types II and III. It is sometimes difficult to categorize an industrial fermentation as one of these types because of the complex nature of the process (e.g., mycelium-producing microorganisms in relation to antibiotic production).

2.1.7.1 Description of a Typical Fermenter

A fermenter is a specially designed vessel built to support the growth of a high concentration of microorganisms. It must be designed so that it is able to provide the optimum environment or conditions to support growth of the microorganisms. Fermenters are commonly cylindrical vessels with hemispherical top and/or bottom, ranging in size from a few liters to cubic meters, and are often made of stainless steel and glass (Fig. 2.20a, b). Some important features are described below.

a

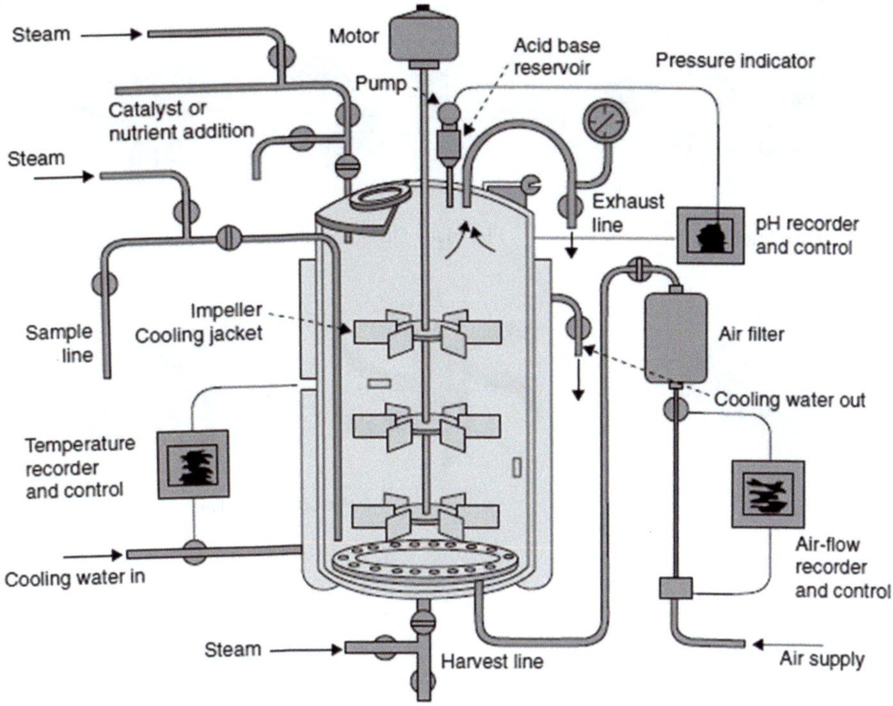

b

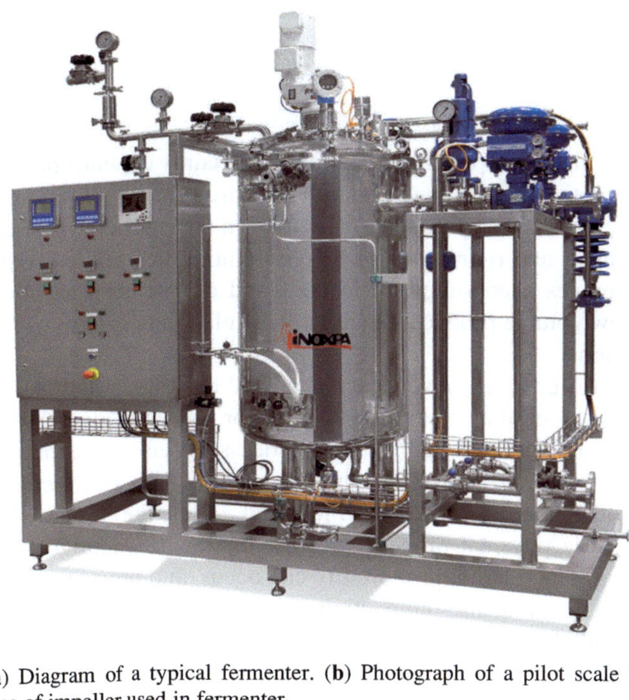

Fig. 2.20 (**a**) Diagram of a typical fermenter. (**b**) Photograph of a pilot scale fermenter. (**c**) Different types of impeller used in fermenter

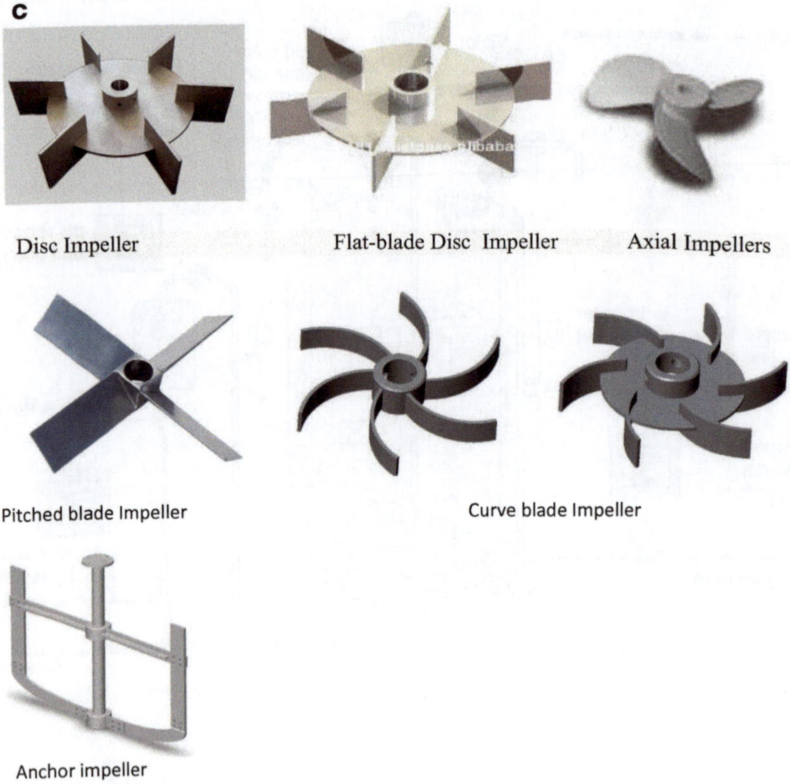

Fig. 2.20 (continued)

Design

The design and mode of operation of a fermenter mainly depends on the production organism, the optimal operating conditions required for target product formation, product value, and scale of production. The design also takes into consideration the capital investment and running costs. Large volume, low value products such as alcoholic beverages need simple fermenters and do not need aseptic conditions. High value, low volume products require a more elaborate system of operation and aseptic conditions.

Bioreactors differ from conventional chemical reactors in that they support and control biological entities. As such, bioreactor systems must be designed to provide a higher degree of control over process upsets and contaminations, as the organisms are more sensitive and less stable than chemicals. Biological organisms will naturally mutate, which may alter the biochemistry of the bioreaction or the physical properties of the organism. Analogous to heterogeneous catalysis, deactivation or mortality can occur, and promoters or coenzymes influence the kinetics of the bioreaction. Although most fundamental bioreactor engineering and design issues are similar, maintaining the desired biological activity and eliminating or minimizing undesired activities often presents a greater challenge than typically found for traditional chemical reactors. Other key differences between chemical reactors and bioreactors are the selectivity and rate. Selectivity is the measure of a system's ability to produce the preferred product (over other outcomes). In bioreactors, high selectivity is of primary importance. In fact, selectivity is especially important in the production of relatively complex molecules such as antibiotics, steroids, vitamins, proteins, and certain sugars and organic acids. Frequently, the activity and desired selectivity occur in a substantially smaller range of conditions than are present in conventional chemical reactors. Construction materials differ according to whether the bioreactor is small scale, pilot scale or large scale. For small-scale processes, glass or stainless steel can be used for vessel construction. For pilot and large-scale processes, stainless steel (>4% chromium), mild steel (coated with glass or epoxy material), wood, plastic, or concrete are used as vessel construction material. Any vessel used should not have any corners, and a smooth surface is essential. The construction material must be nontoxic and corrosion-proof, therefore glass and stainless steel are the most commonly used.

Glass vessels (borosilicate glass) can be classified into two types:

- Type I has a round or flat bottom and a top plate. It can be sterilized by autoclaving and the largest diameter is 60 cm.
- Type II has a flat bottom and top and bottom stainless steel plate. This type is used for in situ sterilization processes and the largest diameter is 30 cm.

Stainless steel is used as vessel construction material with the following modifications:

- >4% chromium (at least 10–13%) may be added
- Film of thin hydrous oxide, which is non-porous, continuous, self-healing, and corrosion resistance
- Inclusion of nickel improves engineering
- Presence of molybdenum confers resistance to halogen salts, brine, and sea water
- Tungsten and silicone improve resistance
- The thickness of the vessel should increase with scale; side plates have lower thickness than top and bottom plates
- Top and bottom plates are hemispherical to withstand pressure

Sealing

Sealing between top plate and vessel is important for maintaining airtight, aseptic, and containment conditions. Three types of sealing have to be made between the two types of surface, that is, glass–glass, glass–metal, and metal–metal. The three types of sealing are gasket, lipseal, and O-ring. Sealing ensures tight joints in spite of expansion of vessel materials during fermentation. The materials used for sealing can be fabric-nitryl or butyl rubbers. The seals should be changed after a finite time.

There are two types of O-ring sealing, simple sealing and double sealing with steam between the two seals.

Baffle

Baffles are metal strips that prevent vortex formation around the walls of the vessel. These metal strips attached radially to the wall for every tenth of vessel diameter. Usually there are four baffles, but six to eight baffles are used when the vessel diameter is over 3 dm^3. There should be enough gap between wall and baffle so that scouring action around the vessel is facilitated. This movement minimizes microbial growth on baffles and fermentation walls. If needed, cooling coils can be attached to the baffles.

Aeration System (Sparger)

A sparger is a device for introducing air into a fermenter. Aeration provides sufficient oxygen for organisms in the fermenter. Smaller bubbles have more surface area than larger bubbles and facilitate oxygen transfer to a greater extent; therefore, fine bubble aerators must be used. Agitation is not required when aeration provides enough agitation, which is the case for airlift fermenters. This type of aeration is only possible for media of low viscosity and low total solids. For aeration to provide agitation, the vessel height/diameter ratio (aspect ratio) should be 5:1. Air supply to the sparger should be supplied through a filter.

The three main types of sparger are porous sparger, orifice sparger, and nozzle sparger:

- Porous spargers are made of sintered glass, ceramic, or metal. They are used only in laboratory-scale non-agitated vessels. The size of the bubbles formed is 10–100 times larger than the pore size. There is a pressure drop across the sparger and the holes tend to become blocked by growth, which is the disadvantage of porous spargers.
- Orifice spargers are used in small stirred fermenters. The sparger is a perforated pipe that is kept below the impeller in the form of crosses or rings. The size should be about 75% of the impeller diameter. Air holes are drilled into the undersurfaces of the tubes and should be at least 6 mm diameter. This type of sparger is used mostly with agitation. It is also used without agitation in some cases such as yeast manufacture, effluent treatment, and production of single-cell protein.
- Nozzle spargers are mostly used in large scale fermenters. The sparger is a single open/partially closed pipe positioned centrally below the impeller. When air is passed through this pipe there is low pressure loss and it does not become blocked.
- Combined sparger agitators supply air via a hollow agitator shaft. The air is emitted through holes in the disc or blades of the agitator.

Exit Gas Cooler
Similar to Liebig condensers, the exit gas cooler condenses moisture from the exhaust gas in the fermenter. This removes as much moisture as possible from the gas leaving the fermenter and prevents excess fluid loss.

Agitation
Agitation provides uniform suspension of cells in homogenous nutrient medium. This agitation provides bulk fluid and gas phase mixing, air dispersion, facilitates oxygen transfer and heat transfer, and maintains a uniform environment throughout the vessel. There are four classes of agitator, namely disc turbine, vaned disc, open turbine of variable pitch, and marine impeller (Fig. 2.20c).

Disc turbines prevent flooding by providing air bubbles. Flooding occurs when air bubbles are not properly dispersed in the air, resulting in formation of a pocket in one area. Flooding occurs only at 120 min/h of air discharge when a disc turbine is used. When an open turbine and propeller are used, the medium is flooded at 21 min/h of air discharge. A Rushton disc turbine with one-third of the fermenter diameter is optimum for some fermentation processes. Recently, new agitator designs have been introduced. Scaba is a new agitator that can handle high flow rates before flooding and has radial flow. However, this is not ideal for top to bottom mixing. The Prochem maxflow agitator has low power consumption with high hydrodynamic thrust. This design has increased downward pumping capacity of blades. In this design, the agitator/vessel diameter ratio is 0.4. Approximately 66% less power is required, even for viscous media, and oxygen transfer efficiency is improved. The Intermig agitator has two units; unlike the earlier design, the agitator/vessel diameter ratio is 0.6–0.7. For this agitator, a larger air sparger is used and top to bottom mixing is not efficient. New turbine designs with dual impellers have also been introduced, one for gas dispersal and the other for aiding circulation with multirod mixing.

Stirrer Glands and Bearings
The entry point of the stirrer into the fermenter can be from top to bottom or from the side. Mostly, entry from the bottom is used, which leaves more space for entry ports on top. There are four types of stirrer glands and bearings:

1. **Stuffing box**
   - Sealed by several layers of packing rings of asbestos or cotton yarn pressed against the shaft by a gland follower
   - At high speeds, the packing wears and pressure should be applied to ensure tightness
   - Difficult to sterilize
   - Satisfactory heat penetration
   - Sufficient for good industrial large-scale practice (GILSP) containment

2. **Mechanical seal**

   - Two parts, one of which is stationary in the bearing housing, whereas the other rotates on the shaft
   - The two parts are pressed together by springs or expanding bellows
   - Steam condensate is used to lubricate and cool seals
   - Safe for containment
   - Double mechanical seal for level 2
   - At levels 2 and 3, the condensate is piped to a kill tank
   - Disinfectants are flushed through the seal
   - Steam condensate outlet monitoring indicates any seal failure

3. **Magnetic drive (for some animal cell cultures)**

   - Shaft does not pierce the vessel
   - Two magnets, one driving, held in a bearing in housing on the outside of the head plate, and one driven, placed on one end of the impeller shaft held in a bearing in suitable housing
   - Ceramic magnets – magnetic power cross, 16 mm gap
   - Rotation of 300–2000 rpm is possible

4. **Simple bush seal**

   The disadvantages of double seals is that they are more difficult to assemble and it is difficult to detect failure of the seals, leading to contamination. Hence, a simple bush seal is preferred in some cases.

Valves

The four types of valves commonly used are addition valves, check valves, pressure control valves, and safety valves:

1. **Addition valves**

   There are four types of addition valve:

   a. Simple on and off
   b. For coarse control
   c. For accurate adjustment
   d. For safety valve-flow in one direction

   There are four different models of valves:

   a. Opening and closing, raising or lowering of blocking unit

      - Gate valve: a sliding disc moves in/out of flow path by a turn of the stem
      - Globe valve: horizontal disc/plug—raised/lowered
      - Piston valve: similar to globe valve except a piston controls flow

- Needle valve: similar to globe valve except the disc is replaced with a tapered plug/needle

b. Drilled sphere/plug turn closes/opens flow path

- Plug valve: parallel/tapered plug with an orifice
- Ball valve: similar to plug valve except a ball with an orifice replaces the plug

c. Disc rotating between bearings

- Butterfly valve: a disc rotates about a shaft and closes against the seal to stop flow

d. Rubber diaphragm/tube pinching

- Diaphragm valve: similar to pinch valve except it does not pinch, but pushes from one side against a diaphragm
- Pinch valve: flexible sleeve closed by a pair of pinch bars (rubber, neoprene, etc.)

Based on the four types of applications, the valves are chosen accordingly:

- On/Off application: globe or butterfly valve
- Crude flow control: gate valve
- Accurate control: needle valve
- Very sterile operation: pinch or diaphragm valve

2. **Check valves**

Check valves used to prevent accidental reversal flow of liquid or gas as a result of breakdown. There are three types, swing check, lift check, and combined stop and check.

3. **Pressure control valves**

These types of valves are used for two purposes:

- Pressure reduction
- Pressure retaining

4. **Safety valves**

There are two types of safety valve by which the increase in pressure is released:

- A spindle lifts from its seating against the pressure to release pressure
- Bursting or rupturing of discs to release pressure

In the case of releasing gas, the escaping gas must be treated before release.

Steam Traps

A steam trap is important for removing any steam condensate. There are two components, the valve and seat assembly and the open/close device. Operation of the components can be based on:

- Density of fluid: The float (ball or bucket) floats in water and sinks in steam. When it floats it closes the valve and when it sinks it opens the valve.
- Temperature of fluid: The water–alcohol mixture senses the change in temperature. This mixture expands in hot steam and closes the valve. When it contracts in cool water the valve opens.
- Kinetic effect of fluid in motion: If low density steam is flowing it has high velocity. Likewise, high density steam flows with low velocity. The conversion of pressure energy into kinetic energy controls the opening and closing of the valve.

2.1.7.2 Types of Fermenter

Stirred Tank Fermenter

Microbial fermentations received prominence during the 1940s, namely for the production of life-saving antibiotics. The stirred tank reactor is the choice for many (more than 70%) although it is not the best. Stirred tank reactors have the following functions: homogenization, suspension of solids, dispersion of gas–liquid mixtures, aeration of liquid, and heat exchange. Stirred tank reactors (Fig. 2.21) are provided with a baffle, and a rotating stirrer is attached either at the top or bottom of the bioreactor. Typical decision variables are the type, size, location, and number of impellers and sparger size and location. These determine the hydrodynamic pattern in the reactor, which in turn influences mixing times, mass and heat transfer coefficients, shear rates, etc. Conventional fermentation is carried out in a batch mode. Because stirred tank reactors are commonly used for batch processes with slight modifications, these reactors are simple in design and easy to operate. Many industrial bioprocesses are still carried out in batch reactors even though significant developments in reactor design have taken place in recent years. The industry still prefers stirred tanks because loss is minimal in cases of contamination or other substandard product formation. Batch stirred tanks generally suffer from low volumetric productivity. The downtimes are quite large and non-steady-state fermentation imposes stress on the microbial cultures as a result of nutritional limitations. The fed-batch mode adopted in recent years eliminates this limitation. Stirred tank reactors offer excellent mixing and reasonably good mass transfer rates. The cost of operation is relatively low and the reactors can be used with a variety of microbial species. Because stirred tank reactors are commonly used in the chemical industry, the mixing concepts are well developed. Stirred tank reactors with immobilized cells are not generally favored due to attrition problems; however, by separating the zone of mixing from the zone of cell culture the system can be successfully operated.

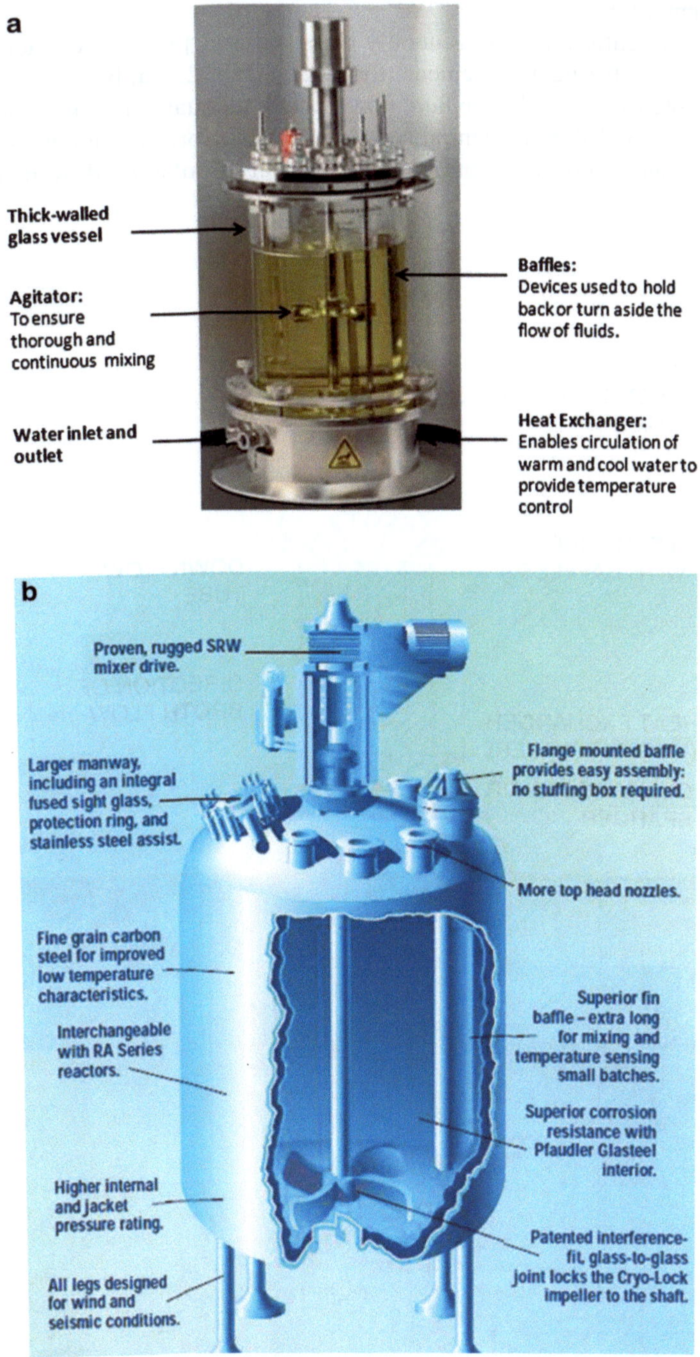

Fig. 2.21 (**a**) Photograph of a typical stirred tank fermenter. (**b**) CSTRs consist of a tank, usually of constant volume, and a stirring system to mix reactants together. Also, feed and exit pipes are present to introduce reactants and remove products. Pictured above is a CSTR that has had a portion of its side removed to show the interior

Airlift Fermenter

The airlift fermenter (ALF) is generally classified as a pneumatic reactor without any mechanical stirring arrangements for mixing (Fig. 2.22a, b).

The turbulence caused by the fluid flow ensures adequate mixing of the liquid. A draft tube is provided in the central section of the reactor. Introduction of the fluid (air/liquid) causes upward motion and results in circulatory flow in the entire

a

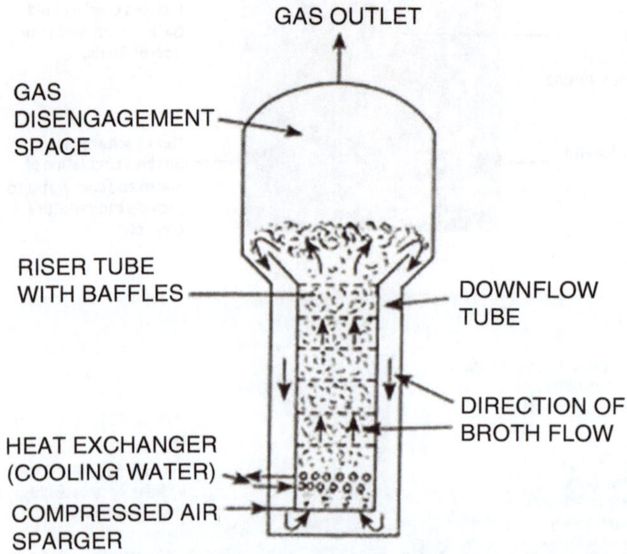

b

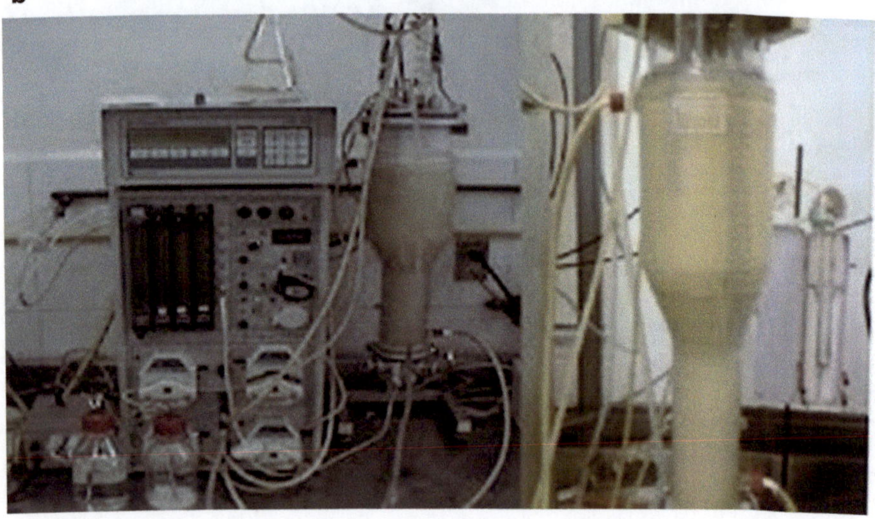

Fig. 2.22 (a) Typical airlift bioreactor. (b) Photograph of an airlift bioreactor

reactor. The air/liquid velocities are low and, hence, energy consumption is also low. ALFs can be used for both free and immobilized cells. There are very few reports on ALFs for metabolite production. The advantage of airlift reactors is elimination of the attrition effects generally encountered in mechanically agitated reactors. They are ideally suited for aerobic cultures because oxygen mass transfer coefficients are quite high in comparison with stirred tank reactors. This is ideal for production of single-cell protein from methanol as carbon substrate. ALFs are used mainly to avoid the excess heat produced during mechanical agitation.

Fluidized Bed Bioreactor

Fluidized bed bioreactors (FBB) (Fig. 2.23) have received increased attention in recent years because of their advantages over other types of reactors. Most of the FBBs developed for biological systems involving cells as biocatalysts are three phase systems (solid, liquid, and gas).

The fundamentals of three-phase fluidization phenomena have been comprehensively covered in the chemical engineering literature. FBBs are generally operated in co-current upflow, with liquid as continuous phase. Other more unusual configurations such as the inverse three-phase fluidized bed or gas–solid fluidized bed are not of much importance. Usually, fluidization is obtained either by external liquid re-circulation or by gas fed to the reactor. In the case of immobilized enzymes, the usual situation is a two-phase system involving solid and liquid, but the use of an aerobic biocatalyst necessitates introduction of gas (air) as the third phase. Differentiation between the three-phase fluidized bed and the airlift bioreactor can be made on the basis that the latter has a physical internal arrangement (draft tube) that provides aerating and non-aerating zones. The circulatory motion of the liquid is induced as a result of the draft tube. Basically, the particles used in FBBs can be of three different types: (1) inert core on which the biomass is created by cell attachment. (2) porous particles in which the biocatalyst is entrapped, and (3) cell

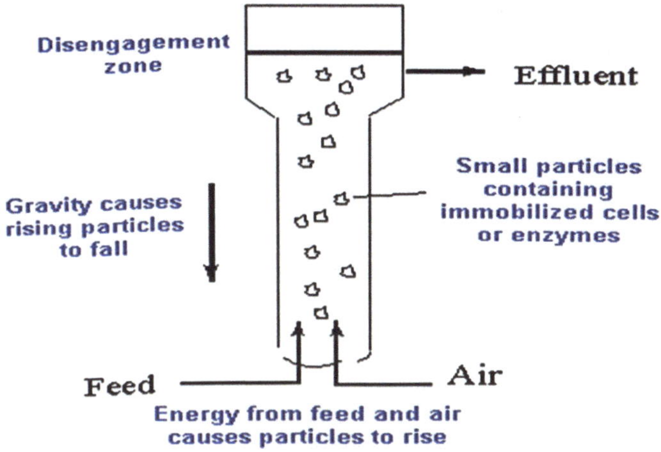

Fig. 2.23 Fluidized bed bioreactor

aggregates/flocs (self-immobilization). In comparison with conventional mechanically stirred reactors, FBBs provide a much lower attrition of solid particles. The biocatalyst concentration can be significantly higher and the washout limitations of free cell systems can be overcome. In comparison with packed bed reactors, FBBs can be operated with smaller sized particles without the drawbacks of clogging, large liquid pressure drop, channeling, or bed compaction. The smaller particle size facilitates higher mass transfer rates and better mixing. The volumetric productivity attained in FBBs is usually higher than in stirred tanks and packed bed bioreactors. There are several successful examples of FBB use in bioprocess development.

Packed Bed Bioreactor
Packed bed or fixed bed bioreactors (Fig. 2.24) are commonly used with attached biofilms, especially in wastewater engineering. The use of packed bed reactors gained importance after the potential of whole-cell immobilization techniques was demonstrated. Immobilized biocatalyst is packed in the column and fed with nutrients either from the top or bottom. One of the disadvantages of packed beds is the changing flow characteristic as a result of alterations in the bed porosity during operation. Working with soft gels such as alginates or carrageenan, bed compaction results in a large pressure drop across the bed. In many cases, bed compaction during fermentation is so severe that gel integrity is severely hampered. In addition, channeling can occur as a result of turbulence in the bed. Although packed beds belong to the class of plug flow reactors, in which back-mixing is absent in many of the packed beds, a slight amount of back-mixing does occur, which changes the characteristics of fermentation. Packed beds are generally used when substrate inhibition governs the rate of reaction. Packed bed reactors are widely used with immobilized cells. Several modifications such as tapered beds to reduce the pressure drop across the length of the reactor, inclined beds, horizontal beds, and rotary horizontal reactors have been tried with limited success.

Bubble Column Fermenter
The bubble column fermenter (BCF) (Fig. 2.25) is the simplest type of tower fermenter and consists of a tube that is air sparged at the base. It is an elongated

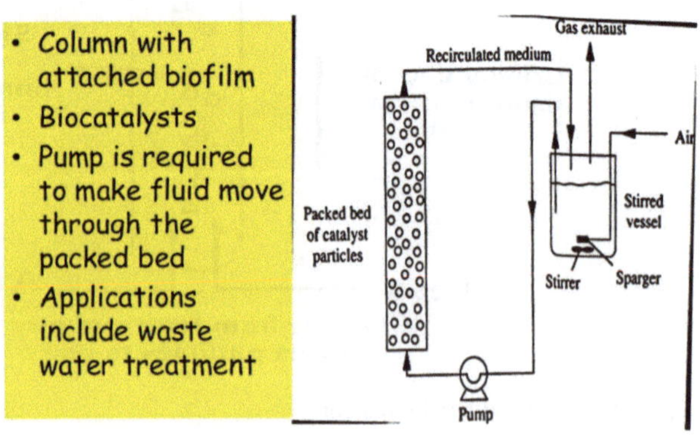

Fig. 2.24 Packed bed bioreactor

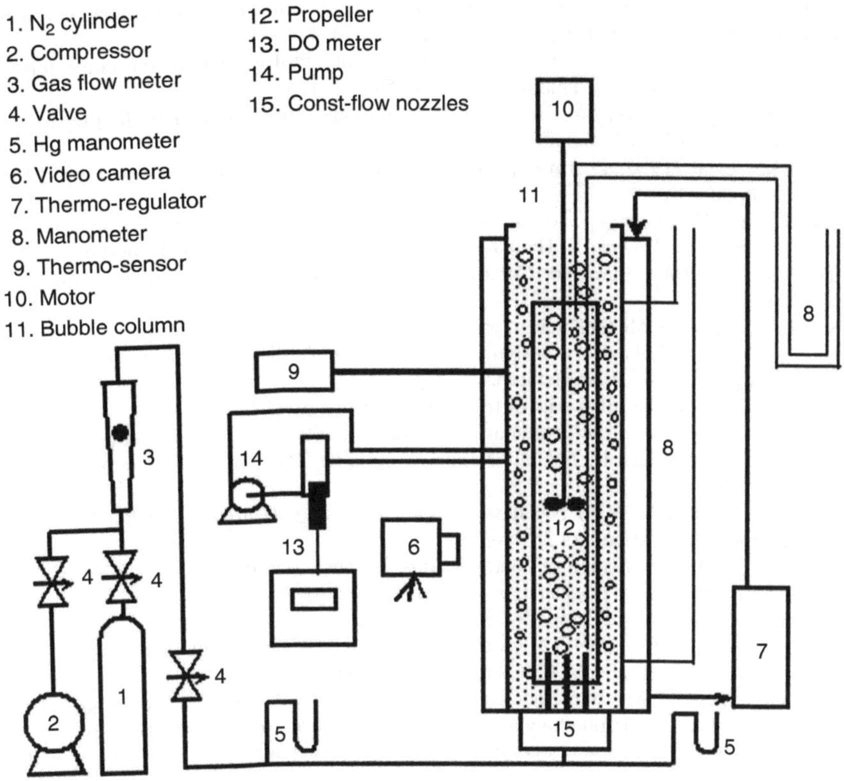

Fig. 2.25 Bubble column fermenter (BCF) with draft tube circulated forcibly liquid

non-mechanically stirred fermenter with an aspect ratio of 6:1. This type of fermenter is used for citric acid production.

References

1. Bruggink A (1996) Biocatalysis and process integration in the synthesis of semi-synthetic antibiotics: biotechnology for industrial production of fine chemicals. Chimia 50:431–432
2. Borowitzka MA (1999) Pharmaceuticals and agrochemicals from microalgae. In: Cohen Z (ed) Chemicals from microalgae. Taylor & Francis, London, pp 313–352
3. Brain JBW (1998) Microbiology of fermented foods, vol 1 and 2, 2nd edn. Blackie Academic and Professional, London
4. http://www.howstuffworks.com/dictionary/biology-terms/fermentation-info.htm
5. Sandle T (2010) Selection of microbiological culture media and testing regimes. In: Saghee MR, Sandle T, Tidswell EC (eds) Microbiology and sterility assurance in pharmaceuticals and medical devices. Business Horizons, New Delhi, pp 101–120
6. Bridson E, Brecker A (1970) Design and formulation of microbiological culture media. In: J.R. Noris, D.W. Ribbons (eds) Methods in microbiology, 3A. London, Academic

7. Cundell A (2002) Review of media selection and incubation conditions for the compendial sterility and microbial limits tests. Pharm Forum 28:2034–2041
8. Sutton SVW (2005) Activities of the USP analytical microbiology expert committee during the 2000-2005 revision cycle. J Pharm Sci Technol 59:157–176
9. Barry AL, Fay GD (1972) A review of some common sources of error in the preparation of agar media. Am J Med Technol 38:241–245
10. Baird RM et al (1986) Pharmacopoeia of culture media for food microbiology. Elsevier Science, London
11. Sandle T (2003) Selection and use of cleaning and disinfection agents in pharmaceutical manufacturing. In: Hodges N, Hanlon G (eds) Industrial pharmaceutical microbiology standards and controls. Euromed Communications, England. (chapter revised on several occasions)
12. Sandle T (2014) The Media Kitchen: preparation and testing of microbiological culture media. In: Sutton S (ed) Laboratory design: establishing the facility and management structure. Parenteral Drug Association, Bethesda, MD, pp 269–293
13. Booth C (2006) Media fills—trial or triumph. Lab News Aug:16–17
14. Evans GL et al (1996) Quality assurance for commercially prepared microbiological culture media: approved standards, 2nd edn. National Committee for Clinical Laboratory Standards, Villanova, PA
15. Nagel JG, Kunz LJ (1973) Needless retesting of quality-assured commercially prepared culture media. Appl Microbiol 26:31–37
16. Snell JJS (1995) Preservation of control strains. In: Snell JJS, Brown DFB, Roberts C (eds) Quality assurance: principles and practice in the microbiology laboratory. Public Health Laboratory Service, London, pp 69–76
17. Brown MRW, Gilbert P (1995) Microbiological quality assurance: a guide towards relevance and reproducibility of inocula. CRC Press, Boca Raton, FL
18. Hunt GR, Stieber RW (1986) Inoculum development. In: Demain AL, Solomon NA (eds) Manual of industrial microbiology and biotechnology. ASM, Washington, DC, pp 32–40
19. Standbury PF, Whitakar A (1984) Principles of fermentaion technology. Pergamon Press, Oxford
20. Chang LT, Elander RP (1986) Long-term preservation of industrially important microorganisms. In: Demain AL, Solomon NA (eds) Manual of industrial microbiology and biotechnology. ASM, Washington, DC, pp 49–55
21. Monaghan RL et al (1999) Culture preservation and inoculum development. In: AL Demain, NA Solomon (eds-in-chief) Manual of industrial microbiology and biotechnology, 2nd ed. ASM, Washington, DC, pp 29-48
22. Lincoln RE (1960) Control of stock culture preservation and inoculum build-up in bacterial fermentation. J Biochem Microbiol Technol Eng 2:481–500
23. Webb C, Kamat SP (1993) Improving fermentation consistency through better inoculum preparation. World J Microbiol Biotechnol 9:308–312
24. McDaniel LE, Bailey EG (1968) Liquid nitrogen preservation of standard inoculum: gas-phase storage. Appl Microbiol 16:912–916
25. Gapes JR et al (1983) A note on procedures for inoculum development for the production of solvents by a strain of Clostrodium butylicum. J Appl Bacteriol 55:363–365
26. Gutierrez NA, Maddox IS (1987) The effect of some culture maintenance and inoculum development techniques on solvent production by *Clostridium acetobutylicum*. Can J Microbiol 33:82–84
27. Cheng K-K, Zhang J-A, Liu D-H, Sun Y, Liu H-J, Yang M-D, Xu J-M (2007) Pilot-scaleproduction of 1,3-propanediol using *Klebsiella pneumonia*. Process Biochem 42:740–744
28. Papagianni M (2004) Fungal morphology and metabolite production in submerged mycelia processes. Biotechnol Adv 22:189–259
29. Möller J et al (1992) The influence of preculture on the process performance of penicillin V production in a 100-1 air-lift tower loop reactor. Appl Microbiol Biotechnol 37:157–163

30. Gutiérrez-Correa M, Villena GK (2003) Surface adhesion fermentation: a new fermentation category. Revista Peruana de Biología 10:113–124
31. Villena G et al (2001) Cellulase production by fungal biofilms on polyester cloth. Agro Food Ind Hi Tec 12:32–35
32. Tengerdy RP (1992) Solid state cultivation of lignocelluloses. In: Doelle HW, Mitchell DA, Rolz CE (eds). Elsevier Science Publisher, London, pp 269–282
33. Crueger W, Crueger A (1984) Biotechnology: a textbook of industrial microbiology. Sinauer Associates, Sunderland
34. Ferenci T (2008) Bacterial physiology, regulation and mutational adaptation in a chemostat environment. Adv Microb Physiol 53:169–229
35. Trilli A (1986) Scale-up of fermentations. In: Demain AL, Solomon NA (eds) Manual of industrial microbiology and biotechnology. ASM, Washington, DC, pp 277–307
36. Ollis DF, Chang H-T (1982) Batch fermentation kinetics with (unstable) recombinant cultures. Biotechnol Bioeng 24:2583–2586
37. Atkinson B, Mavituna F (1983) Biochemical engineering and biotechnology handbook. Macmillan, Surrey
38. Martens J-H et al (2002) Microbial production of vitamin B12. Appl Microbiol Biotechnol 58:275–285
39. Bailey JE, Ollis DF (1986) Biochemical engineering fundamentals, 2nd edn. MaGraw-Hill, New York
40. Greasham R, Inamine E (1986) Nutritional improvement of processes. In: Demain AL, Sollamon NA (eds) Manual of industrial microbiology and biotechnology. ASM, Washington, DC, pp 41–48
41. Nichols HW, Bold HC (1965) *Trichosarcina polymorpha* gen Et sp. Nov. J Phycol 1:34–38
42. Nichols HW (1973) Growth media—freshwater. In: Stein J (ed) Handbook of phycological methods culture methods and growth measurements. Cambridge University Press, Cambridge, pp 7–24
43. Stanier RY et al (1971) Purification and properties of unicellular bluegreen algae (order Chroococcales). Bacteriol Rev 35:171–205
44. http://www.marine.csiro.au/microalgae/methods/Media
45. Annan-Prah A et al (2010) Afr J Microbiol Res 4:2626–2628
46. Tharmila EC et al (2011) Thavaranjit. Arch Appl Sci Res 3:389–393
47. Bader FG (1986) Sterilization: prevention of contamination. In: Demain AL, Solomon NA (eds) Manual of industrial microbiology and biotechnology. ASM, Washington, DC, pp 345–362
48. Cooney CL (1983) Bioreactors: design and operation. Science 219:728–733
49. Van't Riet K, Tramper J (1991) Basic bioreactor design. Dekker, New York
50. Kapelli O (1986) Regulation of carbon metabolism in Saccharomyces cerevisiae and related yeasts. Adv Microb Physiol 28:181
51. Gaden EL Jr (1959) Fermentation process kinetics. J Biochem Microbiol Technol Eng 1:413–429
52. Catherine LW et al (2011) TARDIS-based microbial metabolomics: time and relative differences in systems. Trends Microbiol 19:315–322

Chapter 3
Downstream Processing

3.1 Introduction

Downstream processing (DSP) refers to the recovery and purification of biosynthetic products generated during the upstream stage into finished biotechnology products, including whole cells, organic acids, amino acids, solvents, antibiotics, industrial enzymes, therapeutic proteins, vaccines, and gums (Fig. 3.1).

The downstream stage also includes the sale of those products to other businesses, governments, or private individuals. The type of end user varies depending on the finished product. Regardless of the industry involved, the downstream process has direct contact with customers through the finished product. New methods for process development are being developed. These include the establishment of platform technologies and high-throughput methods optimized using experimental approaches based on quality by design and design of experiment [1–3]. Additionally, integration of modeling, simulation of unit operations, and the use of miniplant facilities can be applied in process development (Fig. 3.2).

The complexity of DSP is determined by the required purity of the product at minimum recovery costs. The major factors affecting DSP include microbial morphology, flocculation characteristics, size, and cell wall rigidity. These factors affect filterability, sedimentation, and homogenization efficiency. The presence of fermentation by-products, media impurities, and additives such as antifoaming agents can interfere with DSP steps. Therefore, a holistic approach is required when developing a new industrial purification strategy. For the whole process, both upstream and downstream factors need to be considered; for example, a cheap carbon and energy source containing many impurities could provide initial cost savings but necessitate increased DSP costs [4–7]. Hence, overall cost savings could be achieved with a more expensive but purer substrate. The overall scheme for upstream and downstream processes is given in Fig. 3.3. In this chapter we consider the following stages: (1) microbial biomass separation, (2) release of intracellular products, (3) concentration and purification, and (4) formulation.

B. Kumara Behera, A. Varma, *Microbial Biomass Process Technologies and Management*, DOI 10.1007/978-3-319-53913-3_3

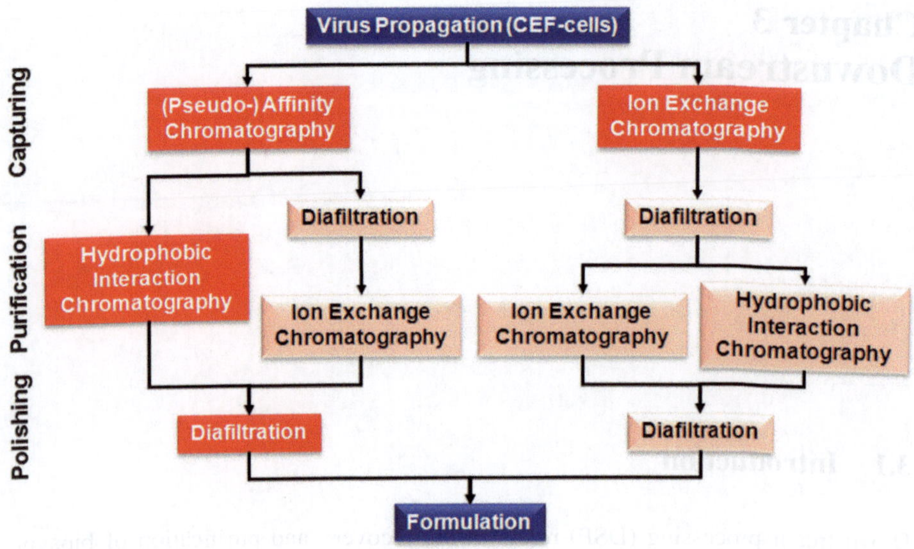

Fig. 3.1 Options for downstream processing in smallpox vaccine manufacturing

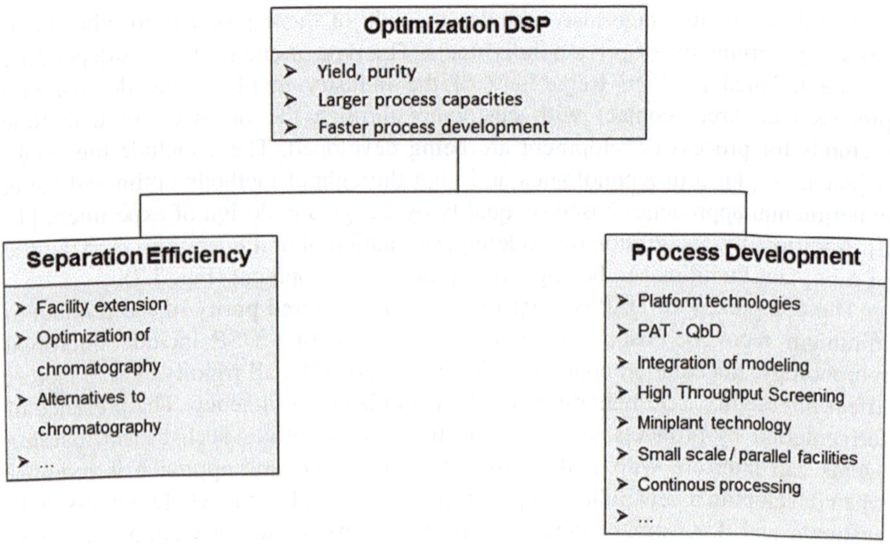

Fig. 3.2 Optimization fields in downstream processing

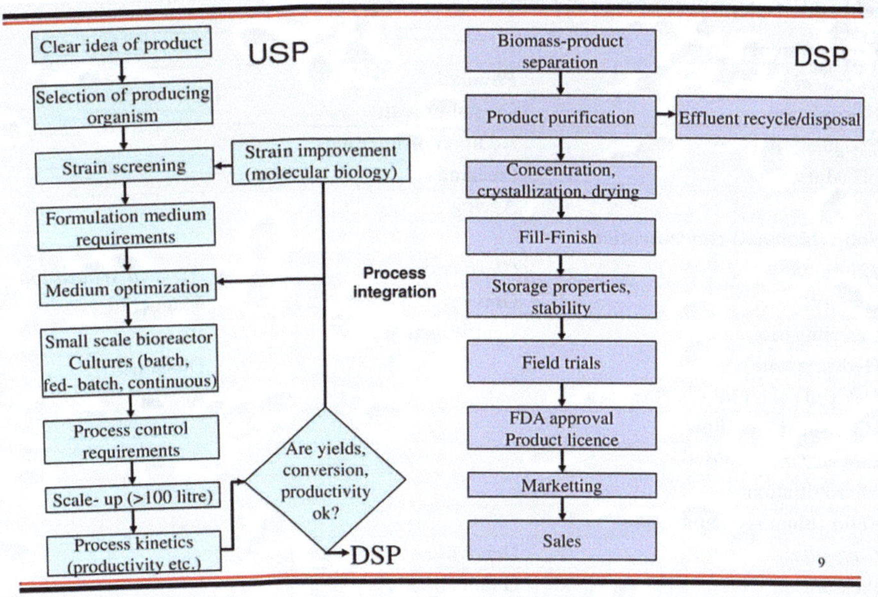

Fig. 3.3 The overall scheme for upstream (*USP*) and downstream (*DSP*) processes

3.2 Microbial Biomass Separation

Liquid–solid separation of fermented broth involves the separation of two phases, solid (microbial biomass) and liquid (extracellular product and left-over unused nutrients). This type of separation is used in many processes for (1) recovery of biomass carrying valuable target metabolite (solid component), with the liquid being discarded; (2) liquid recovery (the solids being discarded); (3) recovery of both solids and liquid; or (4) recovery of neither phase (e.g., when a liquid is being cleaned prior to discharge, as in the prevention of water pollution).

In downstream fermentation processes, the design of any separation system must consider all stages of pretreatment, biomass concentration, biomass separation, and post-treatment. This encompasses a wide range of equipment and processes, as summarized in Table 3.1. Pretreatment is used primarily with difficult-to-filter slurries, enabling them to be filtered more easily. It usually involves changing the nature of the suspended biomass by physical means or by adding a solid (filter aid) to the suspension to act as a bulking agent to increase the permeability of the cake formed during subsequent filtration. In the case of concentrated biomass, part of the liquid may be removed by thickening (using gravity or centrifugation) or by hydrocycloning to reduce liquid volume throughput load on the filter.

Table 3.1 Components of the solid–liquid separation process

Pretreatment		
Chemical	*Physical*	
Flocculation	Crystal growth	
Coagulation	Addition of filter aids	
pH adjustment	Freezing	
	Ageing	
Solid (biomass) concentration		
Thickening	*Clarification*	*Assisted separation*
Gravity	Gravity	Magnetic
Centrifugation	Centrifugation	Electric
Hydrocyclone		Dielectric
Delayed cake filters		Acoustic
High-shear crossflow		
Low-shear crossflow		
Microfiltration		
Solid (biomass) separation		
Cake filter	*Depth filter*	
Pressure	Granular beds	
Vacuum	Cartridges	
Centrifugal	Precoat	
Gravity	Membrane	
Post-treatment		
Filtrate polishing	*Cake treatment*	*Deliquoring*
Microfiltration	Washing	Gravity
Ultrafiltration	Displacement	Displacement
Ion exchange	Re-slurry	Mechanical expression
Reverse osmosis		Thermal drying
Electrodialysis		

3.2.1 Separation and Concentration Techniques

A number of new "assisted separation" techniques (Table 3.1) are making their way into the list of technical alternatives. These use magnetic, electric, or sonic force fields (or combinations) to provide more effective separation.

Separation of solids involves filters, which are classified in many different ways. For the present purposes, a division into those in which cakes are formed and those in which the particles are captured in the depth of the medium is adequate. Cake filtration can be further divided into pressure, vacuum, centrifugal, and gravity operations.

Post-treatment processes involve making improvements to the quality of the solid or liquid products. In the case of the filtrate, these operations are often referred to as polishing processes and may involve micro- or ultrafilters to remove fine substances. Further purification could involve removal of ionic and

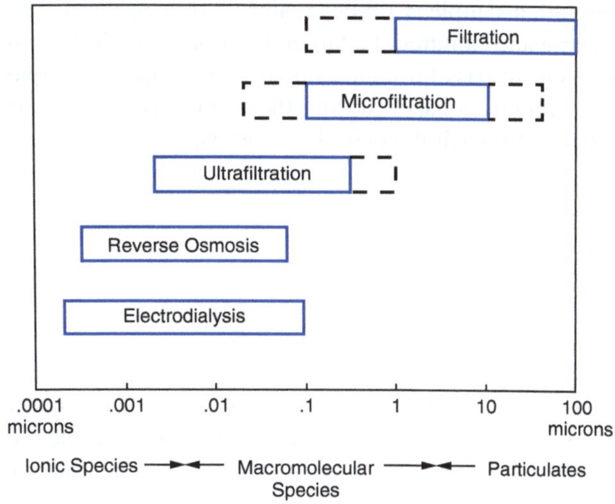

Fig. 3.4 General techniques for removal of contamination from liquids relative to the size of the species to be removed

macromolecular species by, for example, reverse osmosis, ion exchange, or electrodialysis. The relative position of these separation processes in the spectrum of the size of "particle" to be removed from the liquid is shown in Fig. 3.4. Cake post-treatment processes include washing soluble impurities from the cake voids and removing excess liquid from the voids. Thermal drying is often the final stage of liquid removal.

3.2.2 Principles of Separation Techniques

3.2.2.1 Single-Stage Process

Single-stage systems can come in many designs, including continuously stirred tank reactors and plug-flow digesters, each with different modes of operation and differences in design and operation. Generally, single-stage systems are simpler than two-stage systems, and are cheaper to construct and operate.

Potential limitations of single-stage systems are that conditions within the reactor are not optimal for the various trophic groups of microbes. Although two-stage systems can offer more optimal conditions for methanogenic bacterial populations, it does not mean that single-stage systems are unreliable. Methanogenic bacterial populations can be managed by controlling the feeding rate; thoroughly mixing incoming wastes to avoid peak concentrations of potentially harmful contaminants; co-digesting with other organic wastes to provide essential water content, buffering, nutrients, and trace elements; or adding these requirements through the use of chemicals and nutrients.

By means of mathematical models we can consider a single-stage concentration process as the operation of passing a capture element (in reality, the final volume) through a sample N times greater in volume, until it has passed through the whole sample and captured the target microbes from it. If it takes t seconds for the capture element to pass through its own volume in the sample (specific sweep time), and this remains constant during the capture pass, the time T required for the capture element to reach the end of its pass is:

$$T = t(N - 1)$$

3.2.2.2 Two-Stage Process

The sample can be concentrated in two stages, each of which is less efficient (yields a smaller increase in analyte concentration) than the above. Instead of just one, use n capture elements for the first stage and, after concentrating the sample into these, concentrate these n elements into one in the manner above. The total time T for concentration is:

$$T = (T_1 + T_2) = t(n + (N/n) - 2)$$

It is minimal when $n = N^{1/2}$, that is, when the overall concentration factor is achieved in two stages of approximately equal effectiveness.

3.2.2.3 Three or More Stages

A three-stage process is a two-stage process with a stage added at the front end. If t is the same for all three processes, the total time needed is:

$$T = t\left[2n + \left(N/n^2\right) - 3\right]$$

It is minimal when $n = N^{1/3}$; that is, three stages of approximately equal effectiveness yield the minimum concentration time. Similarly, p multiplexed processes execute in a minimum time when $n = N^{1/p}$.

The disadvantage of multistage systems is that they are often more complex, and therefore usually more expensive, than single-stage systems. The implementation of digestion systems with "support material" is, however, only advised for effluent-type feedstocks with low suspended solids; otherwise digesters are prone to clogging.

Some advantages of multistage systems are:

- Greater biological stability
- Greater ability to cope with fluctuating feedstock volume and quality
- Potentially higher throughputs due to optimal conditions

3.2.3 *Filtration*

Filtration is the most commonly used technique for separating biomass from culture filtrate [8–11]. The efficiency of filtration depends on many factors, such as the size of the organism, presence of other organisms, viscosity of the medium, and temperature. The filter medium constitutes the separating agent and retains particles according to size while allowing liquid to pass through the filter. In cake filtration, the particles are retained as a cake on the filter medium. The flow through the filter layers is dependent on the area of the filter and flow resistance provided by the filter medium and the cake. If the particles do not penetrate the filter medium, then its flow resistance remains unchanged. However, as the cake layer thickens it provides increasing resistance. The cake layers, especially in biomass separation, are compressible and the changing effective pressure difference influences flow through the filter. Filter media can be perforated sintered metal, cloth, synthetic fibers, cellulose, glass wool, ceramics, or synthetic membranes. Several types of filters, such as depth filters, absolute filters, rotary drum vacuum filters, and membrane filters, are commonly used and are described next.

3.2.3.1 Filter Press

A filter press is a tool used in separation processes, specifically in solid–liquid separation using the principle of pressure-driven membrane filtration. History recounts that the first form of a filter press was contained in a British patent from 1853 for obtaining seed oil through the use of numerous pressure cells. The main advantages of the filter press are high filtration velocities and low moisture content in the formed filter cakes, as a result of the high filtration pressures that can be applied. Consequently, filter presses have long been used for a wider range of applications than any other type of filter. By the middle of the twentieth century, the requirements of mass production in industry resulted in increasing attention being paid to the automation of filter press operations. A fully automatic horizontal filter press and various other new types were developed in Japan [12–22].

Filter presses are sometimes called “plate-and-frame filters.” This refers to the style of filter element that was most prominent in the mid-1800s until the late 1960s. Although this specialized style of the filter press is still being manufactured, the current manufactured units are more accurately referred to as “recessed chamber” and/or “diaphragm (membrane)” filter presses (Fig. 3.5).

Working Principle of a Membrane Filter Press

The basic principle of a filter press is surface filtration driven by pressure. A filter press is built of a sequence of perforated plates alternating with hollow frames. The plates are covered with a suitable filter medium (cloths) that creates a series of

Fig. 3.5 Photo of an industrial plate filter press. Courtesy of Evoqua Water Technologies

chambers through which the slurry can be forced. Solids are retained in the chambers and the filtrate discharges into the hollows on the plate surface and drains out.

The slurry deposits solids on the filter cloth as the filtrate leaves the press via discharge ports. The solids continue to deposit and eventually form a cake in the spaces between the filter plates. The pressure behind the slurry (typically 100 psi, but up to 900 psi; 7–60 bar) is provided by a feed pump, which is sometimes a positive displacement or centrifugal pump. With a gravity drain on the filtrate side of the press, a pressure differential between the feed pressure and the gravity discharge is created across the medium and the filter cake solids as they thicken (Fig. 3.6).

It is the existence of this pressure differential, and not just the feed pump pressure, that causes the filtering action to occur. Solids within the slurry flow to the area of cake development with the lowest pressure differential, resulting in a filter cake that builds uniformly over the drain field on either side of the chamber walls.

This fill cycle continues until the filter cakes forming on the chamber walls bridge at the center, completely filling the press with solids. It is at this point that the filtration process is complete. Once this is achieved, the hydraulic closure of the press is retracted, the individual filter elements are separated, and the filter cakes are discharged, usually by gravity, to an appropriate receptacle.

The filter cake chambers are formed in either of two ways: by mating two recessed chamber plates or by two flush plates with a cake frame (much like a picture frame) sandwiched in between. The two faces of the filter plate have a drainage surface in the form of ribs, grooves, or pips to allow filtrate to drain behind the cloth to the drainage ports located in each corner of the filter plates. These ports, in turn, connect to the corner eyes, which carry the filtrate drainage to the fixed end of the filter press. A filter cloth is mounted over each of the two faces of the filter plate. The cloth is joined at the feed eye by an impervious sleeve or tube, known as a barrel neck.

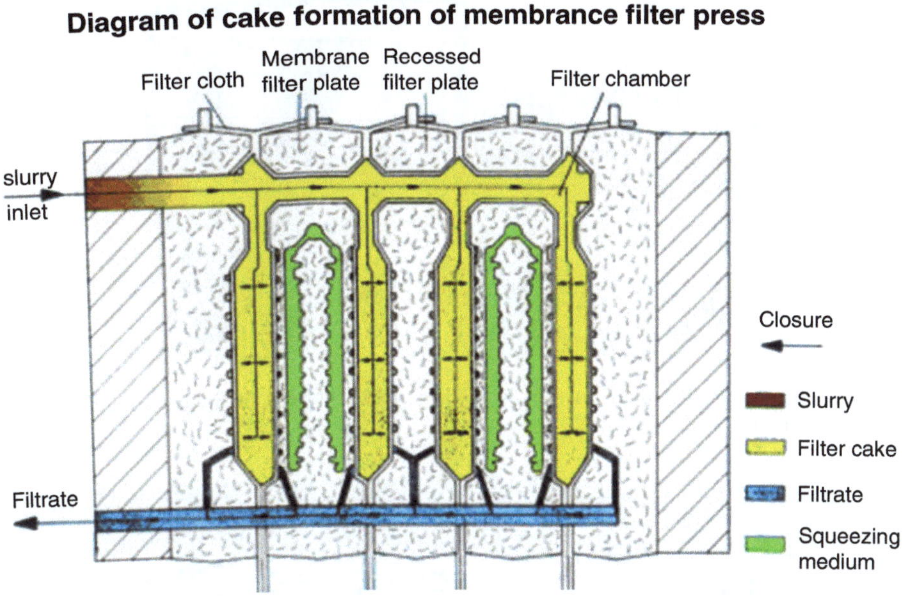

Fig. 3.6 Working principles of a plate filter press (courtesy of Uniwin; www.uniwinfilterpress.com)

Types of Filter Press

Filter presses are used across a wide variety of applications and industries, including wastewater, food and beverage, chemical, minerals, mining, and paper industries. Filter presses are predominately used in the process of solid–liquid separation. The three main types of filter press used in industry today are (1) the plate-and-frame filter press, (2) the recessed or chamber plate filter press, and (3) the membrane plate filter press. These presses and their major components are described below.

Plate-and-Frame Filter Press

This is the most common type of filter used for liquid–solid separation (Fig. 3.5) and mainly consists of a filter frame and filter plate (Fig. 3.7). The plates are metal frames containing large sheets of filter cloth, usually polypropylene. The plates are squeezed together and the slurry is filtered through the press. The slurry deposits solids on the filter cloth as the filtrate leaves the press via discharge ports. The solids continue to deposit and eventually form a cake in the spaces between the filter plates (Fig. 3.8). For maximum filtrate recovery, air can be blown at high pressure through the filter press to dry out the cake almost completely.

Orifices feed the sludge to be filtered under pressure into the filtration chamber. They are usually placed in the center of the plates, allowing proper distribution of flow, right pressure, and better drainage of sludge within the chamber. Solid sludge

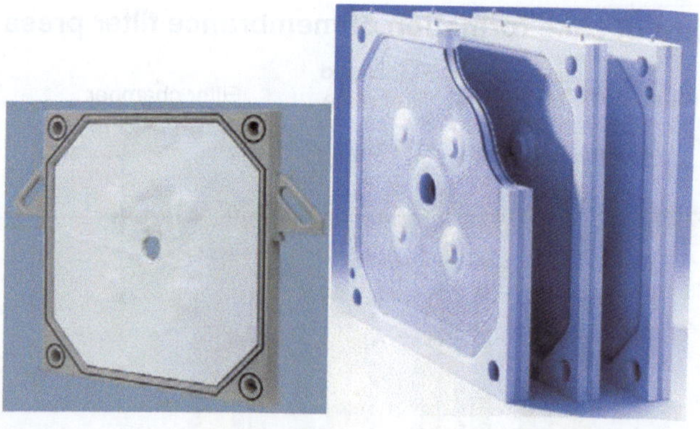

Fig. 3.7 Photo of a filter plate covered with filter cloth (courtesy of Siemens Industry, Inc., Warrendale, PA)

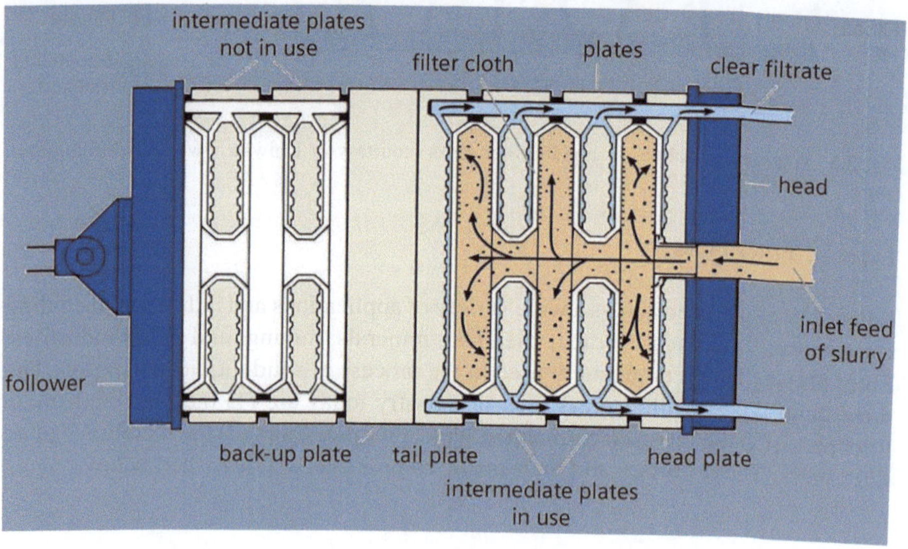

Fig. 3.8 Line diagram of a filter press in operating condition (courtesy of Water Technologies Business Unit of Siemens Industry, Inc., Warrendale, PA)

gradually accumulates in the filtration chamber until the final compacted cake is formed. The filtrate is collected at the back of the filtration support and carried away by internal ducts.

Recessed Plate and Frame Filter Presses

A recessed plate filter press can compact sludge to a dry cake with a density that is dependent on the ultimate compaction pressure and the characteristics of the particular sludge. The recessed plate filter press, or chamber plate filter press as it is also known, is similar in design to the plate-and-frame filter press but has no frame. Each plate serves as both plate and frame because the plates are recessed to form the filtrate chamber when the plates are clamped together. The slurry enters through a large central port, filter cake forms in the recessed hollow chamber, and the filtrate passes through the filter medium. As with most filter presses today, recessed plate filter press plates are constructed from polypropylene.

Membrane Plate Filter Press

The membrane filter press is a development of the standard recessed plate filter press. The filter construction is similar to that in a recessed plate filter press, but the plates incorporate membranes that are inflated at high pressure to squeeze the filter cakes. Normally, the plate pack consists of alternate recessed plates and membrane plates. This design makes it possible to achieve superior results from thin cake slurries and can reduce the cycle times of a batch. Membrane plates usually operate at a feed pressure of up to 7 bar and a squeeze pressure of up to 15 bar. Special plates can be designed to accept higher feed and squeeze pressures.

Use of the Membrane Filtration Technique

The pharmaceutical and cosmetics industries typically focus on monitoring their process water for *Pseudomonas* species. The electronics industry monitors for any and all microorganisms because they must keep their process water free from even the smallest organisms. Microbial monitoring in the food and beverage industry typically employs several types of techniques because of the variety of samples that are encountered. Beverage samples can typically be monitored for microorganisms by the MF technique, but when solid samples cannot be liquefied, alternative methods must be used. MF is an effective and accepted technique for testing fluid samples for microbiological contamination. It involves less preparation than many traditional methods, and is one of a few methods that allows the isolation and enumeration of microorganisms. The MF technique also provides information on the presence or absence of microorganisms within 24 h.

Working Principle

A standard filter press application for metal hydroxide sludge, for example, with plates that provide a 32 mm cake thickness generally takes about 4 h to fill. The unit is usually 80% full after about 2 h. As a result of the high pressure (100 psi from the diaphragm pump, counteracted by 4000 psi of hydraulic pressure), resistance builds and the diaphragm pump begins to slow down considerably. A membrane press is identical to a standard filter press except that the chambers (area between the plates) have membranes to squeeze out excess water from the slurry. A press with all membrane plates can be costly; however, a mixed pack is a more cost-effective alternative. In a mixed pack, each chamber has one side with a membrane that inflates, thereby causing a squeezing action against a solid recessed plate. When the press is filled to about 80% capacity (this is determined by historical information

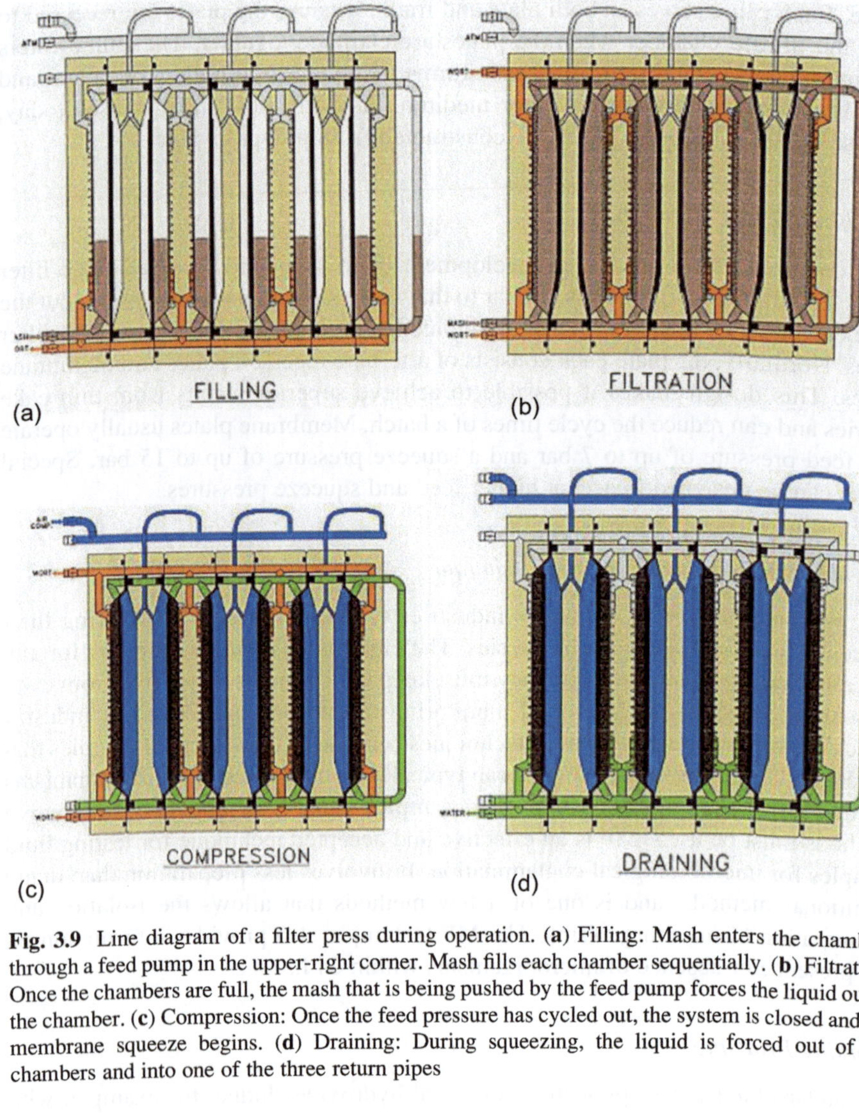

Fig. 3.9 Line diagram of a filter press during operation. (**a**) Filling: Mash enters the chambers through a feed pump in the upper-right corner. Mash fills each chamber sequentially. (**b**) Filtration: Once the chambers are full, the mash that is being pushed by the feed pump forces the liquid out of the chamber. (**c**) Compression: Once the feed pressure has cycled out, the system is closed and the membrane squeeze begins. (**d**) Draining: During squeezing, the liquid is forced out of the chambers and into one of the three return pipes

and/or by the timing of the diaphragm pump), it is the ideal time to inflate the membrane diaphragms to provide extra squeeze (Fig. 3.9).

Major Components of Filter Presses

Filter Cloths

Filter cloths are woven fabrics using monofilament or multifilament synthetic fibers, or a combination of both. The most commonly used materials are polypropylene, polyester, and nylon. Filter cloths are primarily selected for filtration, strength, and cake release properties. With proper cloth selection, good cloth life (typically 1500–1800 cycles) and good cake release from the filter cloth can be expected. Filter cloth style is typically dependent on the type of filter plate selected for the application. Figure 3.10 illustrates two types of filter plates. The caulked and gasket recess filter plate includes a cord sewn into the filter cloth to secure the cloth into the cloth-retaining groove in the filter plate. In the recessed filter plate, the filter cloth is secured by cloth dog pins and/or electrical ties (Fig. 3.10a, b).

Feed Pumps

A proper pumping system that does not impart excess shear, or deteriorate the sludge particles or chemically formed floc, is very important in the design of a filter press dewatering system. The feed pump should be capable of initially delivering high volume/low pressure and then a declining low volume/constant high pressure during the later portion of the filter cycle. The four most common types of pumps employed are:

- Ram or piston pumps, usually hydraulically driven
- Double diaphragm pumps, air driven
- Piston membrane pumps
- Progressive cavity pumps with speed control

Pumps of the first three categories should always include a surge suppression tank mounted on the discharge side of the pump to dampen pressure surges on the filter.

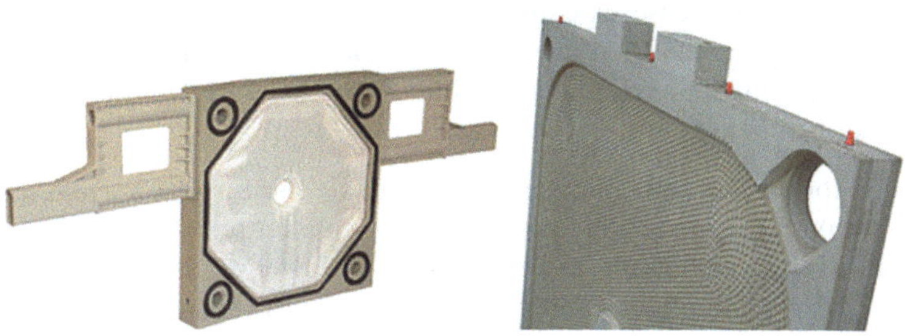

a b

Fig. 3.10 (**a**) Caulked and gasket recess filter plate; a cord is sewn into the filter cloth to secure the cloth into the cloth-retaining groove in the filter plate. (**b**) Recessed filter plate; the filter cloth is secured by cloth dog pins and/or electrical ties

Precoating the Filter Press

A widely accepted remedy for cloth blinding (clogging) is the use of a precoat material on the filtration medium prior to solution feed. Precoat materials that are commonly used are diatomaceous earth and bleached wood pulp. These materials are mixed with water in a slurry and recirculated back through the filter until the water is clear and the filter medium has a precoat on it. The theory is that these porous materials on the filtration medium allow solution to pass through and build a cake. The many "nooks and crannies" of the precoat give the solids places to build while still allowing solution to pass. To precoat a filter, some changes to the plumbing could be required because a tank and pump may be needed. A rule of thumb for precoating is to use approximately 5 kg of material for every 10 m^2 of filtration area in the filter.

Sludge Conditioning

Sludge conditioning is the physical or chemical treatment of sludge to facilitate dewatering. The success of mechanical dewatering is largely due to the development of chemical conditioning, and some of the more recent dewatering processes depend on particular conditioning systems. Effective sludge conditioning or pretreatment is defined as any changes to the raw sludge that effectively reduce viscosity, reduce resistance to flow, increase filtration rates, or increase the concentration of solids in the sludge feed. These changes can be made chemically or physically. Chemically conditioning the sludge involves coagulation and flocculation of the sludge fine particles to produce a filterable floc structure.

Physical means of conditioning sludge include processes such as heat treating and freezing. Anaerobic digestion is sometimes regarded as a form of conditioning but, in fact, does not generally make the sludge any more amenable to dewatering. However, the process of elutriation after digestion is very effective in reducing the amount of chemical required for conditioning. When considering costs such as capital equipment, conditioning agents, and transportation and handling of conditioning agents, in most cases chemical conditioning is the most economical and is, by far, the most generally used method of treatment. Commonly used conditioners for sludge treatment are ferric chloride, alum, lime, and polyelectrolytes, which are briefly described below:

- Ferric chloride and iron chlorosulfate are mainly used in conjunction with lime to condition the sludge before passing through a filter press. They allow a better filterability by coagulating the colloids (thus lowering the content of linked water) and by microflocculation of the precipitates (hydroxides). The dosage for iron salts is between 3% and 15% of the dry content, depending on the quality of the sludge.

- Alum is normally delivered in liquid form (48% solution) and stored in corrosion-resistant storage tanks. Water is normally added to give a diluted solution of approximately 10% strength.
- Lime as a conditioning agent is only used in conjunction with iron salts for filter press applications. It brings a mineral nature to the sludge and strengthens its mechanical properties (higher specific resistance to filtration). The dosages for lime are between 15% and 40% of the dry content. Lime is also used after dewatering to stabilize the sludge. The specific resistance to filtration depends on the size, shape, and degree of agglomeration of the solid particles that make up the cake from a filter press; it is independent of the sludge concentration.
- Polyelectrolytes (polymer) can be provided as a dry powder, liquid solution, or liquid emulsion. The polymer must be diluted and mixed before use. The mixed stock (neat) solution at 0.1–0.2% strength is stored in a corrosion-resistant tank and further diluted at the point of application.

Flocculation

Flocculation, in the field of chemistry, is a process wherein colloids come out of suspension in the form of floc or flake, either spontaneously or as a result of addition of a clarifying agent. Flocculants, with their very high molecular weights (long chains of monomers) and varied ionic charge, can fix the destabilized particles on their chain. Therefore, the particle size in the aqueous phase increases throughout the flocculation step. The formation of flocs induces release of water. This water is easily eliminated during the dewatering step. The origin of destabilized particles varies widely and essentially depends on the nature of the sludge. The charge that the flocculent brings is selected according to the type of destabilized particles present in the sludge to be treated. Generally, most kinds of sludge can be treated with any of the above conditioners, or with a combination, but a particular sludge often responds more readily to a specific conditioner or combination of conditioners to provide ideal dewatering properties (Table 3.2).

Table 3.2 Requirement for a specific flocculation agent or combination of agents for sludge treatment to provide ideal dewatering properties

Type of sludge	Ferric chloride	Alum	Lime	Polymer
Biological waste activated	X		X	
Biological waste activated		X	X	
Biological waste activated				
Oil refinery	X		X	
Water treatment plant—alum			X	
Mineral—clay				X

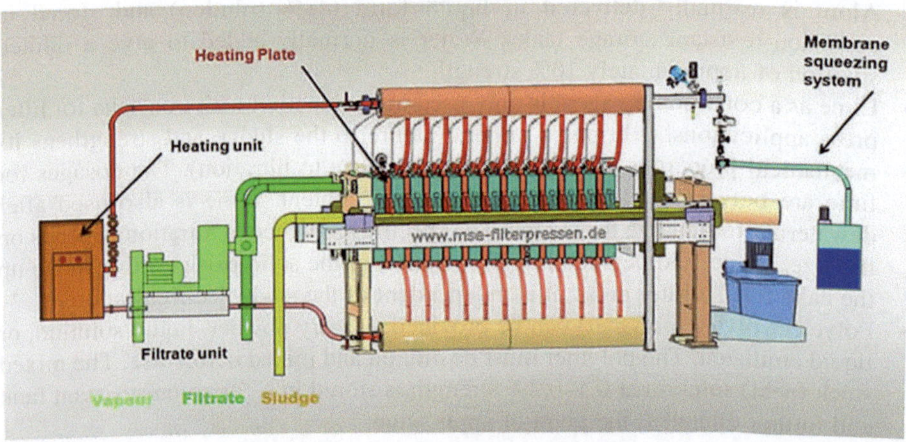

Fig. 3.11 Sludge flow diagram of the thermal cake drying (TCD) filter press system

Basic Sludge-Dewatering System and Performance

A typical filter press dewatering system includes the filter press, sludge feed pumps, sludge storage and/or sludge conditioning tanks with slow speed mixers, sludge transfer pumps, chemical feed, storage equipment, and necessary piping, valves, and controls. Two types of filter press systems are available:

- High pressure (i.e., 225 psi, 16 bar)
- Low pressure (i.e., 100 psi, 7 bar)

Many sludge-dewatering filter press systems are designed for automatic system operation via a main control panel. A typical line-flow diagram is shown in Fig. 3.11.

During the initial step of the filter press dewatering cycle, the filtrate flow from a large press can be as much as 10,000–20,000 gallons per hour. This part of the cycle is often referred to as the fast fill. During this period, the cake chambers of the filter press collect the major amount of sludge solids. As the chambers become progressively more filled with sludge solids, the pressure inside the filter press plate pack rises and the filtrate flow rapidly decreases. Toward the end of the cycle, the filtrate flow gradually drops to virtually nothing. This portion of the cycle is referred to as the "cake consolidation step." During the cake consolidation step, more sludge solids are forced under pressure into the cake chambers, which in turn displaces more water from the loosely formed sludge cake. This enables the filter press to produce harder and drier sludge cakes than other methods of dewatering. A wide range of sludges can be effectively dewatered at pressures as low as 7 bar (100 psi). For more difficult applications such as biological water activated sludge, alum sludge, or where maximum cake dryness is required, high-pressure 16 bar (225 psi) filter presses are generally utilized (Table 3.3).

Table 3.3 Typical results from a filter press are expressed in percentage of cake dry solids by weight

Type of sludge	% Cake solids
Biological waste activated sludge	30–40
Metal hydroxide sludge	30–55
Alum water treatment sludge	25–35
Oil refinery sludge	40–50
Lime sludge	30–60
Brine sludge	60–70

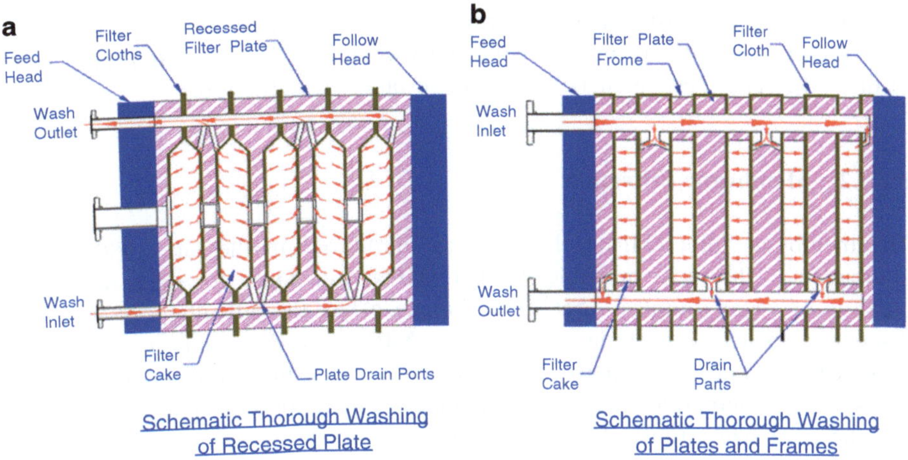

Fig. 3.12 (**a**) Thorough washing using recessed chamber plates (courtesy of FFP Systems, Inc.). (**b**) Thorough washing using recessed chamber plates (courtesy of FFP Systems, Inc.)

Filter Cake Washing and Blowing

Impurities or soluble matter can be extracted or leached from the filter cake by washing the filter cake after it has been formed. The washing medium, water or other liquor/solvent, is forced through the filter cake by pumping at a pressure slightly in excess of the terminating filter pressure. For best washing results, the filter cake should be of uniform thickness and consistency. There are two types of washing methods practiced, simple washing and thorough washing. In simple washing, the wash liquor is introduced to the filter in the same direction as the feed, entering directly into the slurry inlet. Simple washing is sometimes the best method when there is a wide particle size distribution or when the frame or cake chamber is not completely filled with filter cake.

In thorough washing using recessed chamber plates, the washing liquor is fed into one of the filtrate outlet connections on the filter press and enters into the drainage area behind the cloth of the washing plate. It is then forced to flow back through the filter cloth of the wash plate, through the cake and through the filter cloth of the adjacent non-wash plate and then discharges from the non-wash plate (Fig. 3.12a).

A filter using flush plates and frames set up for through washing has separate slurry feed and wash feed ports, and either the same or separated discharge ports for filtrate and wash liquor outlet. During thorough washing, the slurry feed inlet and wash plate outlets are closed, thus forcing the wash liquor entering the washing plates to flow back through the filter cloth, through the cake, and through the cloth of the adjacent non-wash plate and then discharge from the outlet of this plate (Fig. 3.12b).

Blowing the filter cake with air or some other gas can be used to reduce the free moisture retained in the filter cake or in the filtrate outlet channels. As in the washing operation, air blowing can be accomplished either by simple blowing or thorough blowing. Wash plates are normally marked by three buttons on the upper outside corner, or with one button on the non-wash plates. In a plate and frame style of filter, the cake frame is marked with a two-button designation.

3.2.3.2 Membrane Filtration

The membrane filter (MF) technique was introduced in the late 1950s as an alternative to the most probable number (MPN) procedure for microbiological analysis of water samples. The MF technique offers the advantage of isolating discrete colonies of bacteria, whereas the MPN procedure only indicates the presence or absence of an approximate number or organisms (indicated by turbidity in test tubes). Membrane filters or "membranes" are microporous plastic films with specific pore size ratings. Also known as a screen, sieve, or microporous filter, membranes retain particles or microorganisms larger than their pore size primarily by surface capture. Some particles smaller than the stated pore size may be retained by other mechanisms.

Generally, membranes are produced by three different processes. Mixed cellulose esters, cellulose acetate, and nylon are reverse-phase solvent-cast membranes, where controlled evaporation or removal of the complex solvent system forms the porous structure. Both hydrophilic and hydrophobic polytetrafluoroethylene (PTFE) membranes are made by a patented process whereby the membranes are stretched biaxially to form the porous structure. Polycarbonate track-etch (PCTE) membranes are track etched. When filter press cycle time is a concern, a membrane filter press can halve the time needed. Generally speaking, a filter press takes 4 h to fill (actual time depends upon loading capacity and the type of slurry being fed through the press). However, a membrane press can reduce cycle time by approximately 50% on average and, in some cases, even by as much as 75% depending upon the application.

Membrane Inflation Techniques

The safest way to inflate the membranes is to use water instead of air. If air is used to inflate the membrane, tears may appear and there is a chance of plastic breaking, which can be extremely hazardous. If water is used, a rupture results in a water leak, which is much safer and easier to contain. A tank and pump are used to fill the membranes with water via the manifold piping, which is stainless steel, PVC, or steel (depending on the application). Flexible tubing connects the manifold to the plates. The water pumps into the membranes during the squeeze process and, once that is complete, it is drawn back to the water tank for reuse.

3.2.3.3 Vacuum Filters

There are two main types of vacuum filters, rotary drum filters (Fig. 3.13a) and horizontal belt filters (Fig. 3.13b), both of which operate continuously. Vacuum filters are widely used for clarification of fermentation broths (containing 10–40% solids by volume with particle sizes of 0.5–10 μm) because of their simplicity of operation and low cost.

Rotary Drum Vacuum Filters

The best known vacuum filters are the rotary drum vacuum filters (RDVFs), which are used for filtration of filamentous fungi and yeast cells. They are used to clarify large volumes of liquid, with automatic solids discharge. Tangential flow (crossflow filtration) is an effective method for separating cells from liquid where high-value product is involved. The parallel motion of fluid to the membrane helps reduce the thickness of the cell layer on the filter surface. RDVFs consist of a rotating drum lined with filter medium that has approximately one third of its outer wall submersed in the slurry. The filtrate is pulled through the wall of the drum by a vacuum, toward the center of the drum. The drum rotates such that the cake is dumped to the discharge outlet. That section of the drum wall is then washed before performing filtration again. The process of the RDVF is continuous. Each revolution of the drum consists of cake formation, cake washing (if needed), dewatering or drying, and cake discharge (Fig. 3.14). Drum filters are designed for all types of liquid–solid separation including medium concentrate sludge and low-density solids, which do not require efficient washing.

The RDVF, supported by two bearing blocks, is composed of successive cells, which form a horizontal cylinder. The cylinder, in turn, supports the filter cloth, which is partially immersed in the tank. The cells are sealed and interconnected by pipes leading to the distributor, which is divided into several zones, each gathering the various filtrates (mother liquor, washing water, and drying water) coming from the cell suction pipes as they pass during rotation of the drum. The slurry passes through the trough-shaped tank to be filtered. The tank, which is part of the filter frame, also supports the drum, the cake discharge system, and the slurry agitation device. For cake discharge, several methods can be used such as string, belt, scraper, or roller.

a

b

Fig. 3.13 Photographs of (**a**) huge rotary vacuum filter (courtesy of Komline-Sanderson) and (**b**) horizontal belt filter (courtesy of Komline-Sanderson)

Horizontal Belt Vacuum Filters

Horizontal vacuum belt filters are an ideal solution for applications with very large flow rates, stringent low cake moisture requirements, and the need for highly efficient cake washing as a process step. Horizontal vacuum belt filter presses are used to remove water from liquid wastewater residuals and produce a nonliquid

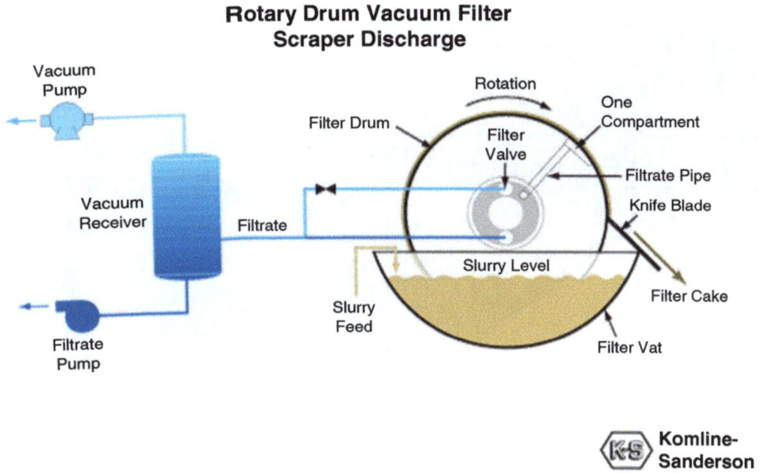

Fig. 3.14 Rotary drum vacuum filter in working condition (courtesy of Komline-Sanderson, Peapack, NJ)

material referred to as "cake." Dewatered residuals, or cake, vary in consistency, from that of custard to moist soil. Dewatering serves the following purposes:

- Reduces the volume, thus reducing storage and transportation costs
- Eliminates free liquids before landfill disposal
- Reduces fuel requirements if residuals are to be incinerated or dried
- Produces a material with sufficient void space and volatile solids for composting when blended with a bulking agent
- Avoids the potential of biosolid pooling and run-off associated with land application of liquids
- Optimizes subsequent processes such as thermal drying

Horizontal belt filters provide high extraction efficiency, low cake moisture, increased production, and reduced operating costs while achieving maximum filtration area compared with other filter options. Horizontal belt filters maximize cake purity at minimum specific cost. They are especially suitable for applications requiring low cake moisture and multistage washing. Low energy consumption and high filtration rate make horizontal belt filters an excellent choice for a wide range of processes (Fig. 3.15a, b).

Feed slurry is uniformly distributed over the full width of the filter by a top feed arrangement. This eliminates cake formation problems that can be associated with fast-settling material. Gravity helps reduce vacuum energy and cake formation time. Several types of feed box arrangements are available, tailored to the process requirements. Cake travels with the filter medium, which is supported by a rubber drainage belt. Dewatering is accomplished by applying vacuum to the bottom of the drainage belt. Differential pressure draws slurry liquor through the filter cloth, along channels in the support belt to drain holes centered over the vacuum pan.

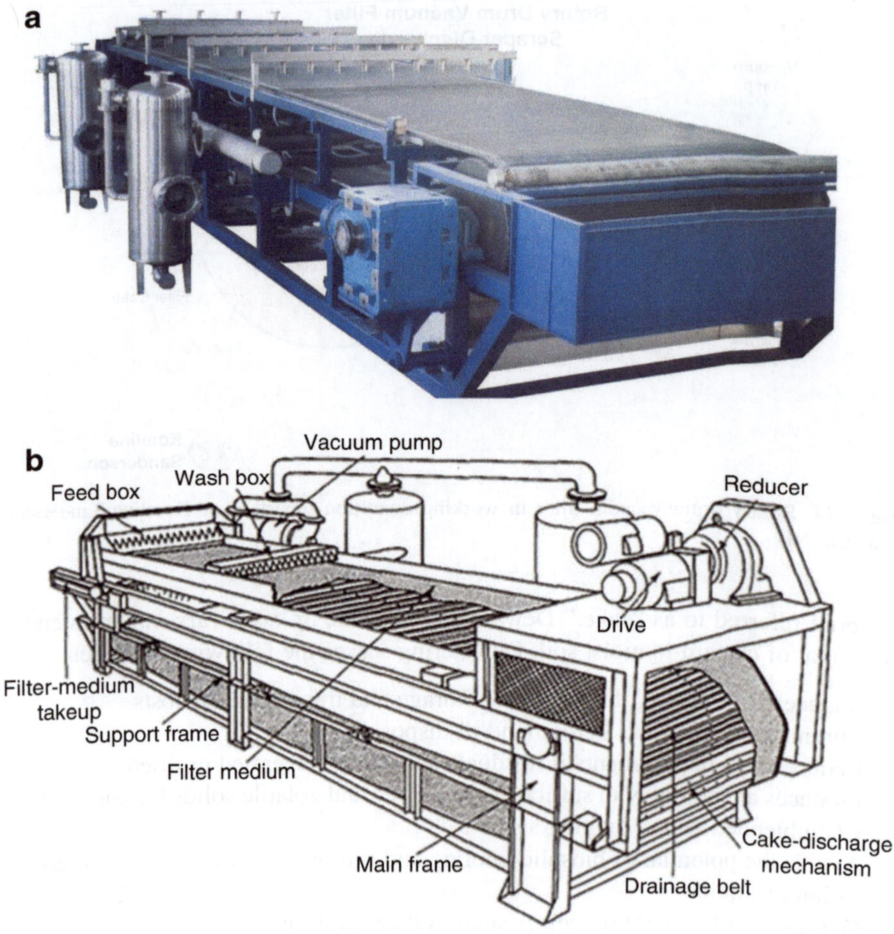

Fig. 3.15 (**a**) Photograph of front view of horizontal vacuum belt filter. (**b**) Diagram showing different functional parts

Channels and drains must be carefully sized to accommodate the hydraulic and pneumatic flows required for efficient filtration. Co-current or counter-current wash systems are designed to meet the specific process requirements, employing one or more independent washing zones to optimize product recovery. Filtrate and air enter the vacuum receiver(s), where the liquid drops out and is pumped away. Air exits at the top of the receiver as a result of the negative pressure developed by the vacuum pump. Cake is discharged as the filter medium travels around a small roller, after separating from the drainage belt. Multiple wash sprays clean the drainage belt and filter medium independently to extend the service life of both. Continuous tension is maintained on the belt and filter medium. Positive, automatic tracking and alignment of the filter medium is controlled by sensors. Cleaning, tensioning, alignment, and support from the drainage belt contribute to improved filter life.

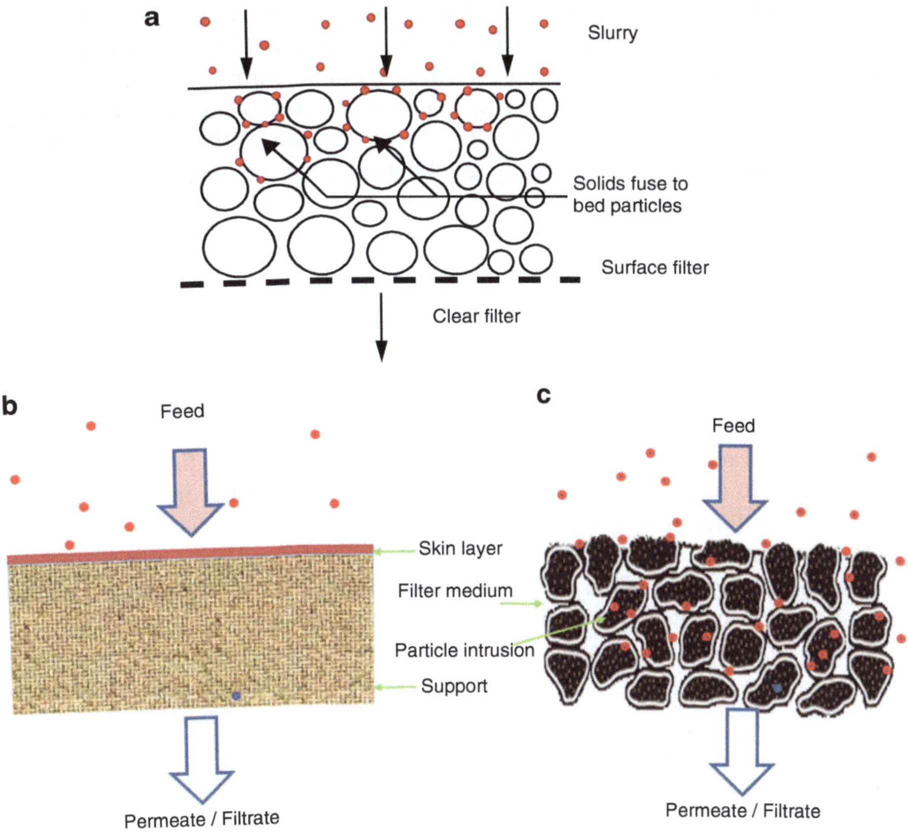

Fig. 3.16 (**a**) Depth filtration works by collecting particulates within the filter media and passing clean fluid to the outlet flow. Surface filtration (**b**) compared with depth filtration (**c**)

3.2.3.4 Depth Filters

Depth filters consist of a matrix of randomly oriented, bonded fibers that capture particulates within the depth of the filter, instead of on the surface (Fig. 3.16a).

A true depth filter allows particles to penetrate the filter matrix and become captured throughout the depth of the medium. As with surface filtration, this only holds true when the particle size/pore relationship is conducive to the process for which the cartridge was designed. The depth filter matrix has a broad pore size distribution; hence, depth cartridges rely on adsorptive retention for a portion of their dirt-holding capacity. Some depth filters, such as the ARD, Nexis, and DFT Classic, have a gradient pore structure, with tighter pores near the center core, to maximize mechanical retention. In some depth cartridges, such as the string-wound type, the medium is not a fixed pore matrix, as with chemically or thermally affixed pleated media. For this reason, depth cartridges should not be subjected to flows as high as those that are possible for pleated cartridges. Most depth filters are made from extruded melt-blown fibers or twisted yarn fibers. Melt-blown depth filters are generally made from polypropylene, polyester, or nylon and can be made with both absolute and nominal retention ratings. These types of cartridges can be made to filter particles sizes from less than 1 μm to over 100 μm. Yarn-wound cartridges, made with fibrous materials, are often brushed to maximize the tortuosity of flow through the filter. They are nominally rated but offer the advantage of being made from a variety of materials.

Depth Filtration Versus Surface Filtration

Filtration can also be categorized according to where particle rejection occurs (as shown in Fig. 3.16a, b):

- Surface filtration: Particles are supposedly rejected on the filter surface and generally do not intrude into the filter medium. All the filter mediums used in membrane filtration fall into this category.
- Depth filtration: Particles can intrude filter medium and are captured by the medium through collision. Because particle removal relies on random collision and there is no absolute barrier, some particles can pass the filter. Cartridge filters, sand filters, multimedia filters, and most air filters fall into this category.

Large-Scale Depth Filtration

Large-scale depth filtration systems are common in the bioprocess industry. They use multiround stainless housings with up to 16-in. depth filter modules stacked three or four high to maximize surface area. Figure 3.17 shows an industrial-scale depth-filtration harvest system consisting of multiple housings and cartridges for clarifying cell harvest from a 15,000-L bioreactor. Before harvest, depth filters are flushed with water or an appropriate buffer to remove loose particulates and extractables from the filter manufacturing process. Once a harvest is completed, the filters are again flushed to recover valuable product held in the housings. Harvest yields as high as 95% are achievable using a post-use flush and ensuring minimum product loss [23].

Depth filters are widely used in the bioprocess industry for clarifying cell culture broths. Single-use depth-filtration technologies can be used for purposes other than clarification of cell culture streams. Because the depth-filter media can carry a surface charge, depth filters retain particles smaller than their rated pore size. Studies have shown that, depending on process design, depth filters can also be used in orthogonal virus removal steps, providing up to 4-log viral clearance [9–11, 24]. Other studies have shown that charged depth filters can remove non-enveloped RNA viruses (such as mouse minute virus) at a process capacity of 320 L m^{-2} from a feed containing 2% spiked viruses.

Fig. 3.17 Industrial-scale depth-filtration harvest system consisting of multiple housings and cartridges for clarifying the cell harvest from a 15,000-L bioreactor

3.2.3.5 Absolute Filters

A filter rating indicates the size of the largest particles that can pass through a screen or filter. An absolute filter can cut off 100% (by weight) of solid particles greater than a stated size. The pore size of an absolute filter is the pore size at which any particle of a particular size can be retained with 100% efficiency under strictly defined test conditions. The conditions that should be specified are challenge pressure, detection method, test organism, and concentration methods used for identifying the contaminants.

3.2.4 Centrifugation

3.2.4.1 Centrifugal Field

Centrifugation is a technique used for the separation of particles using a centrifugal field. The particles are suspended in liquid medium and placed in a centrifuge tube. The tube is then placed in a rotor and spun at a specific speed. Rotation of the rotor about a central axis generates a centrifugal force upon the particles in the suspension.

Two forces counteract the centrifugal force acting on the suspended particles:

- Buoyant force: This is the force with which the particles must displace the liquid medium into which they sediment.

- Frictional force: This is the force generated by the particles as they migrate through the solution.

Particles move away from the axis of rotation in a centrifugal field only when the centrifugal force exceeds the counteracting buoyant and frictional forces, resulting in sedimentation of the particles at a constant rate. Particles that differ in density, size, or shape sediment at different rates. The rate of sedimentation depends upon:

- Applied centrifugal field
- Density and radius of the particle
- Density and viscosity of the suspending medium

One revolution of a rotor is 360° or 2π rad (radians), therefore the speed of rotation in revolutions per minute (rpm) is given by:

$$\frac{\omega \times 60}{2\pi}; \quad \omega = \frac{[2\pi \times (\text{rpm})]^2}{60}$$

where ω is the angular velocity in radians per second. The centrifugal field (CF) can be expressed as:

$$\begin{aligned}\omega^2 \times r &= \frac{[2\pi \times (\text{rpm})]^2}{1} \times \frac{1}{60^2} \times r \\ &= \frac{4\pi^2}{1} \times \frac{(\text{rpm})^2}{3600} \times r\end{aligned}$$

where r is the radial distance of the particle from the axis of rotation.

As the centrifugal field acting on the particle is much greater than the Earth's gravitational field, CF is generally expressed relative to the Earth's gravitational field as multiples of g, the acceleration due to gravity ($g = 980$ cm s^{-2}):

$$\text{Relative Centrifugal Field (RCF)} = \frac{\omega^2 \times r}{g}$$

$$\text{RCF} = \frac{4\pi^2\,(\text{rpm})^2}{3600 \times 980} \times r = 1.11 \times 10^{-5} \times \text{rpm}^2 \times r$$

This expression relates RCF to the speed of the centrifuge (rpm) and the radius of the rotor (r). For example, if a rotor with an average radius of 7 cm revolves at a speed of 20,000 rpm, a centrifugal field of 31,300$\times g$ is created. The sedimentation rate of velocity (v) of a particle can be expressed in terms of its sedimentation rate per unit centrifugal field. This is termed the sedimentation coefficient (s). The sedimentation rate is proportional to $\omega^2 r$, the centrifugal field:

$$v = s\omega^2 r$$

$$s = \frac{V}{\omega^2} \times \frac{1}{r}$$

Sedimentation velocity depends upon the mass, density, and shape of the particles and also on the density and viscosity of the medium in which the particles are suspended.

Despite being cumbersome, centrifugation will probably continue to have a major role in microbial separation. Centrifugation at $2000 \times g$ for 10 s prior to estimating biomass by ATP measurement removed virtually all meat particles from a homogenate without loss of bacterial count [25]. Density gradient centrifugation can remove food debris without loss of bacterial count; for example, a method using colloidal silica clarified food suspensions within 15 min [26], and the automated density gradient apparatus of the Bactoscan instrument also allows concentration of food-related microbes [27]. Sedimentation field-flow fractionation separates pure bacterial cultures and is probably applicable to foods; cells are injected into an open, unpacked channel, first sedimented by a low (5–10 RCF) centrifugal field, then fractionated by a parabolic fluid-velocity field as diluent passes through the chamber.

3.2.4.2 Types of Centrifuge

On the basis of application, centrifuges can be categorized into four types: bench centrifuges, large capacity refrigerated centrifuges, high speed refrigerated centrifuges, and ultracentrifuges.

Bench Centrifuge

A bench-top centrifuge is a piece of laboratory equipment, driven by a motor that spins liquid samples at high speed (Fig. 3.18a). There are various types of centrifuges, depending on size and sample capacity. Like all other centrifuges, laboratory centrifuges work by the sedimentation principle, whereby centrifuge acceleration is used to separate substances of greater and lesser density.

On the basis of application, two types of heads are used in this centrifuge. In one type, the tubes placed are placed in the cups of the rotor to assume a horizontal plane when the rotor is in motion and a vertical position at rest (Fig. 3.18a). During centrifugation, particles travel uniformly and constantly along the tube while the tube is at right angles to the shaft of the centrifuge; thus the sediment is distributed uniformly against the bottom of the tube and remains there when rotor stops, with liquid above it. This liquid can be decanted off, and liquid and sediment separated for analysis. The spinning rotor offers considerable resistance to rotation and generates heat due to air friction (Fig. 3.18b, c). The second type of head has fixed-angle cups in which the centrifuge tubes are placed at an angle of 25–40° to the vertical axis of rotation (Fig. 3.18d). During centrifugation, particles are driven

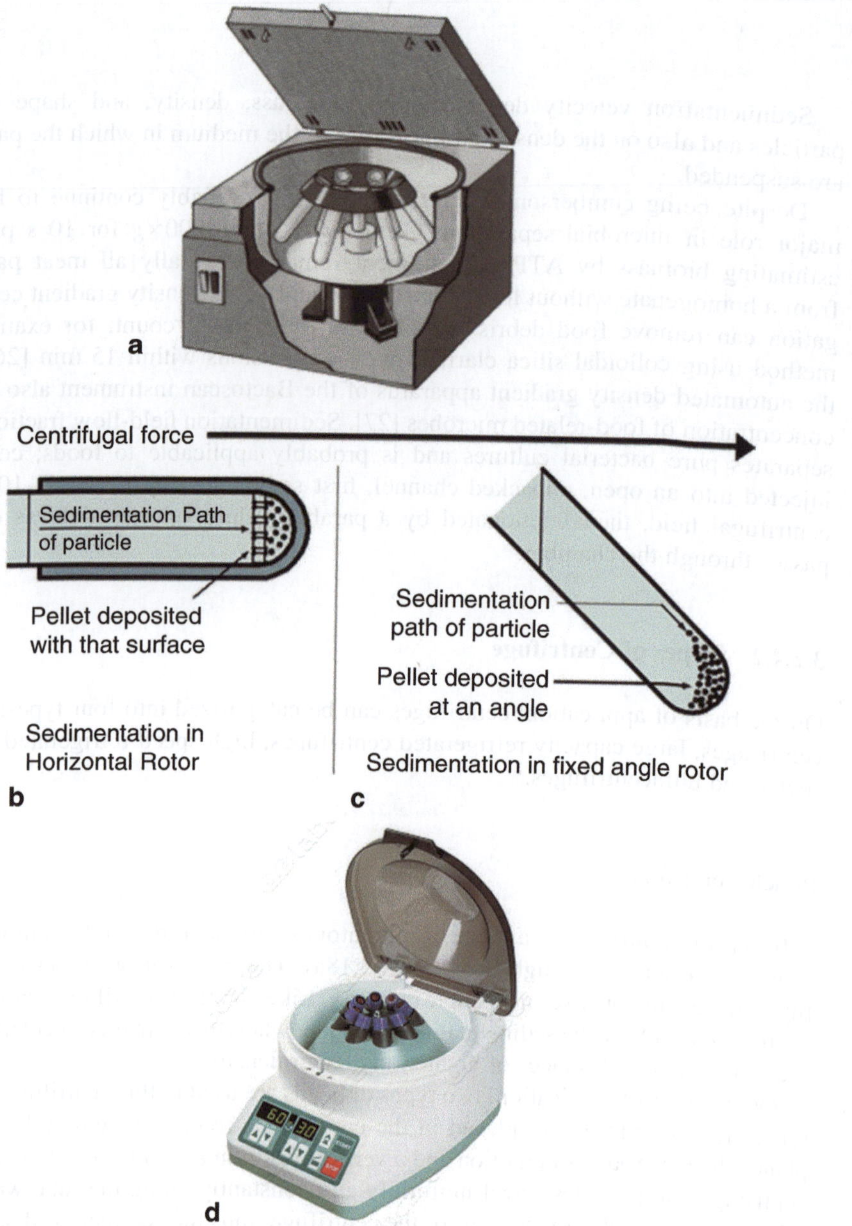

Fig. 3.18 (**a**) Horizontal head or swinging bucket centrifuge. (**b**) Sedimentation of particles evenly on the bottom of the tube in a horizontal rotor. (**c**) Sedimentation of particles fixed at an angle on the bottom of a tube in a fixed angle rotor. (**d**) Fixed angle centrifuge

outward horizontally but strike the side of the tube so that the sediment packs against the side and bottom of the tube, with the sediment surface parallel to the shaft of the centrifuge. As the rotor slows or stops, gravity causes the sediment to slide down the tube and a poorly packed pellet is usually formed.

In order for a rotor to run smoothly and safely at its operating speed, the load it carries must be balanced. A rotor can be properly balanced by following some simple rules (Fig. 3.19):

- A rotor must never be run with buckets missing, although opposing buckets may be left empty
- All opposing loads must balance within a certain weight, as specified by the centrifuge manufacturer's instruction manual

If opposing buckets are run with a partial load of tubes in their adapters, these tubes must be arranged symmetrically, both with respect to the pivotal axis of each bucket and across the center of rotation. With some partial loads, it can be difficult or impossible to achieve the correct symmetry in both sets of opposing buckets. The simplest solution is to fill one or more tubes of the same size with water, or a denser liquid if necessary, and use them to balance the load symmetrically.

Large-Capacity Refrigerated Centrifuge

These centrifuges have a refrigerated rotor chamber. It is possible to use rotor chambers of varying size. They can go up to maximum of $6500\times g$ and are used to sediment or collect substances that sediment rapidly, such as erythrocytes, yeast cells, nuclei, and chloroplasts (Fig. 3.20).

High-Speed Refrigerated Centrifuge

These can generate speeds of about $60{,}000\times g$ and are used to collect microorganisms, cellular debris, larger cellular organelles, and proteins precipitated by ammonium sulfate (Fig. 3.21).

Ultracentrifuge

Ultracentrifuges are a type of centrifuge designed to rotate samples at "ultrahigh" speeds (much higher than conventional centrifuges), with a rotational speed of up to 150,000 rpm, creating a centrifugal force of more than $10^6\times g$. All centrifuges can separate species within a liquid according to density, but the high centrifugal acceleration of ultracentrifuges allows the researcher to focus on the tiny differences between molecules such as proteins or nucleic acids (Fig. 3.22).

Ultracentrifuges are commonly used in molecular biology, biochemistry, and cell biology (Fig. 3.23). Applications for ultracentrifuges include the separation of

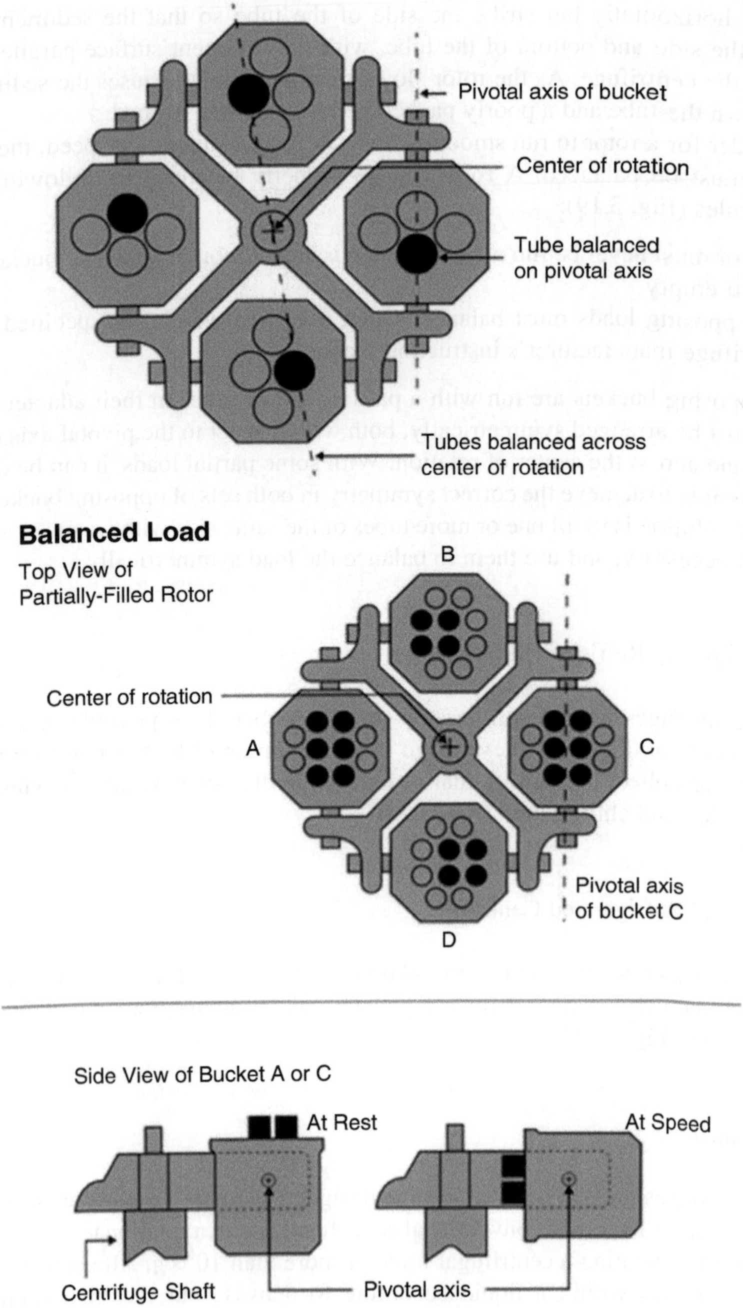

Fig. 3.19 (**a**) Top view shows how to balanced loads on a rotor so that the motor runs smoothly. (**b**) Assuming that all tubes are filled with an equal amount of liquid, the rotor load is balanced. The opposing bucket sets *A–C* and *B–D* are loaded with an equal number of tubes and are balanced across the center of rotation. Each bucket is also balanced with respect to its pivotal axis. (Source: https://www.google.co.in/imgres)

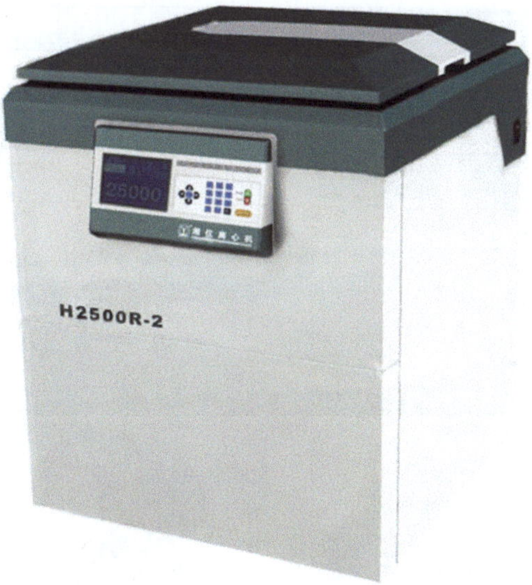

Fig. 3.20 Photograph of a large-capacity refrigerated centrifuge

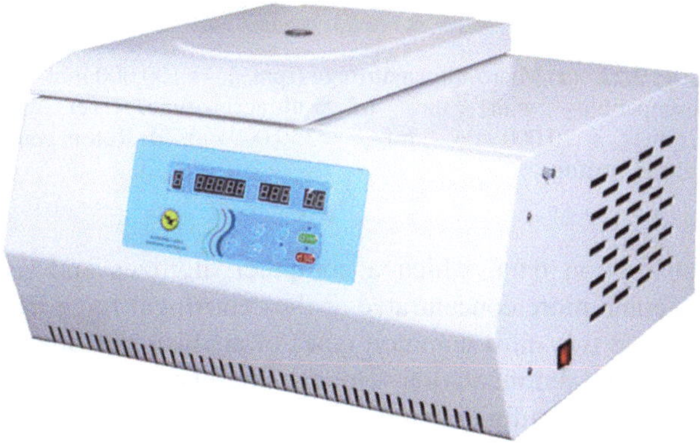

Fig. 3.21 High-speed benchtop refrigerated centrifuge laboratory type (Source: https://www.google.co.in/imgres)

small particles such as viruses, viral particles, proteins and/or protein complexes, lipoproteins, RNA, and plasmid DNA. Ultracentrifuges can be grouped into two types, analytical and preparative. The type chosen depends on the purpose. Ultracentrifuges come in large floor models or more compact bench-top models.

Analytical and preparative ultracentrifuges differ in the way they are used in the laboratory. Analytical centrifugation can yield important information about the molecules in a sample, such as their overall shape, conformational changes, and even the number and stoichiometry of subunits that comprise protein complexes. Analytical ultracentrifuges incorporate a scanning visible/ultraviolet light-based optical detection system for real-time monitoring of the sample's progress during a spin. It is possible to watch the sedimentation process (a detector collects light

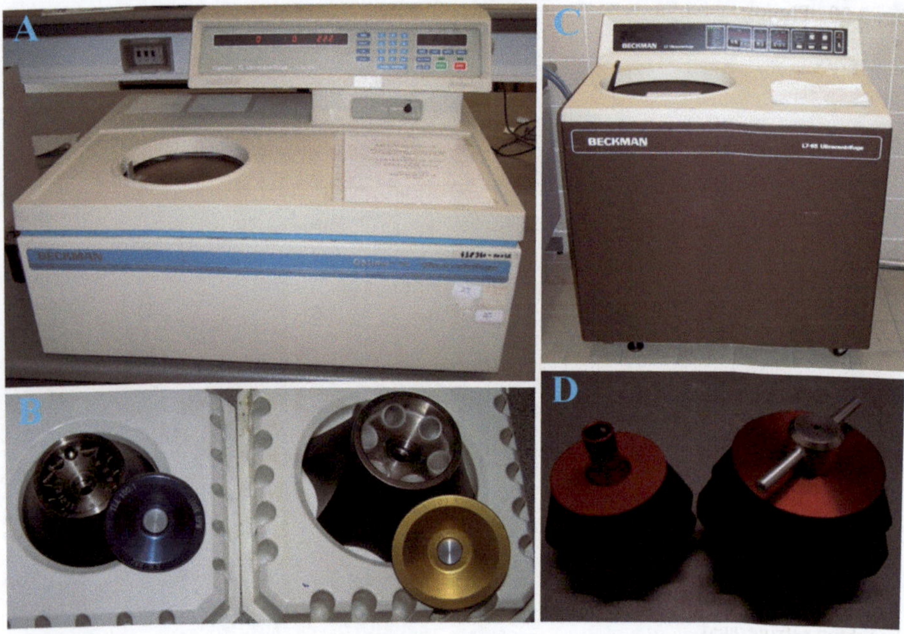

Fig. 3.22 (**a**) Micro-ultracentrifuge ($rpm_{max} = 100{,}000\ min^{-1}$, $rcf_{max} = 300{,}000 \times g$). (**b**) Rotors compatible with the micro-ultracentrifuge. (**c**) Semipreparative ultracentrifuge ($rpm_{max} = 70{,}000\ min^{-1}$, $rcf_{max} = 350{,}000 \times g$). (**d**) Rotors compatible with the semi-preparative ultracentrifuge

absorption data, which a computer digitizes and records) and see the sample become more concentrated as the centrifugal force increases.

The two most common types of analysis performed with analytical ultracentrifuges are sedimentation velocity experiments and sedimentation equilibrium experiments. In sedimentation velocity experiments, the ultracentrifuge, detector, and computer record the time course of the sedimentation process, yielding information about the shape, mass, and size of molecules. Sedimentation equilibrium experiments involve studying the steady-state equilibrium of the sample in solution. Even after sedimentation is completed, the sample exists in steady-state equilibrium with the solute down its concentration gradient. This type of analysis gives information about mass and chemical equilibrium constants.

Preparative ultracentrifuges are often used for separating particles according to their densities, isolating and/or harvesting denser particles for collection in the pellet, and clarifying suspensions containing particles. Sometimes researchers also use preparative ultracentrifuges if they need the flexibility to change the type of rotor in the instrument. Preparative ultracentrifuges can be equipped with a wide range of different rotor types, which can spin different numbers of samples at different angles and at different speeds.

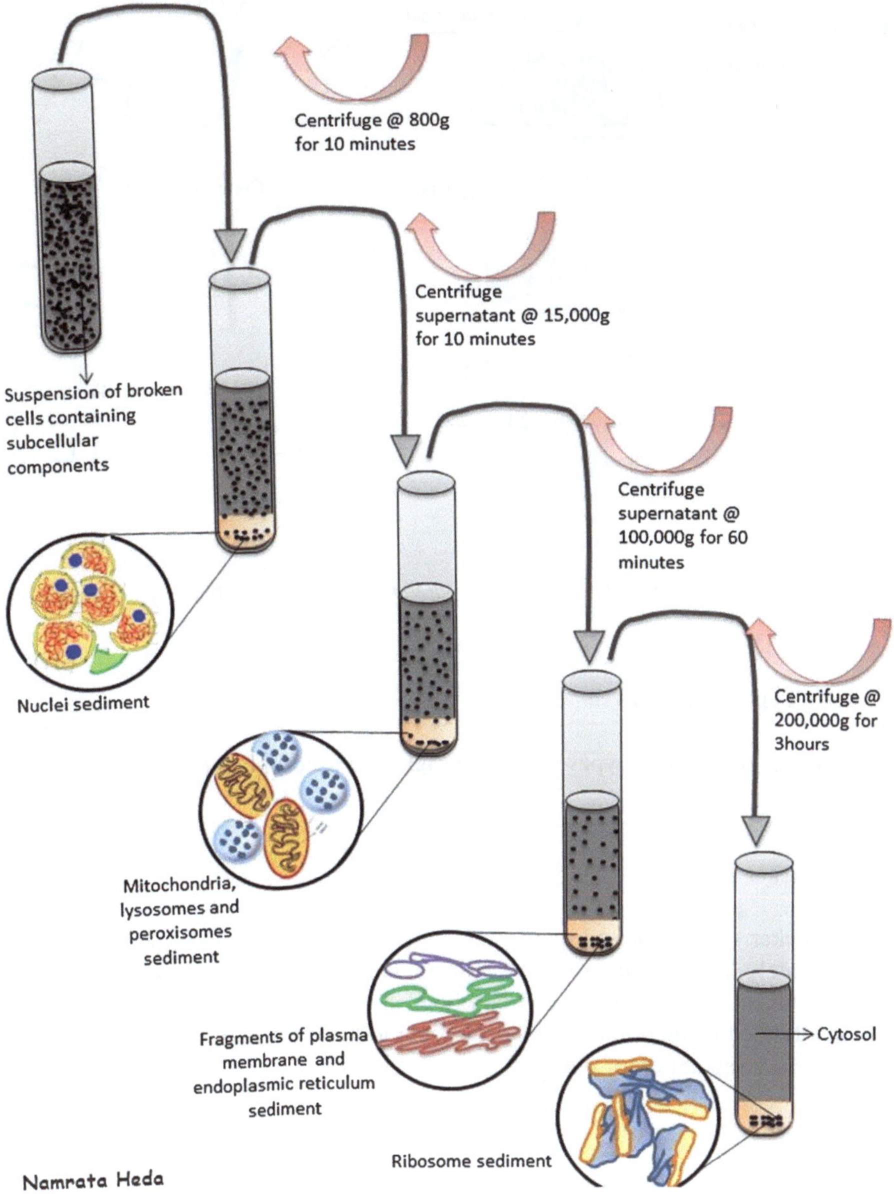

Fig. 3.23 Subcellular fractionation isolation by centrifugation. Source: https://www.google.co.in/imgres

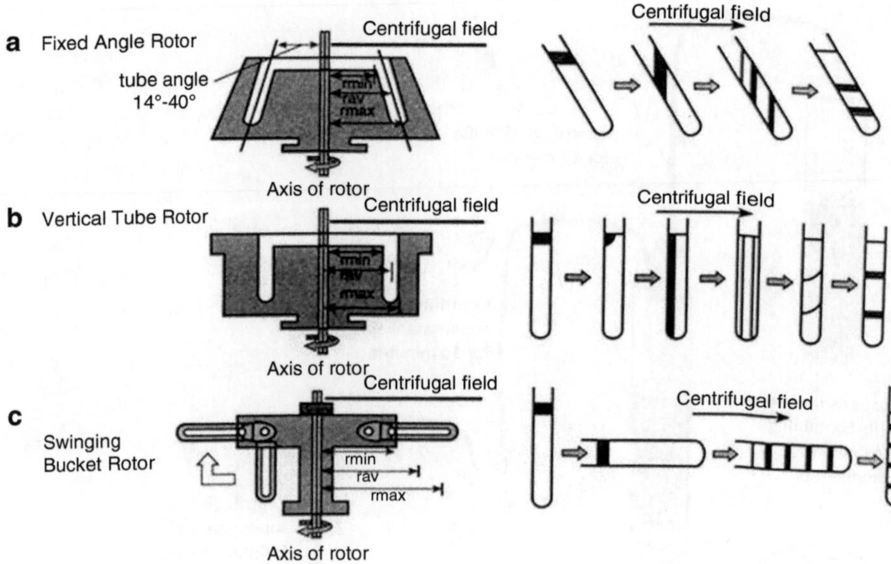

Fig. 3.24 Types of centrifugation rotors, from *top* to *bottom*. Source: https://www.google.co.in/imgres

3.2.4.3 Types of Rotors

Generally three different types of rotor designs are used in centrifuges:

Swing-Bucket Rotor

Swing-bucket rotors, for sedimentation with smaller centrifugal forces of up to approximately $6000 \times g$, are used in medical and research laboratories (Fig. 3.24a).

Advantages: A higher yield can be obtained by filling the centrifuge test tube completely. Sedimentation takes place in the middle of the bottom of the test tube, enabling easy separation of phases.

Drum Rotor

Drum rotors are specialized rotors for higher rotational speeds or (to be precise) for higher centrifugal forces. The vessels (usually reaction vials) are held in cassettes and placed vertically into the rotor. The sedimentation method is similar to that in swing-bucket rotors (Fig. 3.24b).

Advantages: Compared with swing-bucket rotors, drum rotors feature higher rotational speeds and therefor higher centrifugal forces.

Fixed-Angle Rotor

Fixed-angle rotors are used for sedimentation with high centrifugal forces of up to and more than 60,000×g, which is typically needed in research laboratories (Fig. 3.24c).

Advantages: Fixed-bucket rotors allow shorter sedimentation times than swing-bucket rotors because higher rotational speeds and higher centrifugal forces can be achieved. The sedimentation distance is also generally shorter than with swing-bucket rotors, which also has a positive influence on sedimentation time. The capacity of the rotor chamber can be better utilized than with swing-bucket rotors.

3.2.4.4 Industrial Centrifuges

In recent years, continuous flow industrial centrifuges have been developed. These involve continuous feeding of slurry and collection of clarified fluid. The solids deposited are removed intermittently. Five different types of centrifuges are described next.

Tubular-Bowl Centrifuge

A tubular-bowl centrifuge is a type of sedimentation centrifuge. They can be used to separate both solid/liquid and liquid/liquid mixtures. In tubular-bowl centrifuges, feed enters from the bottom of the cylindrical bowl. A distributor and baffle assembly accelerates the incoming liquid to rotor speed. Then, a baffle separates the feed into its components (two liquid layers for liquid–liquid separations, or solid and liquid layers for solid–liquid separations). The outer layer, which consists of the heavier components, becomes concentrated against the wall, while the inner layer, which consists of the lighter components, floats on top. Each layer then travels up the side of the bowl as an annulus. Liquid layers are discharged through overflow ports located on the top of the centrifuge. Solid build-ups remain in the bowl and are recovered manually.

Tubular-bowl centrifuges generally consist of a bowl, a motor, and a drive assembly. The bowl is suspended from an upper bearing and hangs freely. This allows the bowl to find its natural axis of rotation if it becomes unbalanced. The discharge ports at the top of the bowl are located at different radii and elevations depending on the properties of the components to be separated. Figure 3.25 shows the upper bearing suspending the bowl as well as a discharge port.

The inner bowl in tubular-bowl centrifuges can range in capacity from 1 to 15 gallons, and can handle up to 1200 gallons of feed per hour. Tubular-bowl centrifuges can be outfitted with a knife discharge system, which mechanically removes any built-up solids within the bowl. The feed inlet into the bowl of a tubular centrifuge is shown in Fig. 3.25.

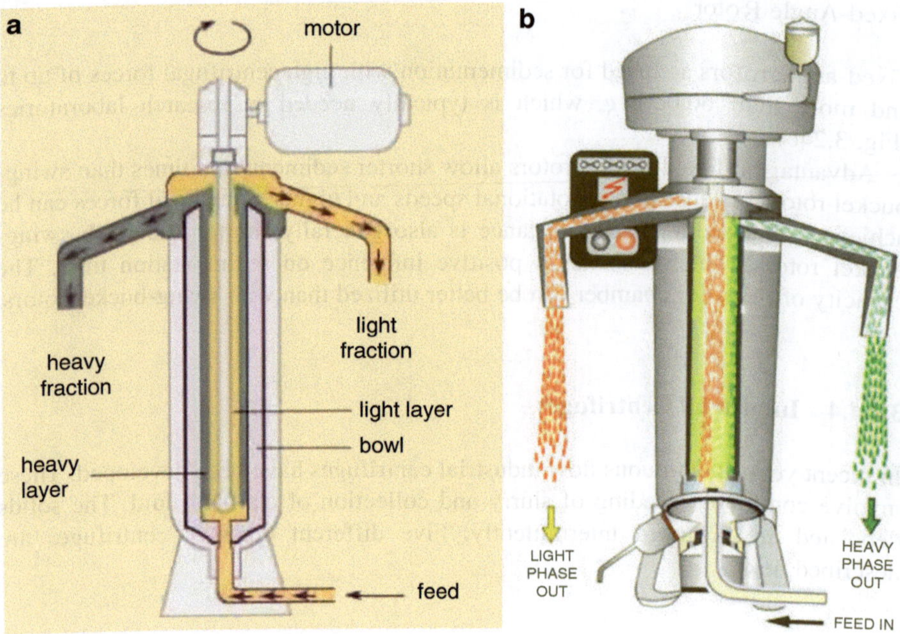

Fig. 3.25 (**a**) Schematic diagram of a tubular bowl centrifuge (Source: https://www.britannica.com/technology/tubular-centrifuge). (**b**) The tubular bowl centrifuge comprises of bow, specially designed slow acceleration motor and starter

Tubular-bowl centrifuges have a wide range of uses in industry. A typical application is the purification of lubricating industrial oils containing 1% or less of sedimentable solids. Tubular-bowl centrifuges also see widespread use in the food, biochemical, and pharmaceutical industries.

Figure 3.25a shows an example of a laboratory-scale tubular centrifuge that is powered by a single-phase motor. These centrifuges can be powered with electric motors, steam turbines, or compressed air. Figure 3.25b shows an example of a tubular centrifuge used in industry.

Disc Nozzle Centrifuge

Disc centrifuges are effective in removing water and finer solids. Feed enters the top of the centrifuge and rotates at high speeds within the bowl, typically at more than 3500 rpm. The high centrifugal forces (over $5000 \times g$) push the water and solids to the outer edge of the disc stack. The lighter bitumen/naphtha mixture remains in the center and is forced out of the top of the centrifuge. The stacked discs provide ample surface area for the fine solids to settle, which are pushed outward along with the heavier water phase. The disk centrifuge shown in Fig. 3.26 is commonly used in the brewing industry. A pressure control valve at the bottom of the centrifuge can

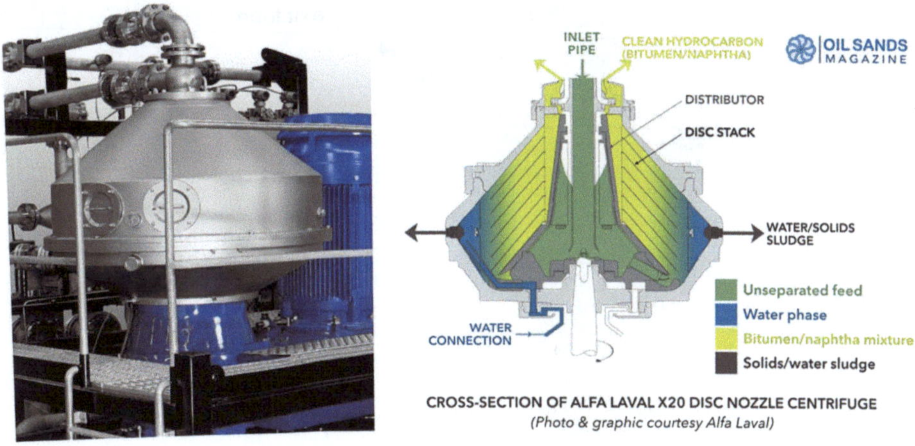

Fig. 3.26 Disc centrifuge (courtesy of Alfa Laval, Richmond, VA)

either let excess water out or allow hot flush water to enter the centrifuge. The space between the rotating bowl and the outer casing is continuously purged with nitrogen or natural gas. After the bitumen/naphtha mixture is processed through the discs, the solids and water content is typically less than 2%. The most common supplier of disc centrifuges is Alfa Laval (Fig. 3.26).

Because of their rotating mechanical parts, centrifuges are susceptible to wear and tear. They are also big consumers of power, which increases the electrical load on the overall operation. Centrifuges are relatively unpopular pieces of equipment and many operators have chosen to add capacity through the addition of more inclined plate settlers to their circuits. Some operators, such as Suncor and Canadian Natural Resources, use cyclones in place of centrifuges to dewater oil and remove fine solids. Despite their shortcomings, centrifuges are very effective in reducing water content as a result of the high *g*-forces imparted onto the diluted froth, which is difficult to achieve with cyclones alone.

Multichamber Centrifuge

A multichamber centrifuge is basically a modified tubular-bowl centrifuge. It consists of several chambers connected in such a way that the feed flows in a zigzag fashion. There is a variation in the centrifugal force in different chambers. The force is much higher in the periphery chambers and, as a result, the smallest particles settle in the outermost chamber (Fig. 3.27). The bowl of a multichamber centrifuge consists of a series of short tubular sections of increasing diameter, linked together to form a continuous tubular passage. Feed is introduced into the smallest diameter tube first and progresses through outer tube diameters as they increase in size. Typically, up to six chambers are connected. Because the centrifugal force

Fig. 3.27 Photo showing short tubular sections of increasing diameter linked together to form a continuous tubular passage in a multichamber centrifuge. Source: https://www.google.co.in/imgres

increases as diameter increases, the heavier particles are deposited in the smallest diameter tube.

Multichamber centrifuges are widely used in the food industry, especially to clarify fruit juices, wort, and beer (Fig. 3.28).

Scroll Centrifuge or Decanter

A scroll centrifuge is composed of a rotating horizontal bowl tapered at one end. The decanter is generally used to concentrate fluids with high solid concentration (biomass content 5–80%). The solids are deposited on the wall of the bowl, which can be scrapped and removed from the narrow end (Fig. 3.29). Solid-bowl scroll centrifuges are used as a first-pass to remove the coarse solids from diluted froth. Sometimes referred to as decanter centrifuges, these units consist of a horizontal revolving shell, or bowl, inside which a screw, or scroll, rotates in the same direction. Feed enters the center of the bowl, which rotates at approximately 1000 rpm, pushing the solids to the outer edge. The inner scroll moves the sludge in the direction opposite to liquid flow, rotating at a slightly lower speed, moving the solids to a discharge hopper. The horizontal bowl has a conical shape at the

Fig. 3.28 Use of multichamber centrifuge to clarify fruit juice, wort, and beer. Adapted from: https://www.google.co.in/imgres

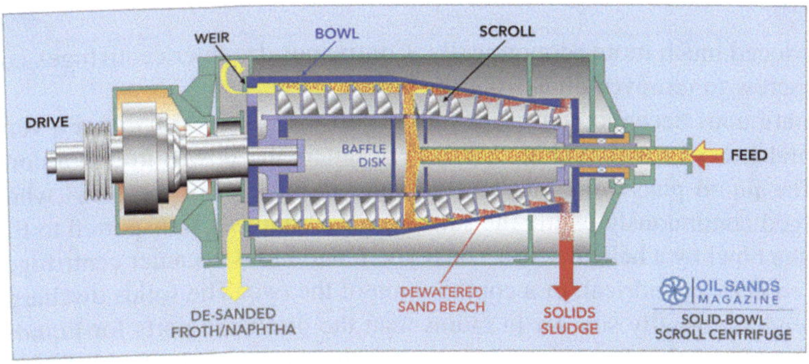

Fig. 3.29 Section view of a scroll centrifuge (courtesy of Oil Sands magazine)

discharge end to encourage beaching and minimize liquid carry-over into the tailings stream. These centrifuge units recover up to 98% of the hydrocarbons but still leave a solids content of 3–5% in the product. A common manufacturer of solid-bowl scroll centrifuges is the Bird Machine Company, now owned by Andritz.

Continuous Decanter Centrifuge

A continuous decanter centrifuges is a horizontal decanter centrifuge for continuous separation of suspended fines from liquids with pasty or slimy sediments. It is used for "soft" products such as yeast, bacteria slurries, precipitated proteins, and other products that are difficult or nearly impossible to process adequately with decanter or disc stack centrifuges. A typical application is the dewatering of soy milk products. By means of this centrifuge, soy milk and soy proteins can be processed

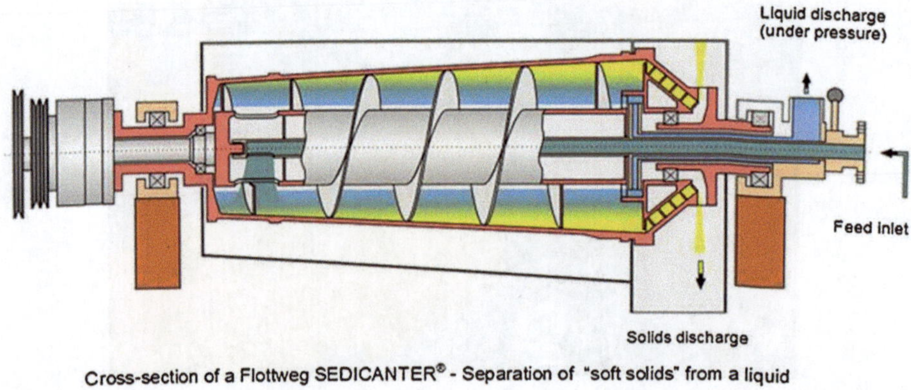

Fig. 3.30 Cross-section of a horizontal decanter centrifuge for continuous separation of suspended fines from liquids with pasty or slimy sediments (courtesy of Schwarz Global Consulting, www.sgconsulting.co.za)

and produced much more economically. Continuous decanter centrifuges contain a helical screw to remove solids from the bowl.

A continuous decanter centrifuge has a solid wall bowl with either a vertical or horizontal axis of rotation (Fig. 3.30). Feed enters the bowl through a concentric tube. The liquid phase migrates to the larger radius end of the bowl where it is discharged continuously, while the solids are continuously transported to the other end of the bowl by a helical screw conveyor. Continuous decanter centrifuge bowls can be conical, cylindrical, or a combination of the two. The solids discharge ports at one end are usually smaller in radius than the discharge ports for liquids at the other end. The helical screw conveyor runs the length of the bowl. Flocculating agents may be added to the centrifuge to collect softer solids (a flocculating agent is a substance that causes soft solids to congeal into small masses).

Another example of a continuous centrifuge is the pusher centrifuge. These are used to separate coarse solids from slurry mixtures. In order to do this, the slurry is deposited on the feed screen where liquors drain through and solids collect. Next, the pusher plate moves the solid film forward, exposing more usable screen space. As the film cake dries, more liquor can be sprayed to remove more impurities. Finally, after the cake has dried it can be pushed out for further processing (Fig. 3.31).

Continuous decanter centrifuges are most typically used for processes in which continuous production is required, such as purifying used lubricating oils. Corn used to make ethanol is shown below, to the left. Ethanol producers use centrifuges in the production of ethanol from corn. The picture to the right is an example of an industrial pusher centrifuge used to separate coarse solids from crystalline slurries continuously with minimum preventive maintenance.

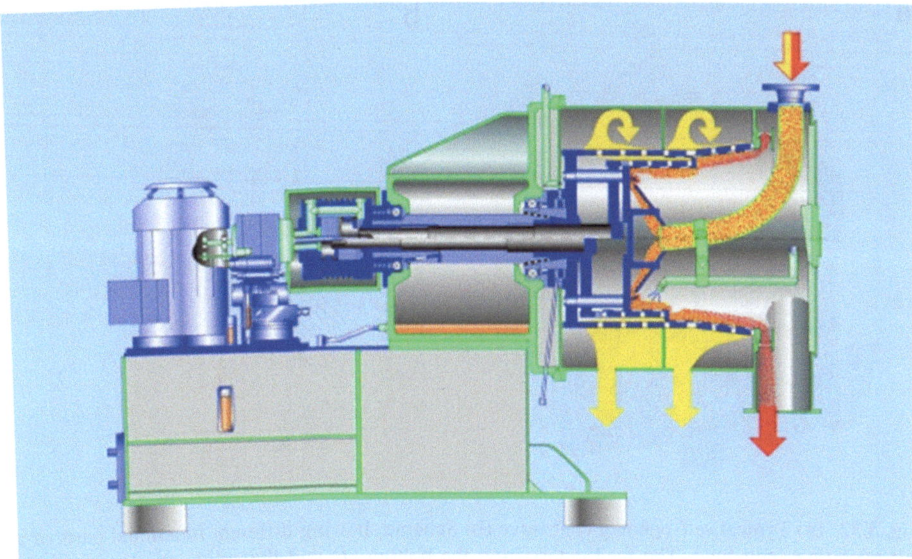

Fig. 3.31 Line diagram of continuous solid/liquid separation, pusher centrifuges

Self Operating Centrifuge

As their name implies, self-opening centrifuges are able to open during operation to discharge solids (Fig. 3.32a). Feed enters the bowl from the bottom, is exposed to centrifugal force, and separates into its components. When the desired amount of solids has accumulated, a sleeve inside the bowl opens and the solids are discharged radially outward. Figure 3.32a shows the sleeve in open and closed positions.

The sleeve in self-opening centrifuges is kept closed by static pressure induced by hydraulic fluid. When the pressure decreases, the sleeve is allowed to open, exposing an annular, ring-like area through which solids are discharged. An elastomeric seal around the top of the bowl ensures that there is no leakage while the sleeve is closed. Self-opening centrifuges can typically handle up to 220 gallons of feed per minute. Figure 3.32b shows a typical self-opening centrifuge.

3.3 Cell Disruption and Release of Intracellular Components

3.3.1 Mechanical Means

The importance of microorganisms as a source of commercially useful chemicals, antibiotics, and enzymes has been recognized for a very long time. Nearly all

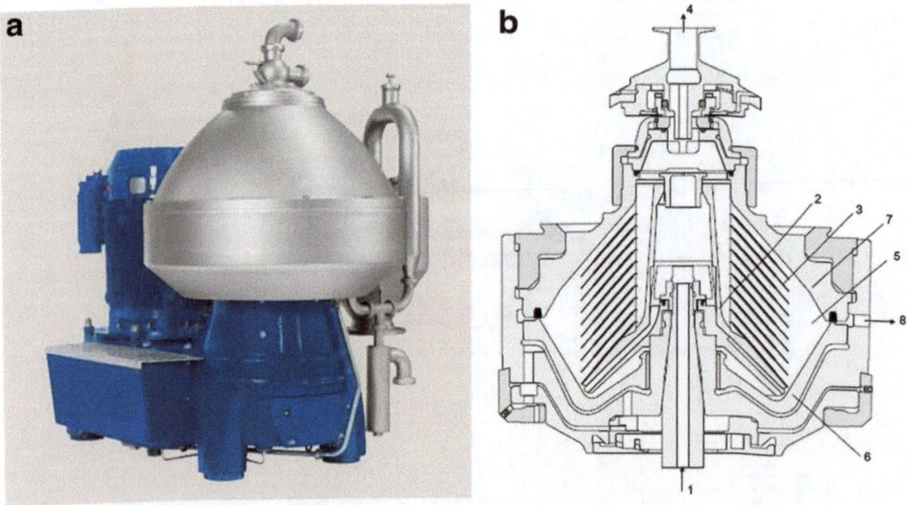

Fig. 3.32 (**a**) Typical self-opening centrifuge (**b**) Scheme showing different functional parts of a self-operating centrifuge. The feed enters from the bottom (*1*) and the solids are separated (*2*) through the disk stack (*3*) and collect in the periphery (*5*), which is surrounded by a cooling jacket (*7*). The concentrate leaves from the top (*4*). The hydraulic system forces the bowl bottom (*6*) to drop down, opening the ports on the side (*8*), allowing the solids to discharge (Source: https://www.google.co.in/imgres)

chemicals of microbial origin produced industrially today are of the extracellular type (Table 3.4); that is, they are produced within the microbial cell but then excreted into the surrounding environment. A much larger proportion of the potentially useful microbial products is retained within the cells. A vast majority of industrial enzymes (Table 3.4) are intracellular in nature [28, 29].

Even greater use of microbial products, many of which are intracellular, can be expected from the predicted surge in biotechnology [29, 30]. The isolation of intracellular material requires that the cell either be genetically engineered so that what would normally be an intracellular product is excreted into the environment, or it must be disintegrated by physical, chemical, or enzymatic means to release its contents into the surrounding medium. The genetic manipulation of microbial cells to make them leaky is limited in scope. Making the cell fully permeable to any significant fraction of intracellular products and enzymes would not only be difficult, but also implies discontinued existence of the cell. It is in this context that microbial cell disruption for intracellular product isolation becomes of increasing importance. However, few studies have been published on the subject. probably because of the high capital and operating costs of pilot plants for large-scale isolation of intracellular products and the requirement of sizeable teams of scientists and technical staff to obtain meaningful biochemical engineering design data [30].

In the last few years, several intracellular enzymes have been produced industrially. Examples include glucose oxidase for food preservation, penicillin acylase

Table 3.4 Some important industrial enzymes and their sources

Enzyme[a]	EC number[b]	Source	Intra/extra-cellular[c]	Scale of production[d]	Industrial use
Animal enzymes					
Catalase	1.11.1.6	Liver	I	–	Food
Chymotrypsin	3.4.21.1	Pancreas	E	–	Leather
Lipase[e]	3.1.1.3	Pancreas	E	–	Food
Rennet[f]	3.4.23.4	Abomasum	E	+	Cheese
Trypsin	3.4.21.4	Pancreas	E	–	Leather
Plant enzymes					
Actinidin	3.4.22.14	Kiwi fruit	E	–	Food
α-Amylase	3.2.1.1	Malted barley	E	+++	Brewing
β-Amylase	3.2.1.2	Malted barley	E	+++	Brewing
Bromelain	3.4.22.4	Pineapple latex	E	–	Brewing
β-Glucanase[g]	3.2.1.6	Malted barley	E	++	Brewing
Ficin	3.4.22.3	Fig latex	E	–	Food
Lipoxygenase	1.13.11.12	Soybeans	I	–	Food
Papain	3.4.22.2	Pawpaw latex	E	++	Meat
Bacterial enzymes					
α-Amylase	3.2.1.1	*Bacillus*	E	+++	Starch
β-Amylase	3.2.1.2	*Bacillus*	E	+	Starch
Asparaginase	3.5.1.1	*Escherichia coli*	I	–	Health
Glucose isomerase[h]	5.3.1.5	*Bacillus*	I	++	Fructose syrup
Penicillin amidase	3.5.1.11	*Bacillus*	I	–	Pharmaceutical
Protease[i]	3.4.21.14	*Bacillus*	E	+++	Detergent
Pullulanase[j]	3.2.1.41	*Klebsiella*	E	–	Starch
Fungal enzymes					
α-Amylase	3.2.1.1	*Aspergillus*	E	++	Baking
Aminoacylase	3.5.1.14	*Aspergillus*	I	–	Pharmaceutical
Glucoamylase[k]	3.2.1.3	*Aspergillus*	E	+++	Starch
Catalase	1.11.1.6	*Aspergillus*	I	–	Food
Cellulase	3.2.1.4	*Trichoderma*	E	–	Waste
Dextranase	3.2.1.11	*Penicillium*	E	–	Food
Glucose oxidase	1.1.3.4	*Aspergillus*	I	–	Food
Lactase[l]	3.2.1.23	*Aspergillus*	E	–	Dairy
Lipase[e]	3.1.1.3	*Rhizopus*	E	–	Food
Rennet[m]	3.4.23.6	*Mucor miehei*	E	++	Cheese
Pectinase[n]	3.2.1.15	*Aspergillus*	E	++	Drinks
Pectin lyase	4.2.2.10	*Aspergillus*	E	–	Drinks
Protease[m]	3.4.23.6	*Aspergillus*	E	+	Baking
Raffinase[o]	3.2.1.22	*Mortierella*	I	–	Food

(continued)

Table 3.4 (continued)

Enzyme[a]	EC number[b]	Source	Intra/extra-cellular[c]	Scale of production[d]	Industrial use
Yeast enzymes					
Invertase[p]	3.2.1.26	*Saccharomyces*	I/E	–	Confectionery
Lactase[l]	3.2.1.23	*Kluyveromyces*	I/E	–	Dairy
Lipase[e]	3.1.1.3	*Candida*	E	–	Food
Raffinase[o]	3.2.1.22	*Saccharomyces*	I	–	Food

[a]The names in common usage are given. As most industrial enzymes consist of mixtures of enzymes, these names may vary from the recommended names of their principal component. Where appropriate, the recommended names of this principal component is given below
[b]EC number of the principal component
[c]*I* intracellular enzyme; *E* extracellular enzyme
[d]+++ >100 ton $year^{-1}$; ++ >10 ton $year^{-1}$; + >1 ton $year^{-1}$; – <1 ton $year^{-1}$
[e]Triacylglycerol lipase
[f]Chymosin
[g]Endo-1,3(4)-β-glucanase
[h]Xylose isomerase
[i]Subtilisin
[j]α-Dextrin endo-1,6-α-glucosidase
[k]Glucan 1,4-α-glucosidase
[l]β-Galactosidase
[m]Microbial aspartic proteinase
[n]Polygalacturonase
[o]α-Galactosidase
[p]β-Fructofuranosidase

for antibiotic conversion, and asparaginase for possible cancer therapy [31]. Other examples of intracellular microbial enzymes produced commercially are given in Table 3.4. The necessity of harvesting the producing cells to subsequently extract an internal constituent is a major economic disadvantage and, in part, results in the present preoccupation with the manufacture of products of very high value. Possible cost reductions, however, can be achieved by simultaneous isolation of a number of intracellular products following cell disruption [32–34]. This would be even more advantageous if extracellular product isolation preceded intracellular product isolation from the same fermentation batch.

It is essential for both research and manufacturing to obtain intracellular contents from cell rupturing in a controlled fashion (Fig. 3.33). Some cells, such as yeasts, are much harder to rupture than others. The method chosen for rupturing the cell also has to take into account the nature of the final product. If the cell is being ruptured to obtain an active protein, the method chosen has to be gentle enough that it does not denature or damage the protein. This can be achieved by either mechanical or non-mechanical methods [35–43].

Microbial intracellular proteins with catalytic or biological activity, or intracellular metabolites of pharmaceutical important, are of great interest to the biotechnology industry. For the large-scale disruption of microorganisms, mechanical disintegrators such high-pressure homogenization or wet milling in high-speed

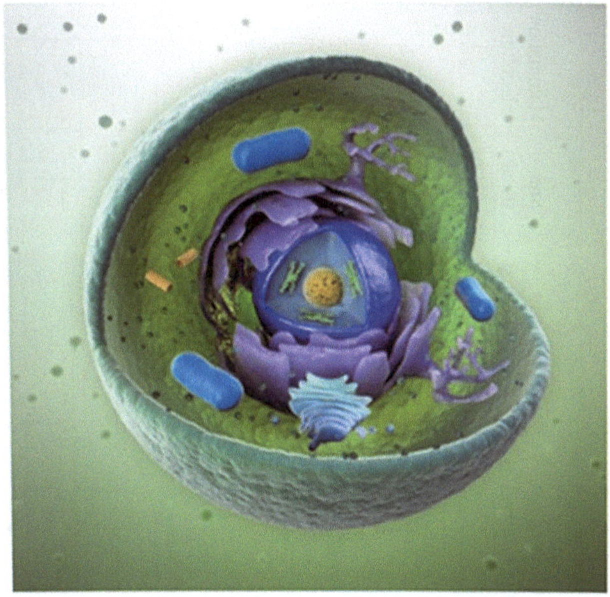

Fig. 3.33 Pictorial view of the 3D structure of a cell, showing intracellular space

agitator bead mills are commonly employed, which are techniques originally designed for the milk and paint industries. The disruption process is quantified by monitoring changes in absorbance, particle size, total protein concentration, or activity of a specific enzyme.

The effectiveness of a cell disruption method determines the accessibility of intracellular proteins for extraction and solubility. Different biological materials require different cell disruption strategies, which can be divided in two categories, gentle and harsher (Table 3.5).

Breaching of the cell wall/envelope and cytoplasmic membrane can pose problems, especially when cells possess strong cell walls (e.g., a pressure of 650 bar is needed to disrupt yeast cells). General problems associated with cell disruption include the liberation of DNA, which can increase the viscosity of the suspension and thus affect further processing. Addition of DNAse can help prevent this problem. If mechanical disruption is used, then heat is invariably generated and cooling measures need to be implemented. Also, products released by eukaryotes are often subject to degradation by hydrolytic enzymes (proteases, lipases, etc.). This damage can be reduced by the addition of enzyme inhibitors, cooling the cell extract, and rapid processing. Currently, intracellular products are released from microorganisms mainly by mechanical disruption of the cells. In this process, the cell envelope is physically broken, releasing all intracellular components into the surrounding medium.

Both solid shear (e.g., bead mill) and liquid shear (e.g., high-pressure homogenizer) methods of cell disruption have proven successful on a large scale. The solid shear methods involve either a grinding action as in a ball mill or extrusion of

Table 3.5 Suitability of cell disruption methods for various sample types

Technique	Description	Bacteria	Yeast, algae, fungi	Seeds	Green plant material	Soft tissues	Mammalian cell culture
Gentle methods							
Osmotic lysis	Suspension of cells in hypotonic solution; cells swell and burst, releasing cellular contents	—	—	—	—	—	•
Freeze-thaw lysis	Freezing in liquid nitrogen and subsequent thawing of cells	—	—	—	—	—	•
Detergent lysis	Suspension of cells in detergent-containing solution to solubilize the cell membrane; this method is usually followed by another disruption method, such as sonication	—	—	—	—	—	•
Enzymatic lysis	Suspension of cells in iso-osmotic solutions containing enzymes that digest the cell wall (e.g., cellulase and pectinase for plant cells, lyticase for yeast cells, and lysozyme for bacterial cells); this method is usually followed by another disruption method, such as sonication	•	•	—	•	—	•
Harsher methods							
Sonication	Disruption of a cell suspension, cooled on ice to avoid heating and subjected to short bursts of ultrasonic waves	•	•	—	—	—	•
French press	Application of shear forces by forcing a cell suspension through a small orifice at high pressure	•	•	—	—	—	•
Grinding	Breaking cells of solid tissues and microorganisms with a mortar and pestle; usually, the mortar is filled with liquid nitrogen and the tissue or cells are ground to a fine powder	•	•	•	•	•	—
Mechanical homogenization	Mechanical homogenization with either a handheld device (e.g., Dounce and Potter-Elvehjem homogenizer), blender, or other motorized device; this approach is best suited for soft, solid tissues	—	—	—	•	•	—
Glass-bead homogenization	Application of gentle abrasion by vortexing cells with glass beads	•	•	—	—	—	•

frozen cells, either alone or as a cell/ice (or other abrasive) mixture, through narrow gaps or orifices under high pressure [44]. Most of the cell disruption equipment in current use was originally designed for the homogenization and size reduction of very different commercial products such as milk and paint. Following Dunnill [31] and Wuytack [45] the application of this hardware to cell disruption has been very successful.

3.3.1.1 Liquid Shear Homogenization

High-pressure homogenization (HPH) is a purely mechanical process, which involves forcing a fluidic product through a narrow gap (the homogenizing nozzle) at high pressure (150–200 MPa, or 350–400 MPa for ultrahigh pressure homogenization, UHPH) [46, 47] (Fig. 3.34). The liquid product is subjected to very high shear stress, causing the formation of very fine emulsion droplets. The origin of this shear is the sudden restriction of flow under high pressure through a restrictive valve [48]. Extreme shear and high-energy input have to be applied to reduce droplets from the micro- to the nanoscale range. The Laplace pressure, which causes the resistance to droplet deformation and breakup, needs to be overcome. This Laplace pressure increases when the droplet diameter decreases and is therefore larger for nanoemulsions than for ordinary emulsions [49]. Nanoemulsions, because of their high surface-to-volume ratio, need high amounts of surfactants (such as proteins) for their stabilization. The interaction of proteins with the newly formed droplets is improved under HPH [50].

A high-pressure homogenizer consists of a high-pressure pump and a homogenizing nozzle. The pump is used to compress the crude emulsion to the required pressure. During depressurization in the homogenizing nozzle, the drops are disrupted. The nozzle is decisive for the efficiency of disruption for emulsions prepared with HPH [46]. The homogenizing pressure is typically between 5 and 50 MPa. Using more recently introduced high-pressure homogenizers (e.g., microfluidizers) it is also possible to produce emulsions at higher pressures (up to approximately 400 MPa) [50]. High-pressure homogenizers can be classified according to the nozzle type or (roughly) the mechanism that disrupts the droplet (Fig. 3.34).

Additional Effects

The high pressures applied cause a temperature increase because of the heat of compression. This temperature increase is enhanced by shear effects and partial conversion of mechanical energy into heat. The result is a total temperature increase of 17–21 °C per 100 MPa. This needs to be controlled in the case of heat-sensitive food compounds. HPH also enhances product texture and mouthfeel. Using nanoemulsions in food products can facilitate the use of less fat without compromising creaminess, thus offering the consumer a healthier option [49–51].

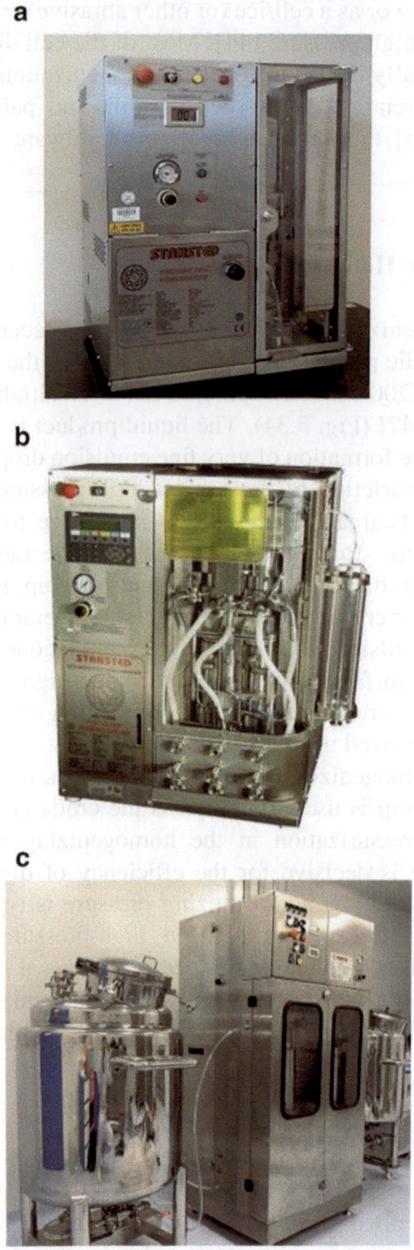

Fig. 3.34 Range of high-pressure homogenizers: (**a**) Continuously adjustable process pressure from 1500 psi (10 MPa) to 60,000 psi (410 MPa). (**b**) For continuous operation at working pressures up to 400 MPa and flow rates up to 16 L/h. (**c**) Hyd-Lok system for outstanding ease of use, cleaning, and service (Adapted from Stansted Fluid Power, UK)

Process Parameters

The mechanism of droplet disruption depends on the operating parameters, including flow rate, pressure, number of cycles, temperature, and device parameters (e.g., nozzle geometry and impingement design). The mechanism also depends on product parameters such as composition (fat, water, protein, etc.) and viscosity (although of less importance than in rotor-stators systems). HPH is a mature technology in milk processing. High-pressure homogenizers of piston-gap type can currently deliver pressures up to 150–200 MPa and even up to 300–400 MPa for UHPH [52–58]. Jet dispersers involve two fluid jets (each from opposite bores) that collide with one another to disrupt particles. Oil emulsification can be performed by direct injection of the oily phase under pressure (up to 350 MPa) into a continuous aqueous phase at low pressure [59]. Microfluidization (up to 200 MPa) produces a tighter particle size distribution than that of traditional techniques. The scalability of process depends on the formulation of the product. Multiple-pass HPH can pose a problem with regard to industrial implementation, but can be overcome by applying higher pressures.

3.3.1.2 Grinding with Glass Beads

Cells mixed with glass beads are subjected to very high speed agitation in a reaction vessel. The cells break as they are forced against the wall of the vessel by the beads. Several factors influence the cell breakage-size and quantity of glass beads required, including concentration and age of cells, temperature, and agitator speed. Under optimal conditions, one can expect a maximal breakage of about 80% of the cells. A cell disrupter employing glass beads is shown in Fig. 3.35. It contains a cylindrical body with an inlet, outlet, and a central motor-driven shaft. To this shaft are fitted radial agitators. The cylinder is fitted with glass beads. The

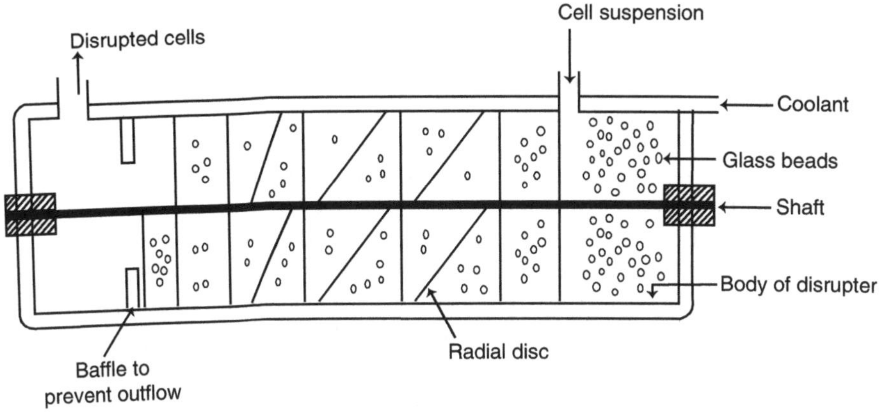

Fig. 3.35 Scheme of a grinding mill

cell suspension is added through the inlet and the disrupted cells leave through the outlet. The body of the cell disrupter is kept cool while in operation.

A ball mill or grinder is used in mines to grind material into a fine powder. There are different sizes of ball mills; large mills rotate on their axis whereas smaller ones consisting of a cylindrical container normally sit on two drive shafts. To transmit the rotary motion, pulleys and belts and sometimes a rock tumbler are also used. The ground materials are, in turn, used in mineral dressing processes such as pyrotechnics, paints, and ceramics. Such devices mix materials such as ores and even ceramic raw materials by rotating them in a round horizontal axis. The mills are filled half way with the material to be ground and the grinding medium added. A ball mill that is commonly used in laboratories is called a planetary bill mill. It is used in demonstrations or for professional use to grind materials to the smallest size.

3.3.1.3 Ultrasonic Disruption

Another widely applied method is cell lysis using high-frequency sound that is produced electronically and transported through a metallic tip to an appropriately concentrated cellular suspension. Because of very high costs it is only used at laboratory scale for lysis of cells with less resistant cell walls, such as bacteria and fungi. Ultrasonic disruption of cells involves cavitations, microscopic bubbles, or cavities generated by pressure waves. It is performed by ultrasonic vibrators that produce high-frequency sound with a wave density of approximately 20 kHz. A transducer converts the wave into mechanical oscillations via a titanium probe immersed in a concentrated cell suspension. However, this technique also generates heat, which can denature thermolabile proteins. Rod-shaped bacteria are easier to break than cocci, and Gram-negative organisms are more easily disrupted than Gram-positive cells. Sonication is effective on a small scale but is not routinely used in large-scale operations because of problems with power transmission and heat dissipation.

Disadvantages of Mechanical Disruption

Mechanical disruption methods suffer from several drawbacks. Because cells are broken completely, all intracellular materials are released. Therefore, the product of interest must be separated from a complex mixture of proteins, nucleic acids, and cell wall fragments. Released nucleic acids can increase the viscosity of the solution and complicate subsequent processing steps or experimental procedures, especially chromatography. The cell debris produced by mechanical lysis often consists of small cell fragments, making the solution difficult to clarify. Complete product release often requires more than one pass through the disruption device, exacerbating the problem by further reducing the size of fragments. Small fragments are difficult to remove by continuous centrifugation because the throughput of the device is inversely related to the square of the particle diameter. Filtration is complicated by the gelatinous nature of the homogenate and by its tendency to foul membranes. Furthermore, mechanical methods expose the cells, and hence the extracted product, to very harsh conditions. Although it is now generally accepted that most proteins can tolerate the high pressure inside a homogenizer or ball mill, most proteins are denaturated by the heat generated unless the device is sufficiently cooled.

3.3.2 *Non-mechanical Means*

Another way to disrupt the cells is to cause their permeabilization. This can be achieved by numerous methods, as outlined below.

3.3.2.1 Chemical Permeabilization

Many chemical methods have been employed to extract intracellular components from microorganisms by permeabilizing the outer-wall barriers. Often used with plant cells (and sometimes in combination with shearing), organic solvents such as toluene, ether, benzene, butanol, chloroform, and methanol can be used to permeate cell walls by creating channels through the cell membrane. Chemical permeabilization can also be achieved using phenylethyl alcohol, DMSO, antibiotics, thionins, surfactants (Triton, Brij, Duponal), chaotropic agents, and chelates. Ethylenediaminetetraacetic acid (EDTA) is widely used to disrupt Gram-negative microorganisms because it chelates cations, causing holes in the cell walls. Its effectiveness is a result of its ability to bond the divalent cations Ca^{2+} and Mg^{2+}. Mg^{2+} stabilizes the structure of outer membranes by bonding the lipopolysaccharides to each other. Once these cations are removed by EDTA, the lipopolysaccharides are released, resulting in areas of increased permeability in the outer walls [38].

Chaotropic agents such as urea and guanidine are capable of bringing some hydrophobic compounds into aqueous solution. They accomplish this by disrupting the structure of water (the hydrogen-bonding network), making it a less hydrophilic environment and weakening hydrophobic interactions between solute molecules.

3.3.2.2 Mechanical Permeabilization

One method of mechanical permeabilization is osmotic shock. Although cells exposed to slowly varying extracellular osmotic pressure are usually able to adapt to such changes, cells exposed to rapid changes in external osmolarity can be mechanically injured. This procedure is typically conducted by first allowing the cells to equilibrate internal and external osmotic pressure in a high-sucrose medium, and then rapidly diluting the sucrose. The resulting immediate overpressure of the cytosol damages the cell membrane. Enzymes released by this method are believed to be periplasmic, or at least located near the surface of the cell.

This can be achieved by equilibrating cells in 20% (w/v) buffered sucrose, rapidly harvesting, and re-suspending in water at 4 °C. It is useful for the release of products from the periplasmic space.

3.3.2.3 Enzymatic Permeabilization

Enzymatic digestion is often used in the laboratory as a very gentle method of cell rupture. The cell walls of yeast and bacteria are digested with special enzymes and then the unprotected cell can easily be ruptured by mild osmotic shock. Cell wall-degrading enzymes such as lysozyme, which degrades β-1,4 glycosidic linkages, is useful for Gram-positive organisms. Addition of EDTA to chelate metal ions improves the effectiveness of lysozyme. Yeast cell walls can be degraded by snail gut enzymes (containing a mixture of glucanases). Enzymatic rupture methods are, in general, too expensive for use in large-scale manufacturing [38].

3.3.2.4 Other Permeabilization Techniques

Basic proteins such as protamine or the cationic polysaccharide chitosan can permeabilize yeast cells. Similarly, mammalian cells can be permeabilized by exposure to several natural substances such as streptolysin or even viruses. Electric discharges have also been used to permeabilize mammalian cells in order to study secretion by exocytosis.

3.3.2.5 Detergents

Detergents that are ionic in nature (e.g., cationic cetyl trimethyl ammonium bromide and anionic sodium lauryl sulfate) can denature membrane proteins and lyse the cells. Nonionic detergents (although less reactive than ionic ones) are also used to some extent (e.g., Triton X-100 and Tween). The problem with the use of detergents is that they affect later purification steps, particularly salt precipitation. This limitation can be overcome by using ultrafiltration or ion-exchange chromatography for purification [59].

3.4 Concentration

Separation of target metabolites after filtration is an important production step. It is considered to be the most expensive part of process production [38, 41]. Conventional techniques such as chromatography, electrophoresis, and precipitation are widely employed. However, these methods involve considerable cost, provide low yields, and are not suitable for large-scale production. A recently developed economic downstream process for biomolecule recovery is the aqueous two-phase system (ATPS) (see Sect. 3.4.3.3 for more details) [60–62].

3.4.1 Evaporation

Water in the broth filtrate can be removed by a simple evaporation process. Evaporators, in general, have a heating device for supply of steam, a unit for separation of concentrated product and vapor, a condenser for condensing vapor, and accessories and control equipment. The capacity of the equipment is variable, ranging from small laboratory scale to industrial scale. Some of the important types of evaporators in common use are briefly described later in this section. The most common solvent in most evaporation systems is water. Evaporation differs from other mass transfer operations such as distillation and drying. In distillation, the components of a solution are separated according to their distribution between vapor and liquid phases, based on the differences in relative volatilities of the substances. Drying is the removal of moisture from a substance in the presence of a hot gas stream that carries away the moisture, leaving a solid residue as the product. Evaporation is normally stopped before the solute starts to precipitate.

An evaporator consists of a heat exchanger for boiling the solution, with special provisions for separation of liquid and vapor phases. Most industrial evaporators have tubular heating surfaces. The tubes can be horizontal or vertical, long or short and the liquid can be inside or outside the tubes. There are many ways to classify evaporators on the basis of how the heat exchanger shell and tubes are set, boiling systems etc. However, a brief description is given on different types of evaporators, in general.

3.4.1.1 Falling Film Evaporators

The process fluid to be vaporized is feed to the evaporator at the top of the tube sheet. A suitable distribution unit is necessary in order to achieve even liquid distribution. It is paramount that a minimal wetting rate is achieved. In general, the process liquid flows inside the tube. However, there are also applications where the liquid is applied on the outside of the tube. Gravity causes the liquid to form a continuous liquid film with hydrodynamic characteristics that are determined by the film Reynold's number and Kapitza number. In general, steam is used as a heating source. The steam condenses on the outside of the tube and flows by gravity to the bottom of the unit. Inside the tubes, the two-phase flow is separated by means of a separator at the bottom of the unit.

In multi-effect falling film evaporators, the vapor produced is used to heat a so-called second effect at a lower temperature and pressure level. This is possible because of the very low required mean temperature difference between heating medium and process liquid. In seawater desalination, up to 12 effects can be used to achieve the most efficient evaporation.

The process of falling film evaporation is illustrated schematically in Fig. 3.36a–c. A liquid film runs down the channel wall and evaporates, with the vapor either flowing co-currently with the film (Fig. 3.36a) or counter-currently to it (Fig. 3.36b). If counter-current flow is used, then there is a limit on the vapor velocity that can be generated before the liquid begins to be carried upward. In falling film evaporators, evaporation occurs directly at the liquid–vapor interface, with no nucleate boiling at the wall. This means that lower temperature driving forces can be used and that the extensive fouling often associated with nucleate boiling of crystallizing solutions is avoided. Falling film evaporators can take a number of different forms. An evaporator in which evaporation occurs on the inside of vertical tubes is shown in Fig. 3.36c. Note that particular care has to be taken to ensure uniform distribution of the liquid at the top of the tubes. In other forms of falling film evaporators, the liquid film may fall over plates, coils, or banks of horizontal tubes.

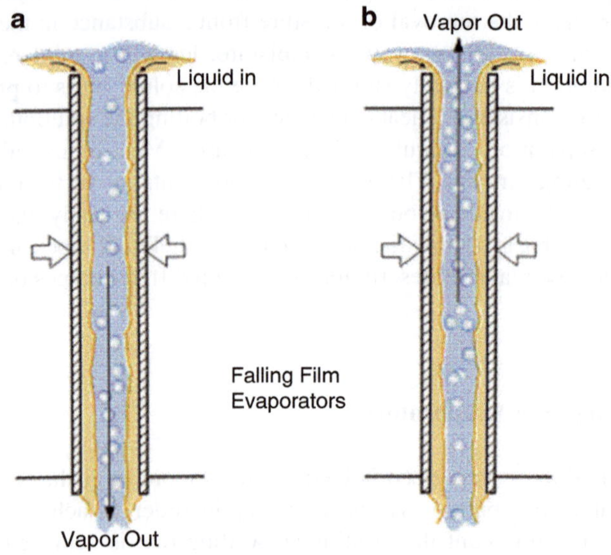

Fig. 3.36 (**a**, **b**) Two modes of falling film evaporation. (**c**) Falling film evaporator unit (courtesy of Thermopedia)

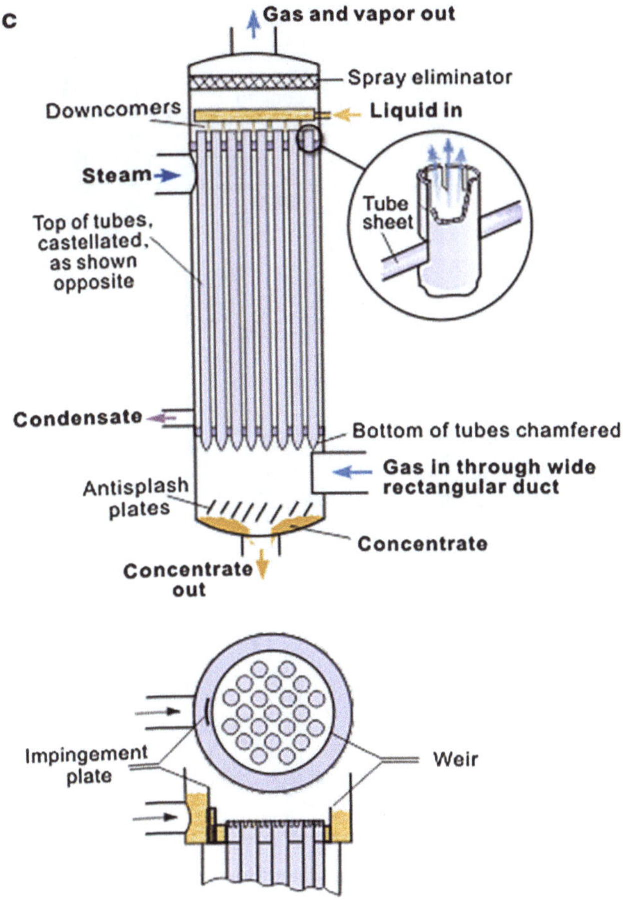

Fig. 3.36 (continued)

3.4.1.2 Forced Circulation Evaporators

Forced circulation evaporators (Fig. 3.37) are usually more costly than natural circulation evaporators. However, natural circulation evaporators are not suitable for some situations, such as:

- Highly viscous solutions, because of their low heat transfer coefficient
- Solutions containing suspended particles
- Heat-sensitive materials

This type of evaporator is suitable for processing liquors that are susceptible to scaling or crystallization. Liquid is circulated at a high rate through the heat exchanger, boiling being prevented within the unit by a hydrostatic head maintained above the top tube plate. As the liquid enters the separator, where the absolute pressure is slightly less than in the tube bundle, the liquid flashes to form a vapor.

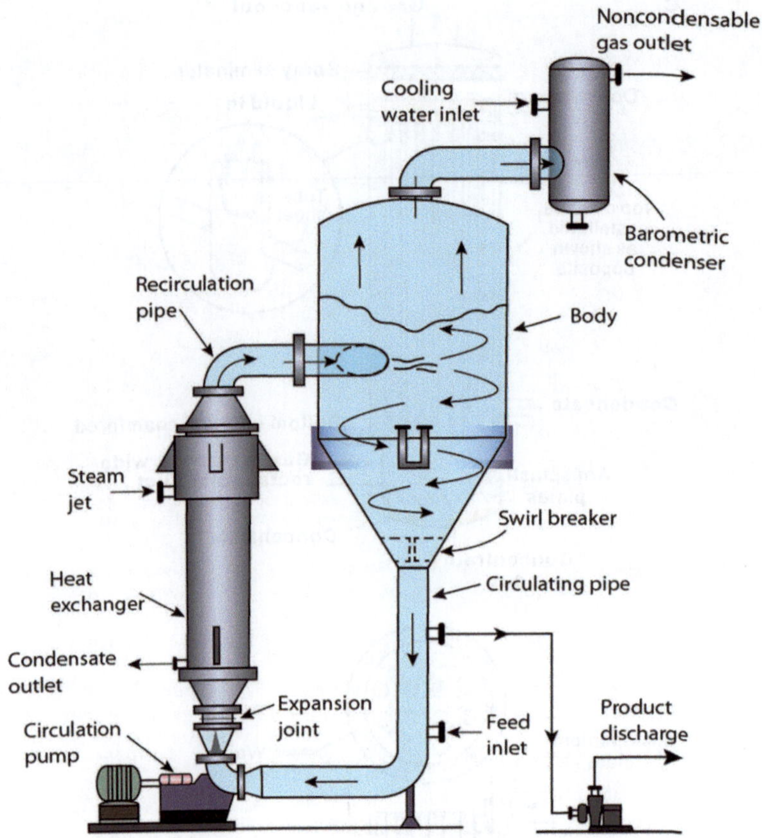

Fig. 3.37 Vertical tube forced-circulation evaporator (Source: https://www.google.co.in/imgres)

Applications of Forced Circulation Evaporators

The main applications for forced circulation evaporators are for concentration of inversely soluble materials, crystallization processes, and concentration of thermally degradable materials, which result in the deposition of solids. In all cases, the temperature rise across the tube bundle is kept as low as possible, often as low as 2–3 °C. This results in a recirculation ratio as high as 100–150 kg liquor per kilogram of water evaporated. This high recirculation rate results in high liquor velocities through the tube, which helps to minimize the build-up of deposits or crystals along the heating surface. Forced circulation evaporators are normally more expensive than film evaporators because of the need for large-bore circulating pipework and large recirculating pumps. Operating costs of such a unit are also considerably higher.

All these problems can be overcome when the liquid is circulated at high velocity through the heat exchanger tubes to enhance the heat transfer rate and inhibit particle deposition. Any evaporator that uses a pump to ensure higher circulation velocity is called a forced circulation evaporator. The main components of a forced circulation evaporator are a tubular shell and tube heat exchanger (either horizontal or vertical), a flash chamber (separator) mounted above the heat exchanger, and a circulating pump. The solution is heated in the heat exchanger without boiling and the superheated solution flashes off (partially evaporates) at a lower pressure in the flash chamber. The pump pumps feed and liquor from the flash chamber and forces it through the heat exchanger tubes back to the flash chamber. Forced circulation evaporators are commonly used for concentration of caustic and brine solutions, and also for evaporation of corrosive solutions.

3.4.1.3 Rising Falling Film Evaporators

The rising falling film evaporator or concentrator (RFC) and the recirculating version (RRFC) work on uniform metered flow entering the bottom of the steam chest and rising through the first pass with rapidly increasing velocity (Fig. 3.38). After leaving the first pass, the flow of the vapor/liquid mixture is reversed and directed downward through the second pass at extremely high velocity. The mixture then enters the separator where the vapor and liquid are separated. When not designed as a circulation unit, the RFC provides once-through evaporation of heat-sensitive materials that require minimum retention time. RFCs are particularly adaptable to multiple-effect systems but provide a lesser turn-down capability than the RRFC configuration. The characteristics of RFCs are as follows:

- High tube velocity minimizes fouling
- Able to achieve high concentrations at low temperatures
- Risk of entrainment is greatly reduced because most of the liquid passes directly out of the steam chest rather than being re-entrained with vapor from the separator
- System may be turned down without loss of efficiency or effectiveness
- Holding time is a matter of seconds (RFC)
- Suitable for evaporation of a wide range of liquid concentrations
- Low headroom requirement

3.4.1.4 Short-Tube Vertical Evaporator (Vertical Tube Evaporator)

Evaporators in which movement of the liquid takes places as a result of a convection current set up by the heating process are called vertical tube evaporators or short-tube evaporators (Fig. 3.39).

Short-tube vertical evaporators consist of a short tube bundle (about 1–3 m in length) enclosed in a cylindrical shell. This is called the calandria. The feed is introduced above the upper tube sheet, and steam is introduced to the shell or steam

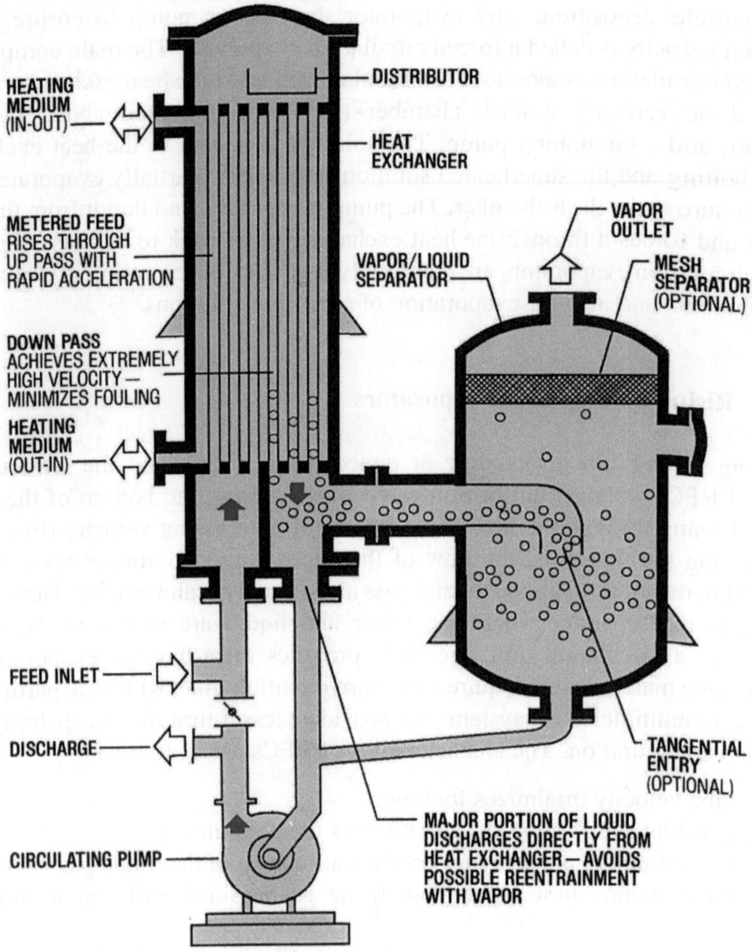

Fig. 3.38 Rising falling film evaporator. https://www.google.co.in/imgres

chest of the calandria. The solution is heated and partly vaporized in the tubes. The central tube in a calandria is of longer diameter. Typically, the "downcomer" area is taken as 40–70% of the total cross-sectional area of the tubes. The circulation rate through the downcomer/downtake is many times the feed rate. The flow area of the downtake is normally approximately equal to the total tubular flow area.

In standard vertical tube evaporators or short-tube evaporators, the liquid passes through the vertical tubes and the steam is supplied from outside the tubes. Heat transfer takes place through the tubes and the liquid inside the tubes is heated. The solvent evaporates and the vapor escapes from the top. The concentrated liquid is collected from the bottom. The main features of vertical tube evaporator construction are listed below:

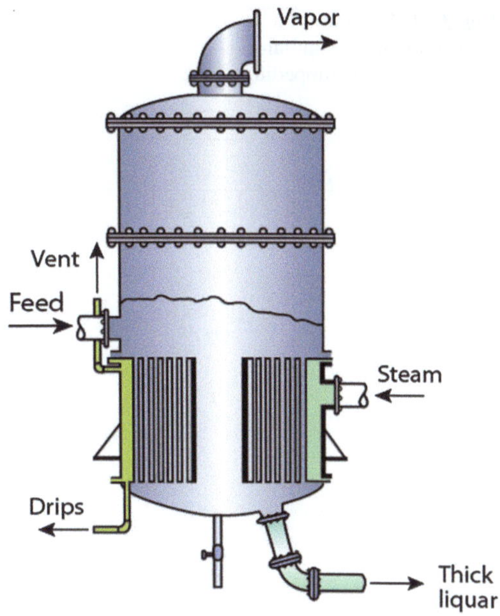

Fig. 3.39 Short-tube vertical evaporator (Source: https://www.google.co.in/imgres)

- Large cylindrical body made of cast iron with dome-shaped top and bottom.
- Inside the body, the calandria is fitted at the bottom. The calandria consists of a number of vertical tubes of 0.05–0.075 m diameter and 1–2 m length.
- About 100 such tubes are fitted in a body measuring 2.5 m or more in diameter. Inlets are provided for the steam and feed.
- Outlets are provided for vapor-concentrated products, noncondensed gases, and condensate.

The vertical tube evaporator functions as follows:

- Steam is introduced outside the tubes. The condensate passes through the corresponding outlet and noncondensed gases escape through the vent.
- The feed is introduced in such a way as to maintain the liquid level slightly above the top of the tubes. The liquid inside the tubes is heated by the steam and begins to boil.
- As the liquid boils, it spouts up through the tubes and returns through the central downtake. Circulation of hot liquid is set up, which enhances the rate of heat transfer.
- The vapor escapes through the top outlets. Steam is supplied until the required concentration of products is obtained.
- Products are withdrawn from the bottom outlet.

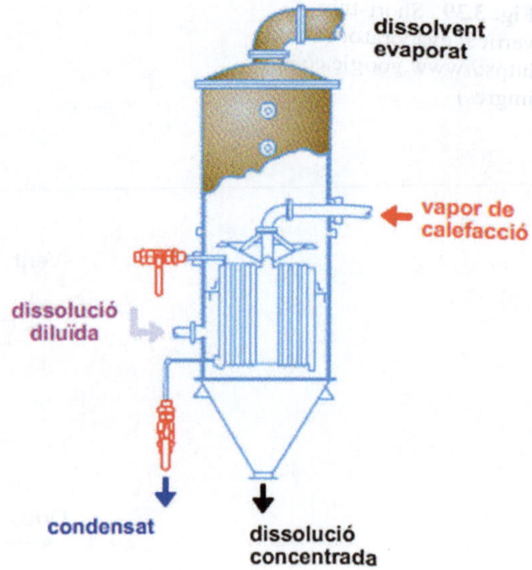

Fig. 3.40 Diagram of a standard basket evaporator (courtesy of Thermopedia)

3.4.1.5 Basket-Type Evaporator

This is a variant of the short-tube evaporator in which the whole tube bank can be easily removed for cleaning. By using vertical rather than horizontal tubes, the natural circulation of heated liquid can be used to give good heat transfer. The standard evaporator shown in Fig. 3.40 is an example of this type. Recirculation of the liquid is through a large downcomer so that the liquors rise through the vertical tubes (of about 5–8 cm diameter), boil in the space just above the upper tube plate, and recirculate through the downcomers. The hydrostatic head reduces boiling in the lower tubes, which are covered by the circulating liquid. The length-to-diameter ratio of the tubes is of the order of 15:1. The basket evaporator is a variant of the calandria evaporator in which the steam chest is contained in a basket suspended in the lower part of the evaporator, and recirculation occurs through the annular space round the basket.

3.4.1.6 Long-Tube Vertical Evaporator

This natural circulation evaporator is widely employed because it is often the cheapest per unit of capacity. The long vertical tube bundle is fixed with a shell that extends into a larger diameter vapor chamber at the top (Fig. 3.41). The long-tube vertical evaporator consists of one pass shell and tube heat exchanger. In this type of evaporator, the liquid flows as a thin film on the walls of a long (about 4–10 m in length) and vertical heated tube. Both rising film and falling types are used. Tube length usually varies from about 6 to 20 m. The main advantage of this type of evaporator is the high heat transfer rate. The feed enters at the bottom and

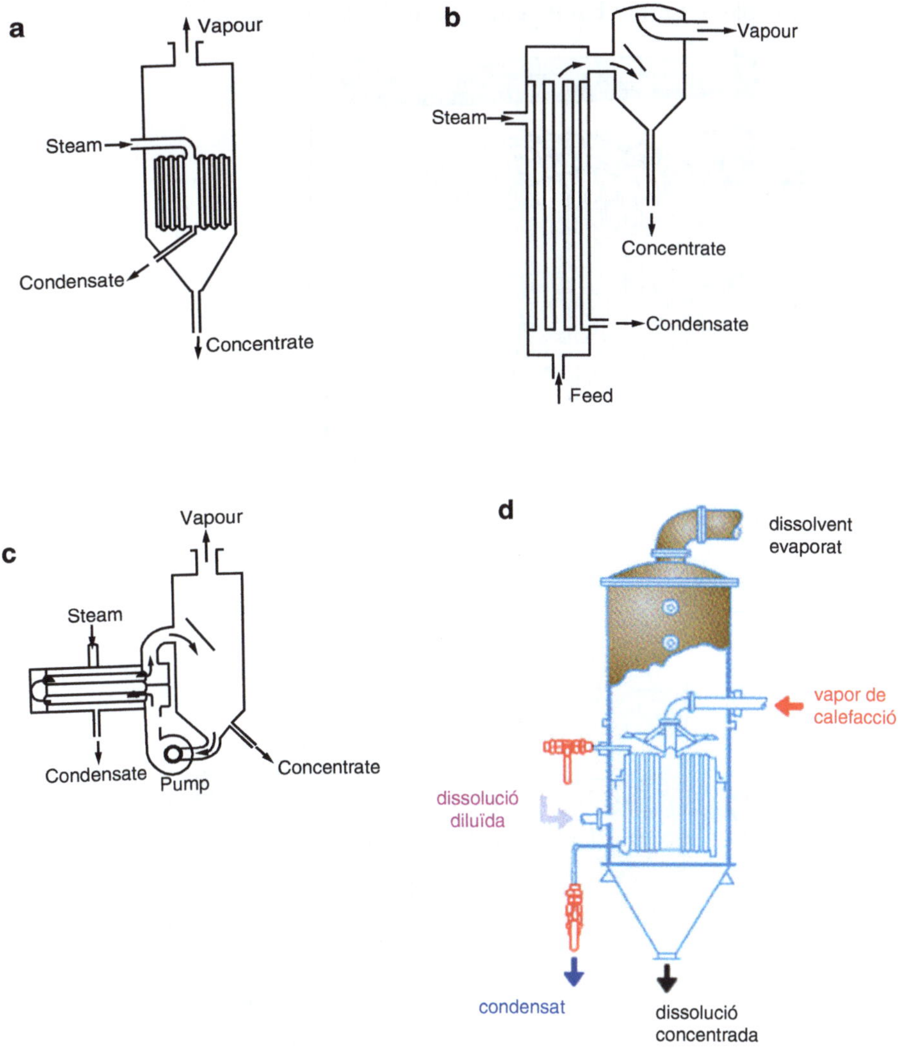

Fig. 3.41 Diagram of a vertical evaporator (courtesy of Thermopedia)

the liquid starts boiling in the lower part of the tube. These evaporators are commonly used for concentrating black liquors in the paper and pulp industries.

3.4.1.7 Plate Evaporator (Gasketed Plate Evaporator)

Gasketed plate evaporators are also called plate evaporators because the design is similar to that of a plate heat exchanger. A number of embossed plates with four corner openings are mounted by upper and lower carrying bars. The gasket is placed

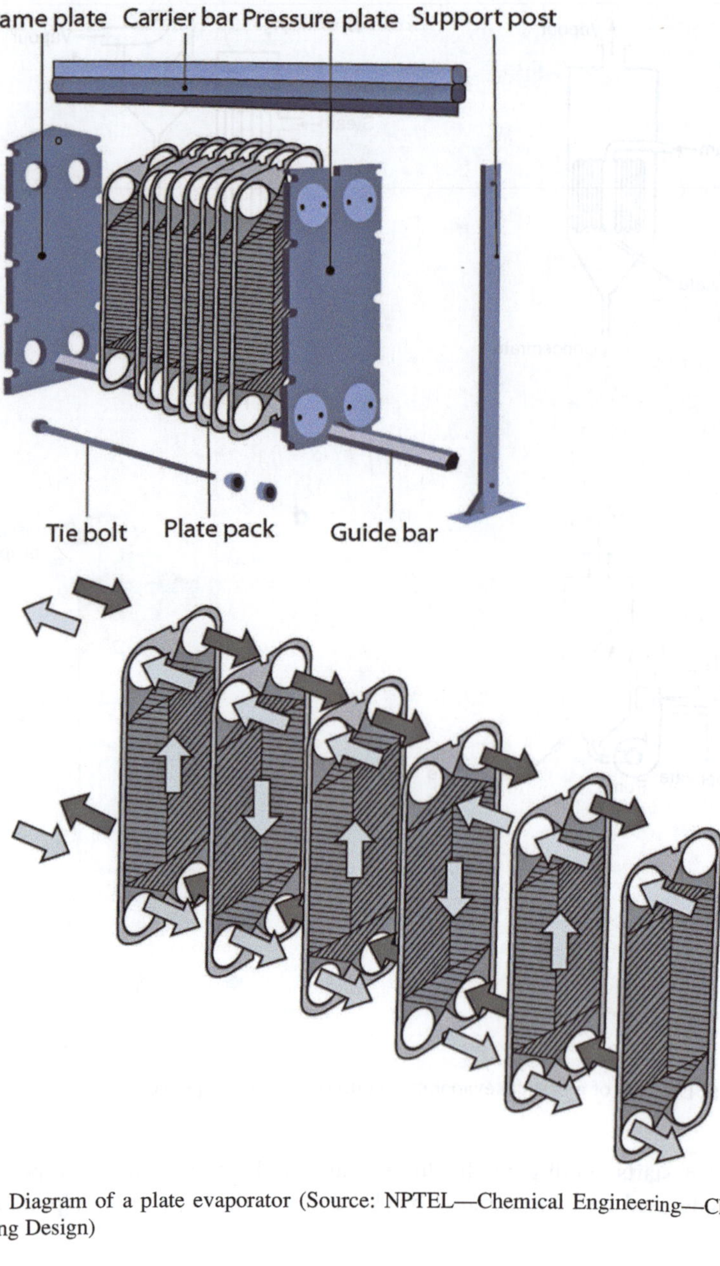

Fig. 3.42 Diagram of a plate evaporator (Source: NPTEL—Chemical Engineering—Chemical Engineering Design)

at the periphery of the plates. The interfering gaskets of two adjacent plates prevent mixing of the fluids and lead the fluid to the flow path through the corner opening (Fig. 3.42). The fluids can either flow in series or parallel depending on the gasket arrangement. The heat transfer coefficient is greatly enhanced as a result of high turbulent flow through the narrow passages. This evaporator is suitable for highly viscous, fouling, foaming, and heat-sensitive solutions. This type of evaporator is mainly used for concentration of food products, pharmaceuticals, emulsions, glue, etc.

3.4.1.8 Horizontal Tube Evaporator

In the case of horizontal film evaporation (Fig. 3.43), the liquid to be vaporized is distributed evenly on the top of a horizontal tube bundle. For this purpose,

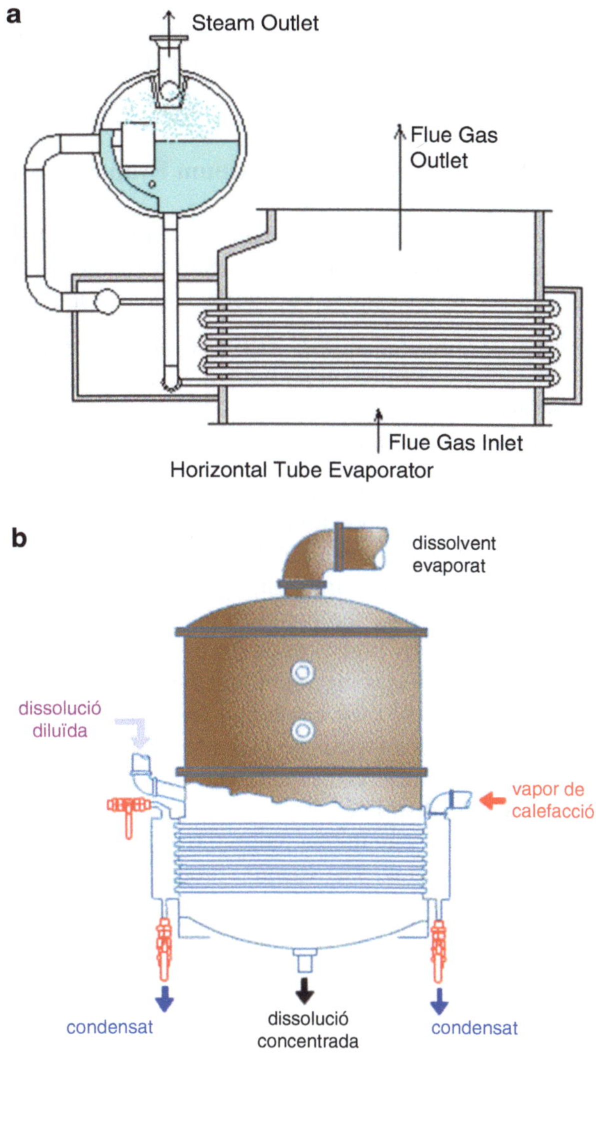

Fig. 3.43 (**a**) Diagram of a horizontal tube evaporator and (**b**) cross-section of a horizontal tube evaporator. Source: https://www.google.co.in/imgres

perforated plates and specially designed jet nozzles are applied. A minimum recirculation rate is required in order to maintain a minimum wetting rate. The liquid characteristics between the tubes are determined by the film Reynold's number and Kapitza number. The fluid can be encountered between the tubes as droplets, jet, or continuous sheet. As a result of liquid impingement, the heat transfer in horizontal units is generally higher than for vertical evaporators.

The horizontal tube evaporator is used for heat recovery from gas turbine exhausts, and also for recovery from flue gases in refinery and petrochemical furnaces.

Horizontal film evaporators have size limitations due to shipping restrictions. They are generally a less expensive unit to manufacture than other configurations. However, for a natural circulation design with large tubes (such as in some CO boilers) or very long tubes, special consideration needs to be given to assure that all tubes are provided with sufficient effluent.

3.4.1.9 Low Temperature Vacuum Evaporator

Low temperature vacuum evaporators (Fig. 3.44a, b) can vaporize water at a lower temperature of around 40 °C, instead of the normal 100 °C, because evaporation takes place in a vacuum. Vacuum evaporators are useful for recovery of:

- Heat-sensitive chemicals such as cyanide plating baths
- Chemicals sensitive to air oxidation, such as cyanide plating baths and stannous tin baths
- Solutions containing volatile components

A low temperature vacuum evaporator (Fig. 3.44b) mainly consists of an evaporator tank, evaporator heat exchanger, process water circulation pump, refrigerant compressor, condenser heat exchanger, distillate tank, vacuum-producing venturi, and distillate circulating pump. The feed water enters the evaporator tank and is circulated by the pump to the evaporator heat exchanger, where the feed water is heated by high-pressure refrigerant from the refrigerant compressor. The hot water flashes as water vapor into the evaporator tank at the top. This water vapor passes through the mist eliminator and the condenser heat exchanger. The refrigerant from the evaporator heat exchanger passes through air cooler and expansion valve and becomes cool liquid. The cool refrigerant cools the hot water vapor to distillate. The distillate is collected in the distillate tank. The distillate is circulated by the distillate circulation pump through a venturi, which produces a vacuum in the evaporator tank. The vacuum produced by the venturi in the evaporation tank causes process water to boil at 40–50 °C. The concentrated process water from the evaporator tank and the distillate from the distillate tank are automatically drained by the electric control system. All necessary controls and switch gears are part of the system.

a

b

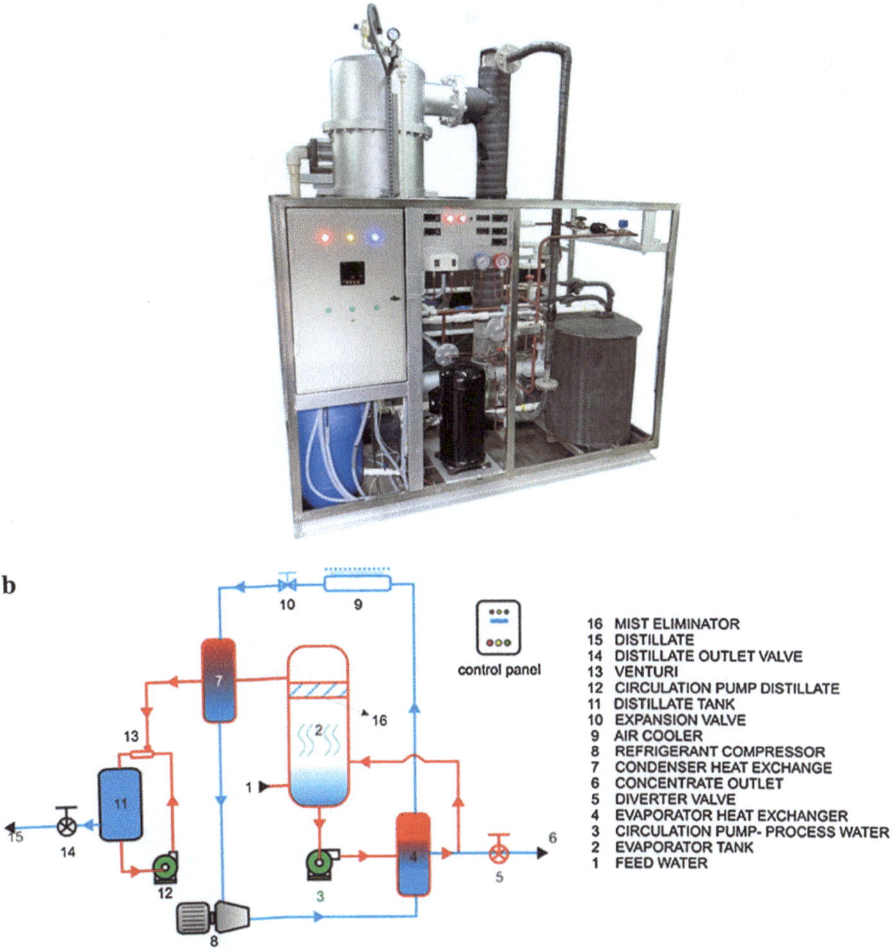

Fig. 3.44 (**a**) Low temperature vacuum evaporator and (**b**) working Principle

3.4.1.10 Flash Evaporation

Flash evaporation is the partial vaporization that occurs when a saturated liquid stream undergoes a reduction in pressure by passing through a valve or other device. If the valve or device is located at the entry into a pressure vessel, so that flash evaporation occurs within the vessel, then the vessel is often referred to as a flash drum (Fig. 3.45a). If the saturated liquid is a single-component liquid (e.g., liquid propane or liquid ammonia), part of the liquid immediately flashes into vapor. Both the vapor and the residual liquid are cooled to the saturation temperature of the liquid at the reduced pressure. This is often referred to as autorefrigeration and is the basis of most conventional vapor compression refrigeration systems. If the saturated liquid is a multicomponent liquid (e.g., a mixture of

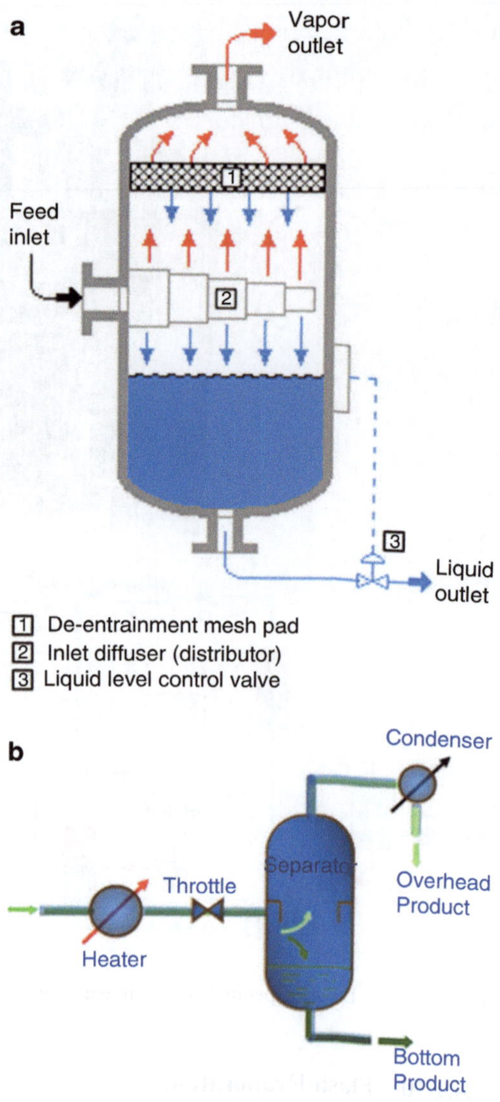

Fig. 3.45 (**a**) Diagram of a typical flash drum. (**b**) Diagram of a flash distillation system

propane, isobutane, and normal butane), the flashed vapor is richer in the more volatile components than the remaining liquid.

Flash distillation (sometimes called equilibrium distillation) is a single-stage separation technique. The process consists of three parts: (1) vaporization of a definite fraction of the liquid, (2) separation of vapor from liquid, and (3) condensation of vapor (Fig. 3.45b). A liquid mixture feed is pumped through a heater to raise the temperature and enthalpy of the mixture. It then flows through a valve and the pressure is reduced, causing the liquid to partially vaporize. Once the mixture enters a large-enough volume (flash drum), the liquid and vapor separate. Because the vapor and liquid are in such close contact up until the flash occurs, the product liquid and vapor phases approach equilibrium. Simple flash separations are very common in industry, particularly in petroleum refineries.

3.4.1.11 Direct Contact Evaporators

In this type of evaporator, a hot gas is injected into a pool of liquid and causes it to evaporate, the vapor being carried away with the efflux gas. Such evaporators are relatively cheap and are useful for the concentration of corrosive fluids, viscous liquids, and slurries, which can be difficult to handle in more conventional heat exchangers. A convenient way of providing the hot gas is to use a submerged combustion system in which the hot gas is produced in situ by the combustion of gas or fuel oil in a combustion chamber mounted in the center of the vessel containing the liquid.

3.4.1.12 Thin-Film Vacuum Evaporator

Thin-film evaporation refers here to thermal separation of products in a mechanically generated, thin and highly turbulent liquid film. After entering the thin-film evaporator (Fig. 3.46), the product comes into contact with the rotor and is uniformly spread on the periphery by the distribution ring. It is then picked up by the first rotor blades and immediately formed in to a film (0.5–3.5 mm) on the heat

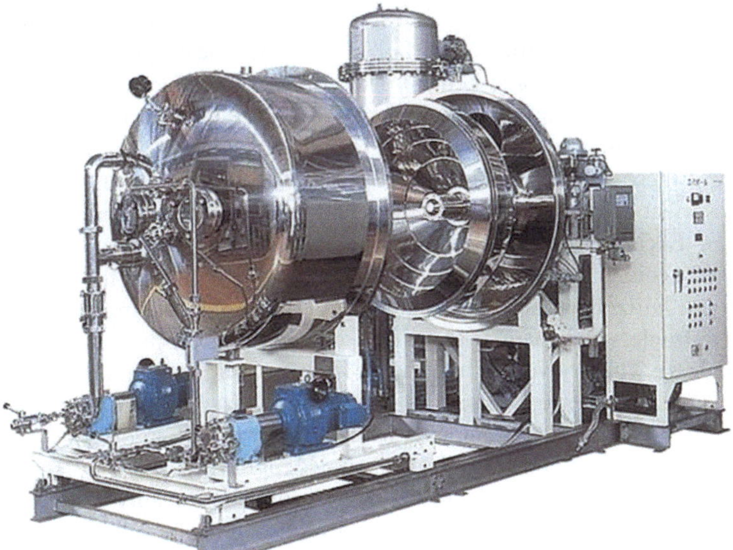

Fig. 3.46 Photo of a thin film evaporator (courtesy of Okawara Mfg. Co., Ltd.)

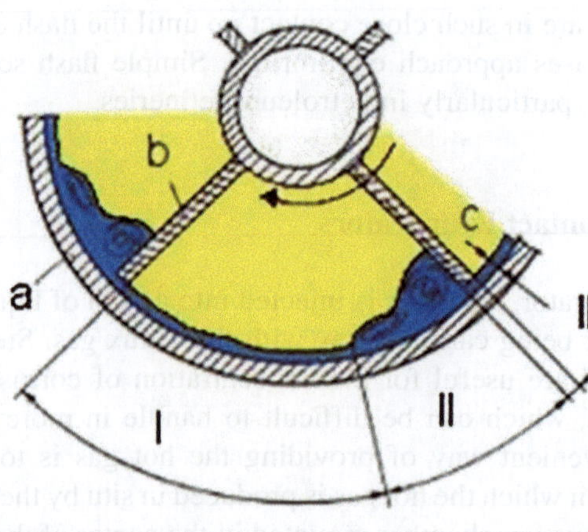

Fig. 3.47 Fluid in front of rotor blades in a thin-film evaporator: (*a*) inner shell (*b*) rotor blade, and (*c*) gap between rotor and inner shell; (*I*) film zone, (*II*) bow wave, and (*III*) gap zone

transfer surface. In front of each rotor blade, the fluid creates a bow wave (Fig. 3.47).

The fluid in the gap between the heat transfer surface and the rotor blade tip is highly turbulent and this leads to intensive heat and mass transfer rates. This turbulence produces high heat transfer coefficients, even with highly viscous products. As a result of the intensive mixing action within the bow wave, temperature-sensitive products are prevented from overheating and fouling on the heat transfer surface can be reduced or eliminated.

The product is fed continuously to the horizontal thin-film evaporator at the larger diameter end, picked up by the rotor blades, and spread immediately in a thin turbulent film on the heat transfer surface (Fig. 3.48). The evaporator's conical form results in a centrifugal force being imparted on the product by the rotor, which effectively has two components: one perpendicular to the heat transfer surface and the other in the direction of the body's larger diameter end. The product hold up created by these forces and that of the incoming product ensures that the heat transfer surface is fully wetted, independently of the evaporation ratio and/or the feed rate. Localized product overheating and thermal degradation are thereby reduced or avoided altogether. The product vapor (low boilers) flows co-currently through the horizontal thin film evaporator and into the rotating separator. Here, entrained droplets and foam are knocked out and pass into the liquid phase outlet (high boilers). The dry vapor then passes into the condensation stage, column, or another downstream stage. The evaporator can be operated for low-temperature concentration (vacuum environment) with extremely short product residence time (less than 1 s). This equipment is the best choice for the processing of heat-sensitive products, at 30–1200 kg h^{-1}, and is widely used for quality improvement of antibiotics, bioactive substances, functional foods, and natural extracts. Its unique features have broadened the application range to food, chemicals, pharmaceuticals, biochemicals, and other products that are difficult to process with other types of evaporators.

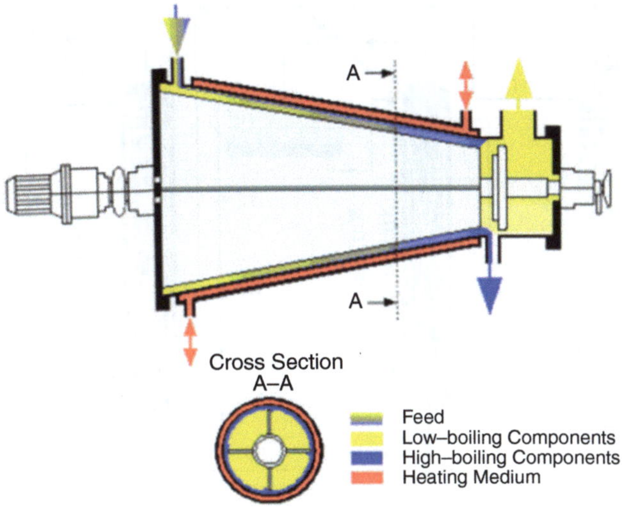

Fig. 3.48 Horizontal conical design of a thin-film evaporator (courtesy of Buss-SMS-Canzler)

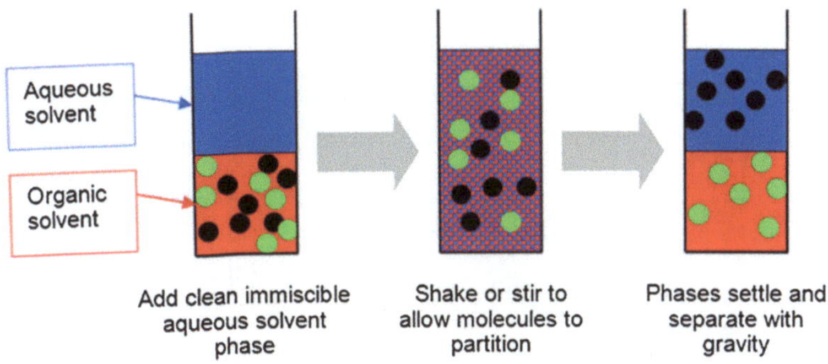

Fig. 3.49 Liquid–liquid extraction process

3.4.2 *Liquid–Liquid Separation*

Liquid–liquid extraction (LLE) consists in transferring one (or more) solute (s) contained in a feed solution to another immiscible liquid (solvent). The solvent that is enriched in solute(s) is called the extract. The feed solution that is depleted in solute(s) is called the raffinate. LLE is an important separation technology, with a wide range of applications in the modern chemical and biochemicals process industries.

The extraction process is based on the different solubilities of components in two immiscible or partially miscible liquids. The components that need to be recovered are extracted from the feed stream with the help of an extractant (often called solvent). Both liquids have to be thoroughly mixed and subsequently separated from each other again (Fig. 3.49). To achieve high purities and yields, it is

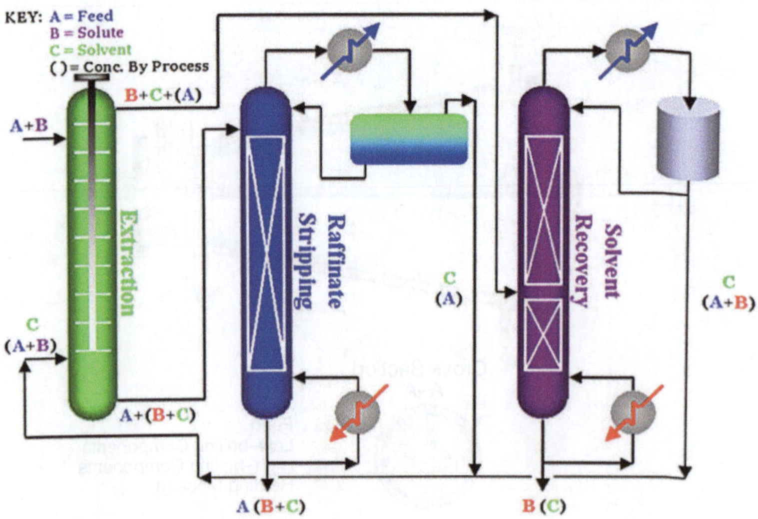

Fig. 3.50 Multiple stage liquid–liquid separation process for achieving high purities and yields. Source: https://www.google.co.in/imgres

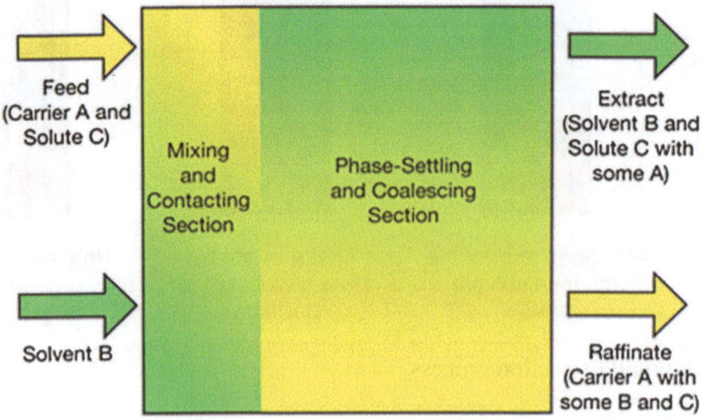

Fig. 3.51 An extraction unit has two inlet streams (liquid carrier containing solute molecules, and solvent) and two outlet streams (raffinate and solute-rich extract). Source: www.aiche.org

necessary to operate with multiple stages and with the liquids flowing countercurrently (Fig. 3.50).

For successful separation by extraction, all components must meet certain specifications. The main criteria for the extractant are favorable partition coefficient, high selectivity, and easy separation from the extracted product. The basic conditions for the pair of liquids are a low mutual solubility and a difference in density, which is the driving force for the motion of droplets. In most multistage extractions, the concentration of biological products can be achieved by transferring the desired product (solute) from one liquid phase to another liquid phase, a phenomenon referred to as LLE (Fig. 3.51).

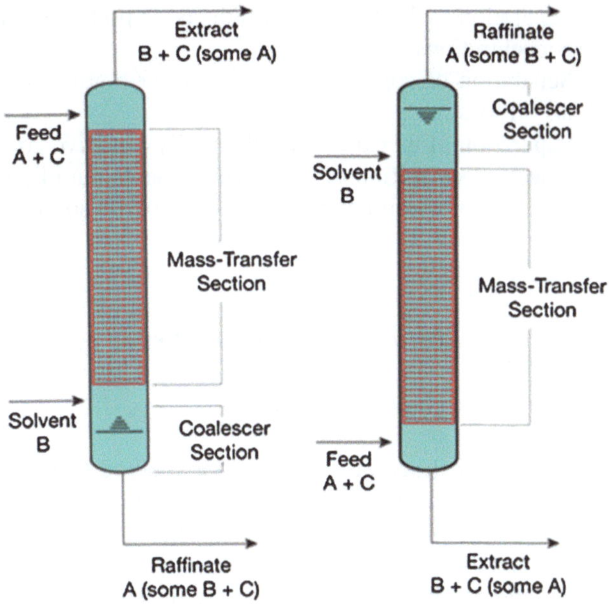

Fig. 3.52 Counter-current extraction units can be set up in one of two ways. *Left*: When the solvent is lighter than the carrier liquid, the solvent is introduced at the bottom of the column and the solute is carried up toward the top of the extractor. *Right*: When the solvent is heavier than the carrier, the solvent is introduced at the top of the column and the solute is carried downward by the solvent toward the bottom of the column. The *red dots* represent the region in which the solute is transferred to the solvent. Source: www.aiche.org

Besides concentration, this technique is also useful for partial purification of a product. The liquids are transported counter-currently (Fig. 3.52). The viscosity and interfacial tension are additional important parameters. In nearly all LLE processes, one of the liquids is dispersed into the second liquid in the form of droplets. The key to high process performance is an adapted droplet size and a uniform hold-up profile throughout the column. This requires specially adapted equipment and in-depth know-how about two-phase liquid flows.

3.4.2.1 Applications of Liquid–Liquid Separation

LLE extractors are successfully applied in every segment of the process industry for the bulk chemical, fine chemical, petroleum, pharmaceutical, biotechnology, and food industries.

By using organic solvents, the lipophilic compounds (low molecular weight products) can be conveniently extracted. However, it is quite difficult to extract hydrophilic compounds. The nonionizing compounds becomes distributed between two liquid phases based on the physical properties.

An additional component has to be introduced as extractant, which makes further separation steps necessary. Therefore, LLE is mostly used when separation of components by distillation is either uneconomical or impossible. Some examples are:

- Simultaneous separation of various components with widely differing boiling points

- Separation of high-boiling products or pollutants that are present in only low concentrations
- Separation of components with similar boiling points or components forming azeotropes
- Separation of mixtures with thermally sensitive components
- Selective separation of single components out of a multicomponent mixture
- Separation of components from electrolytic solutions

3.4.3 *Separation of Different Types of Solute*

3.4.3.1 Dissociation Extraction

Dissociation extraction is a separation method that can be applied to a mixture of two (or more) chemical substances that are each ionizable in solution and have at least a small difference in pK_a values. The mixture is partitioned between two liquid phases and then partially neutralized with a reagent so that the partition coefficients of the components of the mixture are further favorably enhanced. Optimally, the moles of neutralizing agent are equal to the moles of the more reactive component of the binary mixture.

This technique is equally applicable to the concentration of a trace impurity in a substantially pure single substance. The differences between two molecules can be quite remote from the acidic or basic functional group upon which the dissociation depends, but partitioning can occur if the difference can be transmitted electronically or sterically to the ionizing function. Because pharmaceutical substances are most often salt-forming species, this technique provides a powerful method for identification of an impurity that has only minor differences from the drug substance and similar retention time in high-pressure liquid chromatography.

Selection of the solvent for dissociation extraction is crucial. Organic solvents can be classified according to their ability to interact with carboxylic, phenolic, and basic substances. Aliphatic hydrocarbons (e.g., heptane, cyclohexane) are inert solvents with negligible tendency to interact with CO_2H, OH, and NH_2 groups. In order of increasing tendency to interact with the solutes, the aliphatic hydrocarbons are followed by aromatic hydrocarbons, halogenated hydrocarbons, ethers and ketones, esters, and finally alcohols. Knowledge of these particular interactions can help in the selection of the proper organic solvent partner. The following guidelines are generally applicable:

- Use an inert solvent if the stronger component has a relatively free aquaphilic ionizing group.
- Use a polar solvent if the weaker component has a relatively free aquaphilic group. Examples of polar solvents that associate well with free ionizable groups are dibutyl ether and *n*-octanol.

3.4.3.2 Reactive Liquid Extraction

In recent years, process intensification has attracted considerable academic interest as a potential means of process improvement to meet the increasing demand for sustainable production. A reactive liquid extraction process is an LLE process that is intensified through a mechanism involving a reversible reaction between the extracted chemical species and a host chemical species constituting, or present in, the extractant.

Rotating packed beds (RPBs) provide the opportunity for mass transfer processes in high gravity fields as part of the HiGee technology (high "g" for high gravity). This intensification step reduces the required equipment volume and changes operating limits, such as flooding velocities and maximum pressure drops. Although increased mass transfer rates, high loading points, intense shear forces, and high micromixing effects in RPBs result in significantly decreased equipment volumes, the acceptance of this technology in industry is still very low (Fig. 3.53). It is a new energy-efficient recovery process offering substantial improvements over existing recovery technology.

3.4.3.3 Aqueous Two-Phase System

The aqueous two-phase system (ATPS) is a clean alternative to traditional organic–water solvent extraction systems [63]. In ATPS, the distribution of the desired product is based on its surface and ionic character and the nature of the phases (Fig. 3.54) [64]. It has been applied in several fields of biotechnology, such as recovery of proteins, enzymes, biopharmaceuticals, and extractive fermentation. ATPS was developed in Sweden during mid-1950s for the separation of macromolecules, cells, and organelles [65]. These systems were initially applied to the

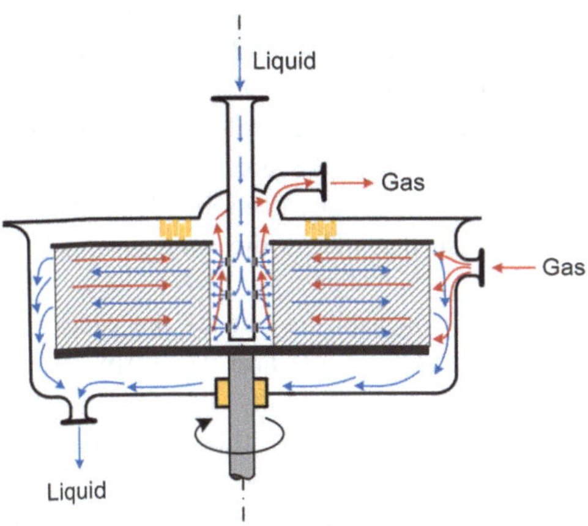

Fig. 3.53 Rotating packed bed evaporator. Source: https://www.google.co.in/imgres

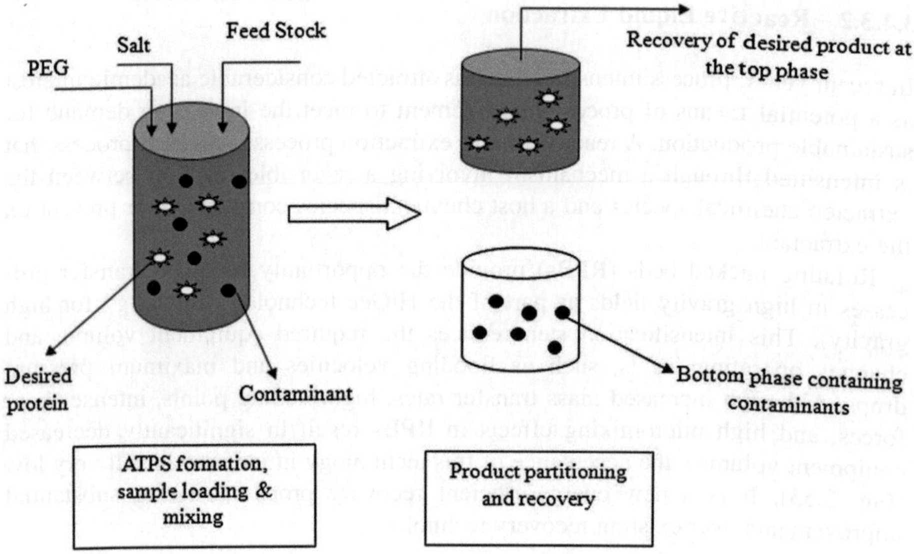

Fig. 3.54 How the target product (e.g., biomolecule) becomes concentrated in one of the phases and contaminants in the other. *ATPS* aqueous two-phase system, *PEG* polyethylene glycol

separation of plant organelles and viruses. Since then, attention has been directed toward widening its scope of application. During the last two decades, much work has been done to develop feasible separation processes using ATPS for various biological materials, including proteins and recombinant proteins [66]. An ATPS is an aqueous, liquid–liquid biphasic system that is obtained by mixing either aqueous solutions of two polymers or solutions of a polymer and a salt. Generally, the former is composed of PEG and a polymer such as dextran [67, 68], starch [69], or polyvinylalcohol [70]. The latter is composed of PEG and phosphate or sulfate salts. This polymer/salt system results in higher selectivity in protein partitioning, leading to an enriched product with high yields in the first extraction step. Water is the main component in ATPS (more than 80%), but the two phases are not miscible. Cells and other solids remain in one phase while proteins are transferred to the other phase. When the limiting concentrations are exceeded, two immiscible aqueous phases are formed. The limiting concentrations depend on the type of phase-forming components and on the pH, ionic strength, and temperature of the solutions.

Polyethylene glycol (PEG) is used as one of the phase-forming polymers in ATPS because it is available at low cost and can form a two-phase system with other neutral polymers as well as with salts. In addition, PEG can significantly enhance the refolding of proteins to recover their activity.

Because of the high cost and high viscosity of the polymer/polymer system, aqueous two phase polymer/salt systems are preferred. Moreover, polymer/salt systems have larger differences in density, greater selectivity, lower viscosity, lower cost, and larger relative size of the drops [69]. Currently, use of citrate salts as a phase-forming component with PEG is preferred because citrate salts are biodegradable and nontoxic [70, 71].

3.4.3.4 Reverse Micellar System

A micelle is an aggregate of molecules possessing both polar and nonpolar regions (Fig. 3.55). Reverse micelles are thermodynamically stable, minute surfactants that hold water in their interior surrounded by organic phase. Only the protein of interest is entrapped in the micelle (Fig. 3.56a, b), whereas other impurities remain in the organic phase. The reverse micellar system is an encouraging strategy for downstream processing. It is ideal for extracting biomolecules from diluted samples. Reverse micellar systems possess high sample loading capacity, are specific, and easy to operate.

Ideally, micelles vary in size from 2 to 20 nm, depending on their number and composition. Besides the critical micelle concentration (CMC), there are other factors that influence the rate of micelle formation, such as temperature. Micelle formation can take place only above a particular temperature. The Krafft temperature is the minimum temperature at which the amphipathic molecules form micelles. Below this temperature, molecules remain in a crystalline form.

In a polar medium like water, the number of counter-ions is always equal to its degree of binding to the micelle. As counter-ions bind to the polar heads of amphipathic molecules, they decrease the repulsion and, thus, increase the stability of micelles. The CMC decreases with an increase in the number of counter-ions.

Detergents are capable of forming micelles that function as emulsifiers, dissolving substances that are usually insoluble in water. The hydrophobic tails of detergents interact with insoluble dirt and oil such that the dirt is entrapped in the

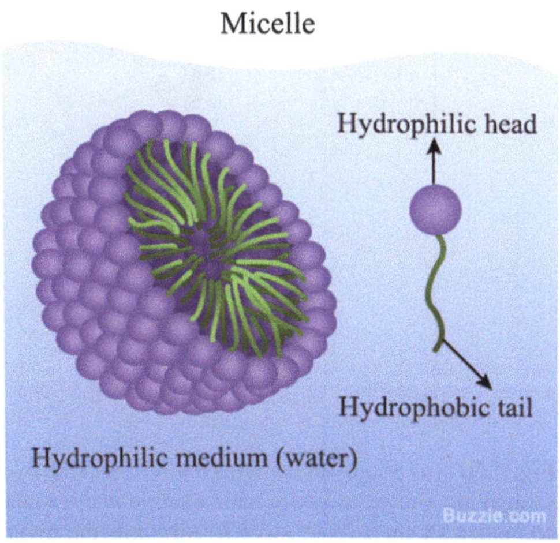

Fig. 3.55 In a micelle, molecules orient themselves in such a way that the hydrophilic heads of the molecules are in close contact with the polar medium (water) and their hydrophobic regions are away from it. Source: http://www.buzzle.com/articles/micelle-function-structure-and-formation.html

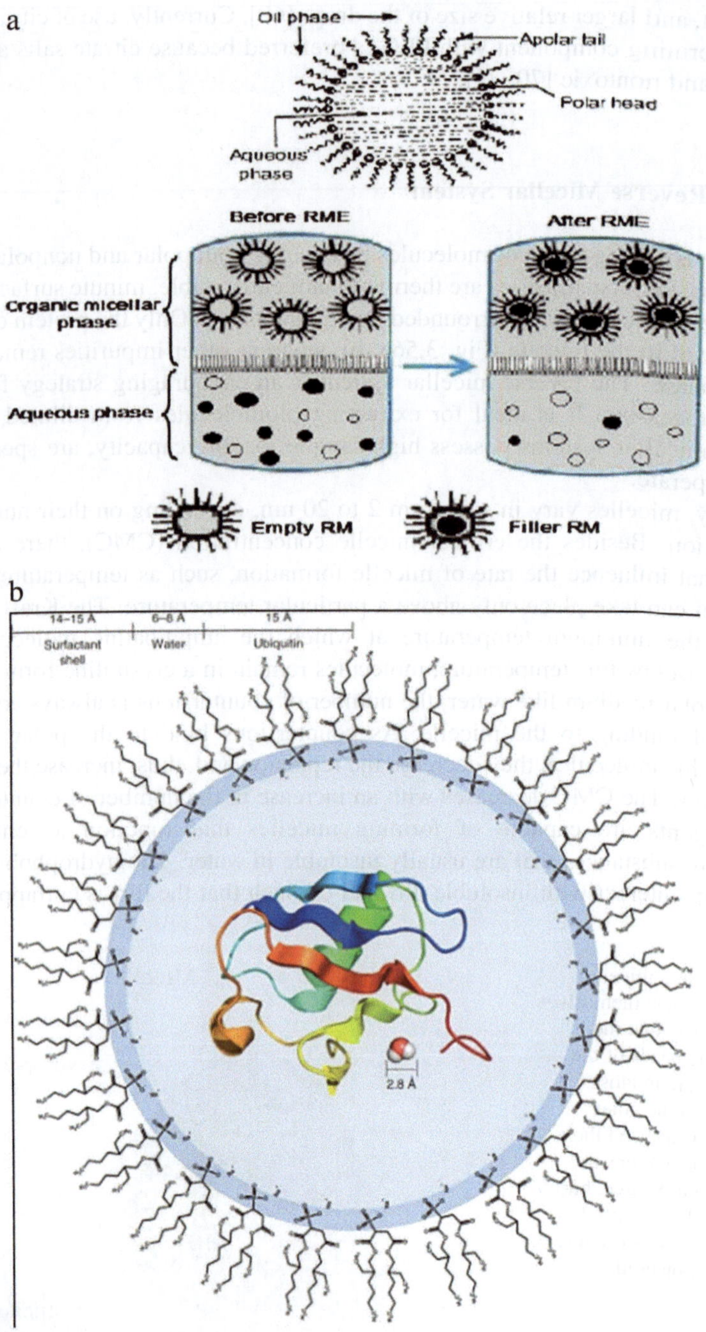

Fig. 3.56 (**a**) Stages in the working of a reverse micellar system. (**b**) Surfactant, water pool, and protein are shown to scale, with a single water molecule pictured (also to scale) for comparison. Nonpolar alkane solvent, which surrounds the reverse micelle (RM) particle, is not shown. The RM diameter was determined on the basis of measurements of the rotational correlation time of ubiquitin in AOT reverse micelles at a water loading comparable to that used here. This value

hydrophobic center of these micelles. Washing clothes with water removes the dirt particles entrapped in micelles and cleans the surface.

A reverse micellar extraction cycle is basically composed of two steps, forward and back extraction. In the forward extraction process, biomolecules are transferred from the initial aqueous phase to the reverse micelles. In the back extraction process, biomolecules are transferred from the reverse micelles back to the aqueous stripping solution. The extractions can be performed batchwise on a bench scale, employing different procedures. Continuous extraction using reverse micellar systems is an efficient process with a reduced number of steps and is used in the purification of biomolecules.

3.4.3.5 Supercritical Fluid Extraction

This technique differs from the above procedures in that the materials used for extraction are supercritical fluids (SCFs). SCFs are intermediates between gases and liquids and exist as fluids above their critical temperature and pressure. Supercritical CO_2, with a low critical temperature and pressure is commonly used in extractions. SCF extraction is relatively expensive and hence not widely used (SCF has been used for the extraction of caffeine from coffee beans, and of pigments and flavor ingredients from biological materials).

Replacing conventional organic solvents with SCFs in extraction procedures is a major advancement in today's pollution prevention programs. SCF extraction allows separation and minimization of waste, as well as solvent recycling. Other advantages of SCF extraction include high efficiency, high extraction rates, and good selectivity. Processes such as supercritical CO_2 extraction [71–73] and microwave-assisted extraction [74, 75] are green processes that have been used to obtain bioactive compounds from plants (phytochemicals). The use of such green processes is attracting attention because of their sustainability and ability to increase target molecule specificity. Supercritical extraction procedures have been used extensively in the past to extract target compounds from a variety of matrices at laboratory and commercial scales. The most well-known examples of large-scale SCF extraction are the decaffeination of coffee beans [74, 76, 77] and the extraction of α-acids from hops to produce hop resins.

CO_2 is the main solvent used in this technique, especially when the target molecule is polar. It is safe, food-grade, non-flammable, recyclable, and readily available at a relatively low cost and high purity. In addition, CO_2 has a relatively low critical temperature of 31.1 °C and a low critical pressure of 7.4 MPa. At

Fig. 3.56 (continued) agrees closely with simple geometric predictions of RM dimensions. The outer, *darker blue* ring approximates the layer of solvent that is dynamically restricted by interactions with the surfactant headgroups, whereas the *lighter blue* region corresponds to the central water pool that hydrates the protein [Source: Nature Structural & Molecular Biology **18**, 245–249 (2011)]

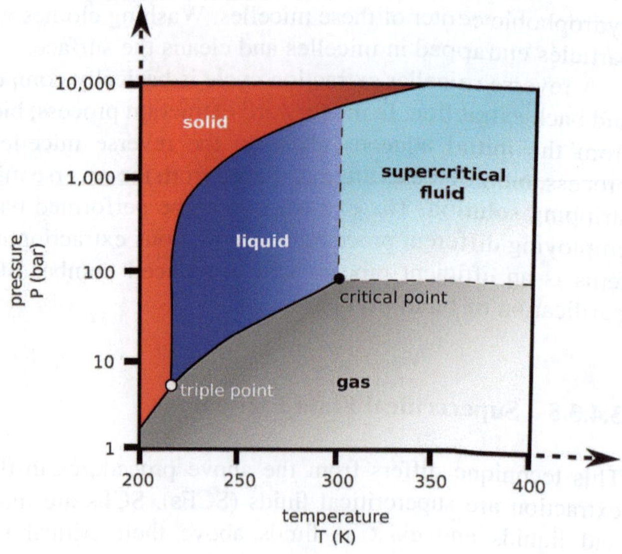

Fig. 3.57 Phase separation diagram for CO_2

supercritical conditions, above its critical temperature and above its critical pressure (Fig. 3.57), CO_2 possesses the properties of both a gas and a liquid. Thus, it acquires very good mass-transfer properties because of the high, more gas-like, diffusion coefficients and low viscosity values. These properties provide effective penetration into solid matrices and help preserve the integrity and stability of more delicate compounds as a result of its relatively low surface tension. Solvents other than CO_2 are not necessarily required and the expense of removing polluting organic solvents can be eliminated. Another advantage is its high selectivity and easy separation from the extracted compounds, without leaving toxic residues in extracts.

The application of supercritical CO_2 in the food industry has been widely developed for extraction of organics. Recent new technologies are emerging for the use of supercritical CO_2 in the extraction of metals and nonorganics. Uses of supercritical CO_2 continue to be explored and expanded because of its many benefits.

Application of Supercritical Fluid Extraction

The food and beverage industry was the first to make commercial use of supercritical CO_2 extraction. Three examples of this are as follows:

- Caffeine extraction from coffee beans; as early as 1974, European companies were extracting caffeine using supercritical CO_2
- Fat extraction from processed snack foods
- Extraction of flavor oils from hops for the brewing industry

3.4.3.6 Microwave-Assisted Extraction

Microwave-assisted extraction is also a novel green method for extracting soluble compounds into a fluid from a wide range of matrices [78]. Microwaves are electromagnetic fields with a frequency of 0.3–300 GHz. They are made up of two oscillating fields, electric and magnetic, that are perpendicular. These waves can penetrate the sample matrix and interact with polar molecules (such as water) to create heat. Consequently, microwaves can simultaneously heat the whole material to the penetration depth. The extraction process involves the separation of solutes from active sites of sample matrix under increased temperature and pressure, diffusion of solvent across the sample matrix, and release of solutes to solvent caused by expansion and breakdown of cell walls (Fig. 3.58). The process thus partitions compounds of interest from the sample matrix into the solvent.

Furthermore, the migration of dissolved ions increases solvent penetration into the matrix and thus facilitates the release of phytochemicals. The effect of microwave energy is strongly dependent on the dielectric susceptibility of both the solvent and the solid plant matrix.

3.4.3.7 Membrane Filtration

Membrane filtration has become a common separation technique in industrial biotechnology. It can be conveniently used for the separation of biomolecules and particles, and for the concentration of fluids. The membrane filtration technique basically involves the use of a semipermeable membrane that selectively retains particles or molecules that are bigger than the pore size, while smaller molecules pass through the membrane pores [79–83].

Membranes used in filtration are made up of polymeric materials such as polyethersulfone and polyvinyl difluoride. It is rather difficult to sterilize membrane filters. In recent years, microfilters and ultrafilters composed of ceramics and steel have become available and are more easily cleaned and sterilized. Three other types of membrane filtration techniques are described briefly below.

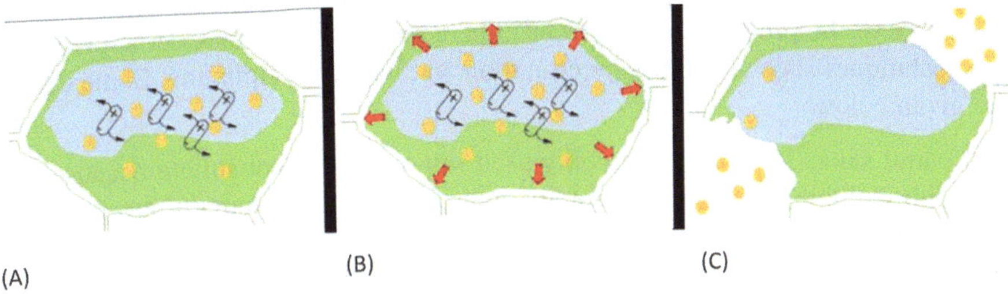

Fig. 3.58 Mechanism of cell disruption caused by microwave irradiation. (**a**) Stimulation of rotation of in-situ water; (**b**) temperature and pressure increase inside the plant cell; and (**c**) expansion and breakdown of cell wall and consequent release of target biomolecules

Membrane Adsorbers

Membrane adsorbers are micro- or macroporous membranes with attached ion exchange groups and/or affinity ligands. Membrane adsorbers can bind to proteins and retain them. Methods and buffers known from conventional gel cromatography can be directly applied to separate target molecule.

Pervaporation

This is a technique in which volatile products can be separated by a process of permeation through a membrane, coupled with evaporation. Pervaporation is useful for the extraction, recovery, and concentration of volatile products. However, this procedure cannot be used for large-scale separation of volatile products because of its relatively high cost.

Perstraction

This is an advanced technique based on the principle of membrane filtration coupled with solvent extraction. Hydrophobic compounds can be recovered or concentrated by this method.

Precipitation

Precipitation is the most commonly used technique in industry for the concentration of macromolecules such as proteins and polysaccharides. Furthermore, precipitation can also be employed for the removal of certain unwanted by-products such as nucleic acids and pigments. Neutral salts, organic solvents, high molecular weight polymers (ionic or nonionic), and alterations in temperature and/or pH are used in precipitation. In addition to these nonspecific protein precipitation reactions (i.e., the nature of the protein is unimportant), there are some protein-specific precipitation techniques such as affinity precipitation and ligand precipitation. Some details are given below:

- Neutral salts: The most commonly used salt is ammonium sulfate because it is highly soluble, nontoxic to proteins and cheap. Ammonium sulfate increases hydrophobic interactions between protein molecules, resulting in their precipitation. The precipitation of proteins is dependent on several factors such as protein concentration, pH, and temperature.
- Organic solvents: Ethanol, acetone, and propanol are commonly used organic solvents for protein precipitation. They reduce the dielectric constant of the medium and enhance the electrostatic interactions between protein molecules

that lead to precipitation. Because proteins are denatured by organic solvents, the precipitation process has to be carried out below 0 °C.

- Nonionic polymers: PEG is a high molecular weight nonionic polymer that can precipitate proteins. It reduces the quantity of water available for protein solvation and precipitates protein. PEG does not denature proteins and is nontoxic.
- Ionic polymers: Charged polymers such as polyacrylic acid and polyethyleneimine are used. They form complexes with oppositely charged protein molecules, resulting in charge neutralization and precipitation.
- Increase in temperature: Heat-sensitive proteins can be precipitated by increasing the temperature.
- Change in pH: Alterations in pH can also lead to protein precipitation.
- Affinity precipitation: Affinity interaction (e.g., between antigen and antibody) can be exploited for precipitation of proteins.
- Precipitation by ligands: Ligands with specific binding sites for proteins have been successfully used for selective precipitation.

Adsorption

The biological products of fermentation can be concentrated by using solid adsorbent particles. In the early days, activated charcoal was used as the adsorbent material. More recently, cellulose-based adsorbents have been employed for protein concentration. The matrices used for concentration of low molecular weight compounds (vitamins, antibiotics, peptides) are generally based on polystyrene, methacrylate, and acrylate. The process of adsorption can be carried out by making a bed of adsorbent column and passing the culture broth through it. The desired product, held by the adsorbent, can be eluted.

3.4.4 Chromatography

Chromatography is a versatile method of separating many different kinds of chemical mixtures or biomolecules according to differences in specific properties (Table 3.6) [84–86].

In any chromatographic technique, the stationary phase is usually a solid, thick liquid, or bonded coating that stays fixed in one place. The mobile phase or eluent (usually a liquid or gas) moves through it or across the stationary phase. A sample to be separated, when placed on the stationary phase, gradually moves along in the same direction as the mobile phase. If a sample compound (or analyte) has no interaction with the stationary phase, it runs right through and comes out of the system (as elute) at the same rate as the mobile phase (Fig. 3.59). On the other hand, if an analyte has no interaction with the mobile phase, it sticks directly to the stationary phase and is never eluted. Neither of these are good outcomes.

Table 3.6 Chromatographic separation of biomolecules on the basis of specific properties

Property	Technique[a]
Biorecognition (ligand specificity)	Affinity chromatography
Charge	Ion-exchange chromatography
Size	Gel filtration (sometimes called size-exclusion chromatography)
Hydrophobicity	Hydrophobic interaction chromatography
	Reverse phase chromatography

[a]Expanded bed adsorption is a technique used for large-scale purification. Proteins can be purified from crude samples without the need for separate clarification, concentration, and initial purification to remove particulate matter

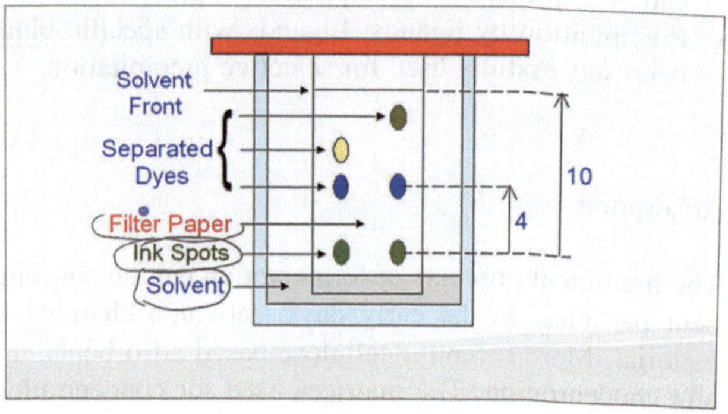

Fig. 3.59 Demonstration of simple paper chromatography: draw some blobs of ink on paper, roll the paper into a cylinder, and place it in a wine glass with a small amount of water. As the water creeps up the paper, the colors separate into their components

In a well-designed chromatography process, precaution is taken so that stationary and mobile phases both have at least some interaction with the analytes. Any individual sample molecule interacts first with one phase and then the other, back and forth repeatedly, but the fraction of each analyte in each phase remains constant. This distribution ratio between the selected phases must differ for each analyte in order for them to be separated. Compounds will not separate chromatographically if they have the same distribution ratio on a particular system. Analytes with greater attraction to the stationary phase still flow through the system because they spend part of their time moving in the mobile phase, but they tend to lag behind those analytes that have greater interaction with the mobile phase and thus spend more time in the mobile phase. Gradually, as the analytes progress through the system, they separate from each other (usually in order of their distribution ratios) and can be captured or at least detected in relatively pure form as they elute [86–88]. There are hundreds of stationary phase, mobile phase, and detector combinations in wide use throughout the chemical, medical, and biotechnology industries, but we can identify six major types, which are briefly described next.

3.4.4.1 Paper Chromatography

Paper chromatography (Fig. 3.60) has proved to be very successful in the analysis of chemical compounds, and lipid samples in particular. In paper chromatography, the sample mixture is applied to a piece of filter paper, the edge of the paper is immersed in a solvent, and the solvent moves up the paper by capillary action [89, 90]. The retention factor, R_f, is a quantitative indication of how far a particular compound travels in a particular solvent. The R_f value is a good indicator of whether an unknown compound and a known compound are similar, if not identical. If the R_f value for the unknown compound is close or the same as the R_f value for the known compound, the two compounds are most likely similar or identical. For accurate work, the distance moved by the solvent is marked carefully with a pencil and the distances moved by each "center" of the colored spots is also measured. These can be compared with known substances but, in this case, identical paper and solvent must be used. R_f, is defined as the distance the solute moves (D_1) divided by the distance traveled by the solvent front (D_2):

$$R_f = D_1/D_2$$

Over the years, different types of paper chromatography have evolved and are classified by the direction of flow of the mobile liquid phase. Some examples are ascending paper chromatography, descending paper chromatography, ascending–descending chromatography, and two-dimensional chromatography.

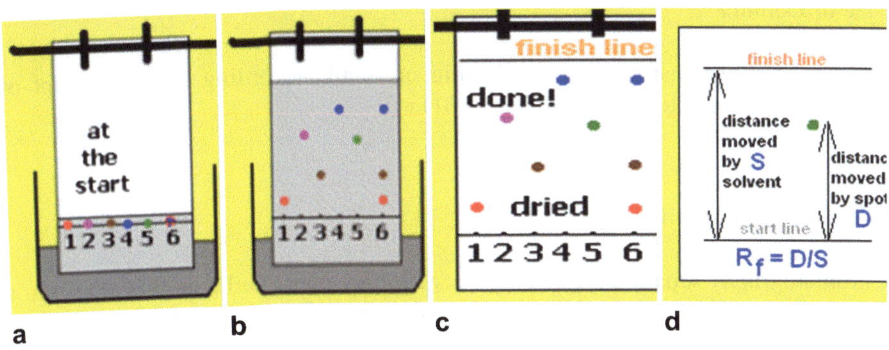

Fig. 3.60 The principle of paper chromatography (Source: www.dochbrown.info). (**a**) At the start, spots are marked on the ***start line*** (baseline), making sure it is above the surface of the solvent, so that spots are not dissolved. We start with an even line of solute chemical spots (e.g., dyes or any coloured compound). (**b**) The **solvent** is allowed to move up the paper. The colors begin to separate. It is best done in a larger covered beaker so that the solvent does not evaporate into the laboratory! (**c**) When the solvent is near the top of the paper, the paper is removed and allowed to dry before examination and measurement of the R_f values of all the dyes. The final result is called the **chromatogram**. (**d**) **How do we measure an R_f value**? It is quite simple to measure a reference value (R_f) for a dye color. For example, for the green dye spot, if S is 7.2 cm and D is 5.3 cm, then $R_f = D/S = 5.3/7.2 = 0.74$

3.4.4.2 Column Chromatography

Column chromatography (CC) is another common and useful separation technique for a mixture of biomolecules or organic molecules. This separation method involves the same principles as thin-layer chromatography (TLC), but can be applied to separate larger quantities than TLC. CC can be used on both a large and small scale. The applications of this technique are wide-reaching and cross many disciplines, including biology, biochemistry, microbiology, and medicine. Many common antibiotics are purified by CC [86, 91–93]. Alumina and silica are the two most popular stationary phases used and the partitioning works in an analogous manner: the more polar sample is retained on the stationary phase longer. Thus, the least polar compound elutes from the column first, followed by each compound in order of increasing polarity.

Although the interactions between the mobile and stationary phases are based on the same principles for CC and TLC, care must be taken when predicting the order of elution. Because the direction of solvent flow in TLC moves up and in CC the solvent flows down, it appears that the order is "upside-down". In TLC, the more polar molecules have lower R_f values, but in CC they are retained longer on the column for both alumina and silica. The types of stationary phase chosen are determined experimentally, or often based on results from a previous TLC experiment. The type of adsorbent, size of column, polarity of the mobile phase, and rate of elution all affect the separation.

Size of Column

Columns for chromatography can be big or small, according to the amount of material that needs to be loaded (Fig. 3.61).

Stationary and Mobile Phases

Chromatography separation consists of a stationary phase and a mobile phase. In a mixture of different types of molecules, the target molecule interacts with the stationary phase while the rest molecules in the mixture remain in the mobile phase. The stationary phase is the solid media or medium through which the unknown sample or solution to be analyzed is passed. As the term implies, it is stationary and substances in the unknown sample that have affinity to it (or to reagents in it) do not travel along the column, but stay with the stationary medium, thereby allowing separation of the substances present in the unknown sample or solution. Alumina and silica are the two most popular stationary phases in column chromatography. Mobile phases generally consist of a water/organic solvent mixture, aqueous buffer, or mixtures of organic solvents with or without modifiers. In isocratic mode, the composition of the mobile phase remains unchanged throughout

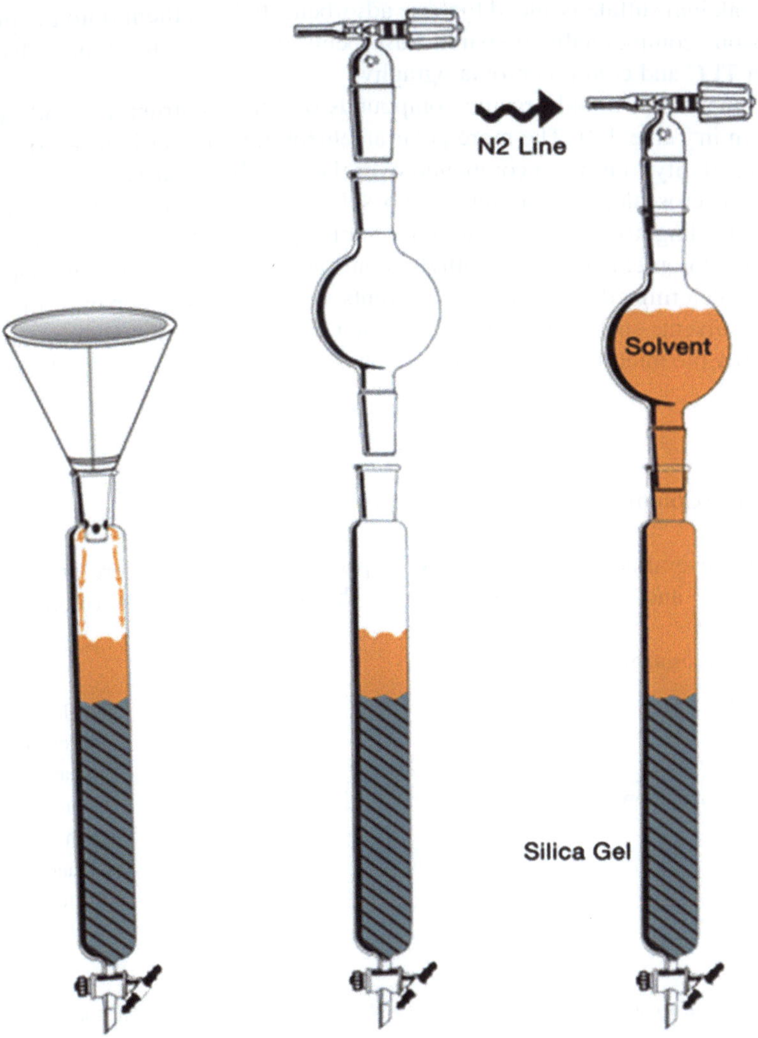

Fig. 3.61 (**a**) Microcolumns can separate about a 1–5 g of material. (**b**) Gravity column chromatography can separate 5–10 g of material. (**c**) Larger chromatography columns have a pressurised device for connection to the N_2 line. The size employed depends on the amount of material that needs to be separated

the analytical run, whereas in gradient elution mode the mobile phase composition changes through programming of the pump. There are many factors related to selection of the mobile phase, such as the solubility and ionization of the analyte, eluent polarity, and whether to use normal or reverse phase mode. An appropriate column should also be selected.

Alumina (Al_2O_3) and silica gel (SiO_2xH_2O) are the most commonly used adsorbents in TLC and column chromatography. However, for use in TLC a binder such as calcium sulfate is added to these adsorbents to hold them onto the plate. For this reason, commercially prepared adsorbents cannot be used interchangeably between TLC and column chromatography.

The eluents are liquid organic compounds of various structures and polarities (as shown in Table 3.7). The more polar an eluent, the greater is its eluting power, that is, its ability to move a compound over the adsorbent surface.

On a macroscale, the mixing of two solvents can create heat and crack the column, leading to a poor separation. Some typical solvent combinations are ligroin/dichloromethane, hexane/ethyl acetate, and hexane/toluene. Often an experimentally determined ratio of these solvents can sufficiently separate most compounds. Solvents such as methanol and water are normally not used because they can destroy the integrity of the stationary phase by dissolving some of the silica gel (Table 3.8).

Packing the Column

There are several acceptable methods for packing a column. There are two main types, slurry and dry packing methods. The slurry method is often used for

Table 3.7 Choosing solvents and adsorbents

Least polar	Cyclohexane
	Petroleum ether
	Hexane
Increasing eluting power	Toluene
	Dichloromethane
	Ethanol
	Acetone
Most polar	Methanol

Table 3.8 Some major stationary-phase classifications and their general applicability

Phase	Ligand	Applicability
C_{18} (alkyl)	$-C_{18}H_{37}$	Most retentive alkyl phase for pharmaceuticals, steroids, fatty acids, phthalates, environmental pollutants
C_{18} polar embedded	$-C16H_{33}$ NO	Carbamate or similar spacer in ligand promotes retention of polar analytes and aids separation based on functional group differences
Cyano	$-(CH_2)_3$ CN	Unique selectivity for polar analytes and solutes with widely different chemistries
Phenyl (diphenyl)	$-C_6H_5(-C_{12}H_9)$	Aromatic and moderately polar compounds
Pentafluorophenyl	$-C_6F_5$	Enhanced selectivity for halogenated, polar and isomeric analytes
Amino	$-(CH_2)_3NH_2$	Weak anion exchanger that can be used with pH control to enhance electrostatic retention

macroscale separations. The solid stationary phase is combined with a small amount of nonpolar solvent in a beaker. They are thoroughly mixed until a consistent paste is formed that is still capable of flowing. This homogeneous mixture is poured into the column as carefully as possible using a spatula to scrape out the solid. The slurry method normally gives the best column packing, but is the more difficult technique to master (Fig. 3.62).

Whether the dry or slurry method is chosen, the most important aspect of packing the column is creating an evenly distributed and packed stationary phase. As mentioned, cracks, air bubbles, and channeling lead to poor separation. Once the column is loaded, the stopcock is opened the solvent level allowed to drop to the top of the packing, but not below this point. Allowing the solvent level to go below the stationary phase (known as letting the column "run dry") should always be avoided because it allows air bubbles and channel formation to occur, leading to poor separation.

Dry packing is the method of choice for a microscale column. The column is first filled with a nonpolar solvent. Slowly, the powdered alumina or silica is added while gently tapping the side of the column with a pencil. The solid should "float" to the bottom of the column. The column should be filled as evenly as possible because cracks, air bubbles, and channels lead to poor separation (Fig. 3.63). For the dry method, the sample is first dissolved in a minimum amount of solvent, with subsequent addition of about 100 mg of silica gel. The mixture is gently swirled until the solvent evaporates and only a dry powder is left. The dry powder is transferred onto a folded piece of weighing paper and added to the top of the

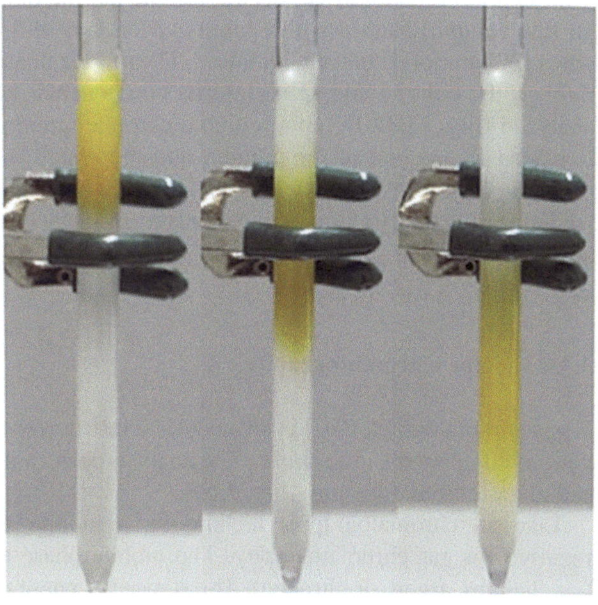

Fig. 3.62 Column packed with silica slurry and sample to be eluted. The colored bands travel down the column as the compound is eluted (the column is loaded using the slurry method for macroscale separations)

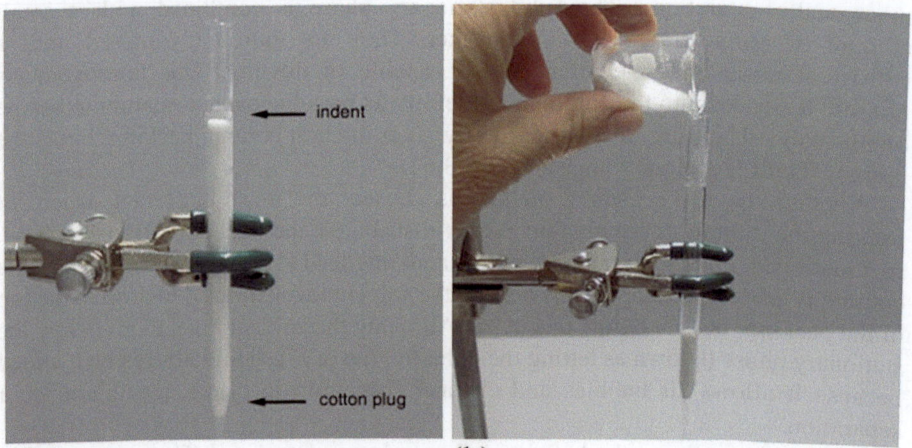

Fig. 3.63 (**a**) Dry packing is the method of choice for a microscale column. (**b**) Another way to fill the column is to pour gel into the column using a 10 mL beaker

prepared column. This method is limited by the fact that it can only separate small amounts of sample. It works best for 20–30 mg of sample.

Stationary phases for CC come in a variety of sizes, activities, and acidic and basic variations for both alumina and silica. The types of stationary phase chosen are determined experimentally, or often based on results of a previous TLC experiment. The type of adsorbent, size of column, polarity of the mobile phase, and rate of elution all affect the separation. The stationary phase is a vertical glass jar (the column) packed with a highly adsorbent solid, such as crystals of silica or silica gel, or a solid coated with liquid. The mobile phase is pumped at high pressure through the column and splits into its components, which are then removed and analyzed (Fig. 3.63). In liquid-column chromatography, the mixture being studied is placed at one end of the column and the eluant is poured in to help it travel through. Thin-film chromatography is a variation of this technique in which the "column" is actually a film of glass, plastic, or metal coated with a very thin layer of adsorbent material.

3.4.4.3 Gas Chromatography

A gas chromatograph (GC) is an analytical instrument that measures the content of various components in a sample. The analysis performed by a gas chromatograph is called gas chromatography (Fig. 3.64).

Like all chromatographic techniques, a mobile and a stationary phase are required for gas chromatography. The mobile phase (carrier gas) is an inert gas (i.e., helium, argon, or nitrogen). The stationary phase consists of a packed column in which the packing or solid support acts as stationary phase or is coated with a

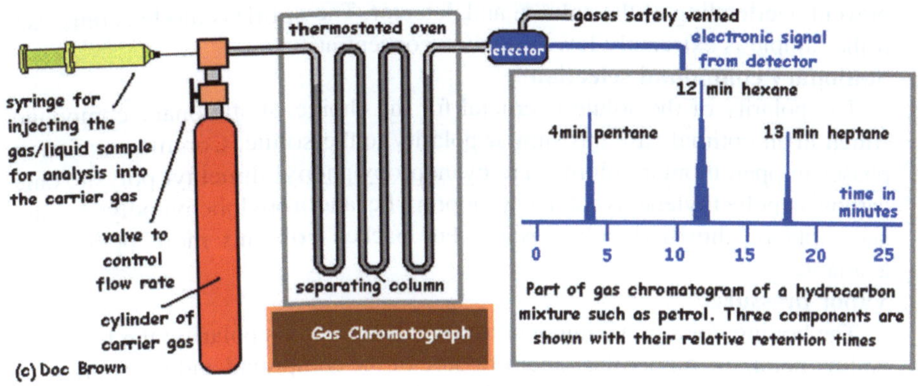

Fig. 3.64 Gas chromatography. Source: www.dochbrown.info

liquid stationary phase (high boiling polymer). Most analytical GCs use capillary columns in which the stationary phase directly coats the walls of the small-diameter tube (i.e., 0.25 μm film in a 0.32 mm tube).

The separation of compounds is based on the different strengths of interaction of the compounds with the stationary phase ("like-dissolves-like" rule): the stronger the interaction, the longer the compound interacts with the stationary phase, and the more time it takes to migrate through the column (longer retention time). In Fig. 3.64, heptane interacts more strongly with the stationary phase and therefore lags behind pentane in its movement through the column. As a result, pentane has a much shorter retention time than heptane.

Practical Procedure

The sample solution injected into the instrument enters a gas stream, which transports the sample into a separation tube known as the "column." The various components are separated inside the column. The detector measures the quantity of the components that exit the column. To measure a sample with an unknown concentration, a standard sample of known concentration is injected into the instrument. The standard sample peak retention time (appearance time) and area are compared with those of the test sample to calculate the concentration. Six factors that influence the separation of components are listed below:

1. **Amount of material injected**
   Ideally, the peaks in a chromatogram display a symmetric shape (Gaussian curve). If too much sample is injected, the peaks show significant tailing, which causes poorer separation. Most detectors are relatively sensitive and do not need a lot of material to produce a detectable signal. Strictly speaking, under standard conditions, only 1–2% of the compound injected into the injection port passes through the column, because most GC instruments are operated in split-mode to

prevent overloading of the column and detector. The splitless mode is only used if the sample is extremely low in analyte concentration.

2. **Stationary compound selection**

   The polarity of the solute is crucial for the choice of stationary compound, which in an optimal case has similar polarity to the solute. Common stationary phases in open tubular columns are cyanopropylphenyl dimethyl polysiloxane, carbowax polyethyleneglycol, biscyanopropyl cyanopropylphenyl polysiloxane, and diphenyl dimethyl polysiloxane. For packed columns more options are available.

3. **Vapor pressure**

   The boiling point of a compound is often related to its polarity: the lower the boiling point, the higher the vapor pressure of the compound and the shorter the retention time because the compound spends more time in the gas phase. That is one of the main reasons why low boiling solvents (i.e., diethyl ether, dichloromethane) are used as solvents to dissolve the sample. The temperature of the column does not have to be above the boiling point because every compound has a non-zero vapor pressure at any given temperature, even solids. That is the reason why we can smell compounds such as camphor (0.065 mmHg/25 °C), isoborneol (0.0035 mmHg/25 °C), and naphthalene (0.084 mmHg/25 °C). However, their vapor pressures are low compared with those of liquids (e.g., water, 24 mmHg/25 °C; ethyl acetate, 95 mmHg/25 °C; and diethyl ether, 520 mmHg/25 °C).

4. **Column temperature**

   The column(s) in a GC are contained in an oven, the temperature of which is precisely controlled electronically. The rate at which a sample passes through the column is directly proportional to the temperature of the column. The higher the column temperature, the faster the sample moves through the column. However, the faster a sample moves through the column, the less it interacts with the stationary phase, and the less the analytes are separated. In general, the column temperature is selected as a compromise between the length of analysis and the level of separation. A method that holds the column at the same temperature for the entire analysis is called "isothermal." However, if the column temperature is increased during analysis, then the initial temperature, rate of temperature increase (the temperature "ramp"), and final temperature are called the "temperature program." A temperature program allows analytes that elute early in the analysis to separate adequately, while shortening the time it takes for late-eluting analytes to pass through the column.

5. **Carrier gas flow rate**

   A high flow rate reduces retention times, but results in poor separation because components have very little time to interact with the stationary phase and are just being pushed through the column.

6. **Column length**

   Longer columns provide more time during which the partitioning and, thus, separation can occur. This is especially needed for separations of compounds with similar functional groups and weights.

3.4.4.4 Gel Filtration Chromatography

Gel filtration is the simplest and mildest of all the liquid chromatography techniques and separates molecules on the basis of differences in size. This is also known as size-exclusion, gel permeation, molecular sieve, and gel-exclusion chromatography. In this technique, the separation of molecules is based on the size, shape, and molecular weight of the sample rather than on chemical interaction with the gel beads. Gel filtration resins are beads contain pores of a defined size range. The sponge-like gel beads with pores serve as molecular sieves for separation of smaller and bigger molecules. Large proteins, which cannot enter these pores, pass around the outside of the beads. Smaller molecules enter the gel beads through their pores and become trapped (Fig. 3.65).

Gel filtration is mainly carried out in columns. The column is filled with beads of gel, which contain pores. The gels are made up of dextrans (Sephadex), agarose, or polyacrylamide. The sample is dissolved in buffer and allowed to flow through the column. Large molecules cannot enter the pores of the beads so they are excluded and pass down the column rapidly. Small molecules can enter into the pores of the beads so their speed is retarded and they move down slowly.

Types of Gels

The separating gels used for gel chromatography can be divided into swelling and non-swelling materials. Additional crosslinking of linear polymers or a weak crosslinking during polymerization produce gels that can take up great amounts

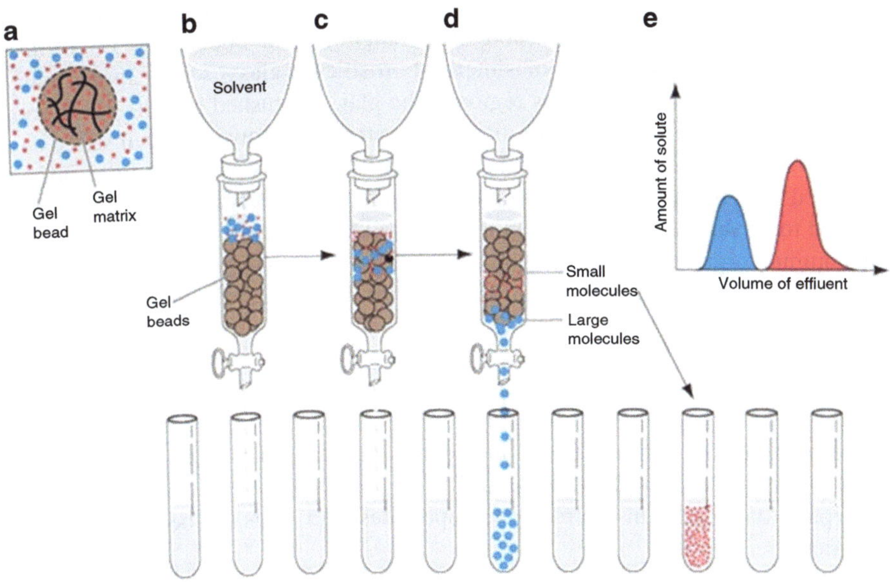

Fig. 3.65 Gel-filtration chromatography (showing how the target product is collected)

of solvent and swell appreciably. Dextran gels belong to this type and have proved very valuable for measuring and separating proteins and other water-soluble polymers. Sephadex is the most popular gel for separation of proteins and most biomolecules.

When *Leuconostoc mesenteroides* indulge in sucrose fermentation, large polymers of glucose are the result. These polymers are known as dextrans and are used to prepare Sephadex, which is a crosslinked polymer. Sephadex gels are insoluble in water and are stable in bases, weak acids, and mild reducing and oxidizing agents. They cannot be used to separate biopolymers larger than 300,000 Da. Some identified gel serial numbers are G-25, G-50, G-75, G-100, and G-200.

Very rigid polymer gels of large pore size are also available, such as those prepared by copolymerization of styrene and divinylbenzene in the presence of different diluents, (e.g., in a mixture of toluene and *n*-dodecane). The pore size of this gel depends on the composition of the reaction mixture, but all gels are rigid and swell only very little. For completely non-aqueous separations, a gel that swells in an organic solvent is required. Styragel provides this option. It is a rigid crosslinked polystyrene gel. The gel structure is unaffected by temperatures as high as 150 °C. The gel can be used with such solvents as tetrahydrofuran, cresol, dimethyl sulfoxide, chloroform, and carbon tetrachloride. Similar gels with large pores and great rigidity are also prepared from vinyl acetate and divinyladipate in suitable diluents.

Silica gels are principally applicable to gel chromatography. Recently it became possible to produce this material as beads with well-defined diameters and pore sizes. It is commercially available as Spherosile or Porasilc and has already been used extensively in gel chromatography. A glass powder of uniform pore size is also available. These fine glass spheres are manufactured from borosilicate glass to contain large number of pores with a very narrow size distribution. The glass spheres have a molecular exclusion limit ranging from 3000 to 9×10^6 Da. Manufacture involves the use of sintered borosilicate glass, which is thermally treated to produce heterogeneous regions. The glass is crushed and the silica-poor phase is leached out to leave the desired pore system. The powder is well suited as a packing for polymers. Polyacrylamide is also a very popular medium and is produced by polymerizing acrylamide into bead form. Polyacrylamide gels can be used to separate molecules of up to 300,000 Da. This gel is insoluble in water, but common organic solvents in the pH range of 2–11 can be used. Some common gels are Bio-gel P 10, Bio-gel P60, Bio-gel P100, Bio-gel P200, and Bio-gel P300.

Applications of Gel Filtration Chromatography

- Separation and purification of molecules, viruses, nucleic acids, hormones, enzymes, proteins, and antibodies
- Separation of vitamins, steroids, neuropeptides, and drugs can be achieved very quickly
- Determination of molecular weight of molecules.

- Study of protein receptor binding
- Separation of 4S and 5S tRNA
- Separation of DNA (from bacteria, usually Gram-positive) from the invariable contaminants, the teichoic acids.

3.4.4.5 Ion-Exchange Chromatography

Ion-exchange chromatography (Fig. 3.66) is a chromatography process that separates ions and polar molecules based on their affinity for the ion exchanger. This technique is applicable for almost any kind of charged molecule, including large proteins, small nucleotides, and amino acids.

Ion-exchangers are of two types: (1) cation exchangers, which have negatively charged groups such as carboxymethyl and sulfonate; and (2) anion exchangers, with positively charged groups such as diethylaminoethyl. The most commonly used cation exchangers are Dowex HCR and Amberlite IR, the most popular anion exchangers are Dowex SAR and Amberlite IRA.

Basic Principle of Function

Selection of a suitable ion-exchange matrix is probably the most important decision in an ion-exchange protocol and is based on factors such as ion exchanger charge

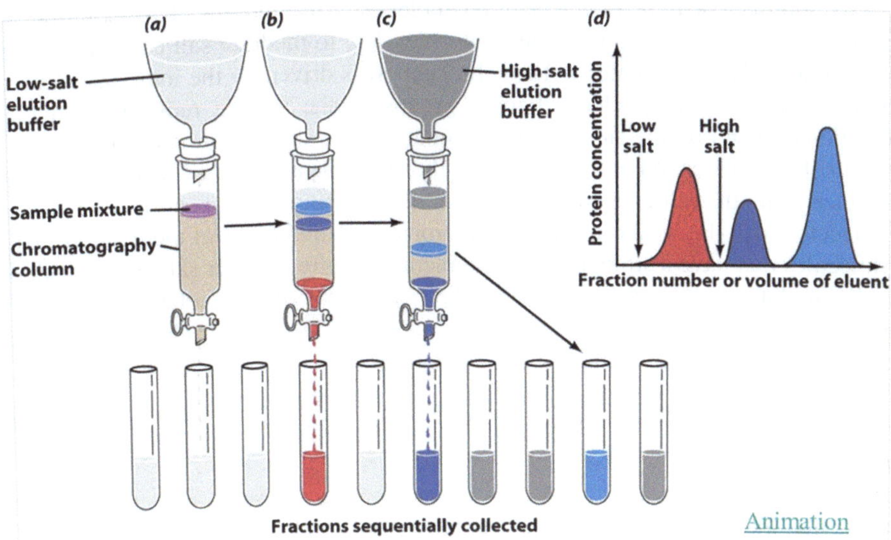

Animation

Fig. 3.66 Ion exchange chromatography

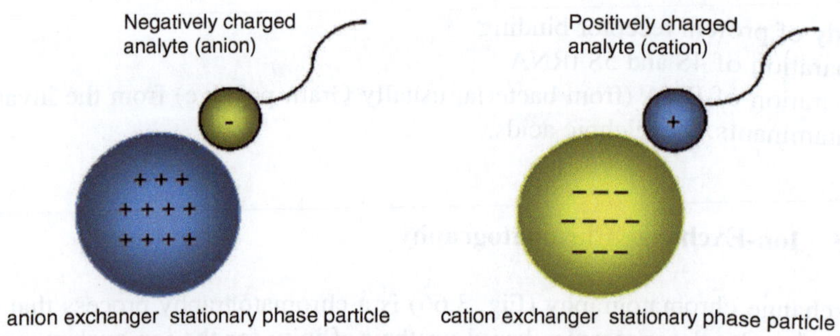

Fig. 3.67 *Left*: In cation exchange chromatography, positively charged molecules are attracted to a negatively charged solid support. *Right*: In anion exchange chromatography, negatively charged molecules are attracted to a positively charged solid support

and strength, linear flow rate, sample volume, and sample properties. In ion-exchange chromatography, numerous stationary phases are available from different manufacturers and vary significantly in chemical and physical properties. Stationary phases comprise two structural elements: the charged groups, which are involved in the exchange process, and the matrix on which the charged groups are fixed. Ion exchangers are characterized both by the nature of the ionic species comprising the fixed ion and the nature of the insoluble ion exchange matrix itself. In cation-exchange chromatography, positively charged molecules are attracted to a negatively charged solid support. Conversely, in anion-exchange chromatography, negatively charged molecules are attracted to a positively charged solid support (Fig. 3.67).

To optimize binding of all charged molecules, the mobile phase solution is generally of low to medium conductivity (i.e., low to medium salt concentration). Adsorption of the molecules to the solid support is driven by the ionic interaction between the oppositely charged ionic groups of the sample molecule and the functional ligand on the support. The strength of the interaction is determined by the number and location of the charges on the molecule and on the functional group. By increasing the salt concentration (generally by using a linear salt gradient), the molecules with the weakest ionic interactions start to elute from the column first. Molecules that have a stronger ionic interaction require a higher salt concentration and elute later in the gradient. The binding capacities of ion-exchange resins are generally quite high. This is of major importance in process-scale chromatography, but is not critical for analytical-scale separations.

Working Conditions

As a rule, the pH of the mobile phase buffer must be between the pI (isoelectric point) or pK_a (acid dissociation constant) of the charged molecule and the pK_a of the charged group on the solid support. For example, in cation-exchange chromatography, using a functional group on the solid support with a pK_a of 1.2, a sample molecule with a p*I* of 8.2 can be run in a mobile phase buffer of pH 6.0. In anion-exchange chromatography, a molecule with a p*I* of 6.8 can be run in a mobile phase buffer at pH 8.0 when the pK_a of the solid support is 10.3. As in most other modes of chromatography, a protein sample is injected onto the column under conditions where it is strongly retained. A gradient of linearly increasing salt concentration is then applied to elute the sample components from the column. An alternative to using a linear gradient is to use a step gradient. This requires less complicated equipment and can be very effective for eluting different fractions if the appropriate concentrations of salt are known, usually from linear gradient experiments.

Many chromatographers also use changes in pH to effect a separation. In cation-exchange chromatography, raising the pH of the mobile phase buffer causes the protein molecule to become less protonated and, hence, less positively charged. The result is that the protein can no longer form ionic interactions with the negatively charged solid support, which ultimately results in the molecule eluting from the column. In anion-exchange chromatography, lowering the pH of the mobile phase buffer causes the molecule to become more protonated and, hence, more positively (and less negatively) charged. The result, as above, is that the protein can no longer form ionic interactions with the positively charged solid support, which causes the molecules to elute from the column.

Applications of Ion-Exchange Chromatography

Ion-exchange chromatography can be applied for the separation and purification of many charged or ionizable molecules such as proteins, peptides, enzymes, nucleotides, DNA, antibiotics, and vitamins from natural sources or of synthetic origin.

3.4.4.6 Affinity Chromatography

Affinity chromatography is a method of separating biochemical mixtures based on a highly specific interaction, such as that between antigen and antibody, enzyme and substrate, or receptor and ligand.

M	+	**L**	®	**ML**
(Macromolecule)		(Ligand)		(Complex)

Affinity chromatography was originally developed for purification of enzymes but was later extended to nucleotides, nucleic acids, and immunoglobulins. For purification of enzymes, the ligand attached to the adsorbent is usually a powerful inhibitor that binds to only one enzyme in a complex mixture. Affinity purification and many assay techniques depend on the reversible binding interactions of target biomolecules in liquid samples with specific chemical groups or macromolecules that are immobilized on a solid material. The latter class of molecules are called affinity ligands and they can be attached (immobilized) to the solid material

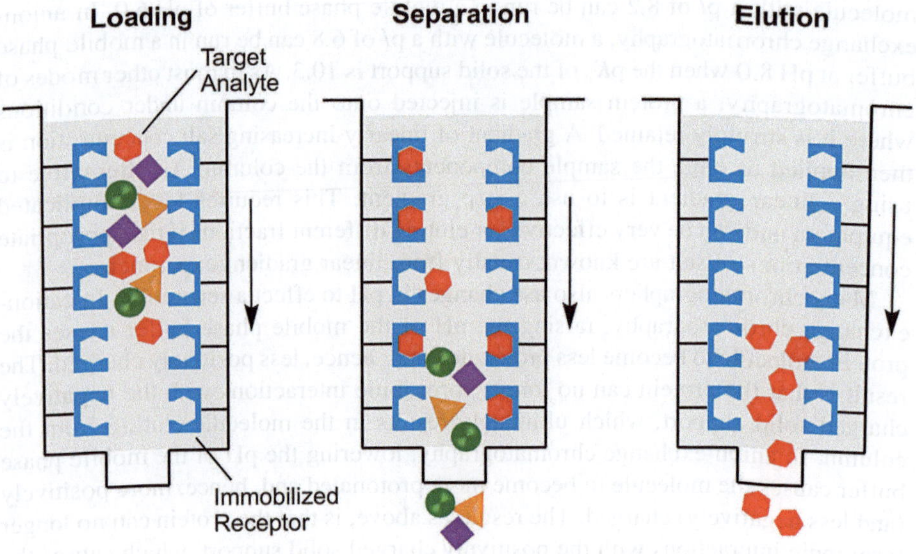

Fig. 3.68 Affinity chromatography involves three major steps: loading, separation, and elution

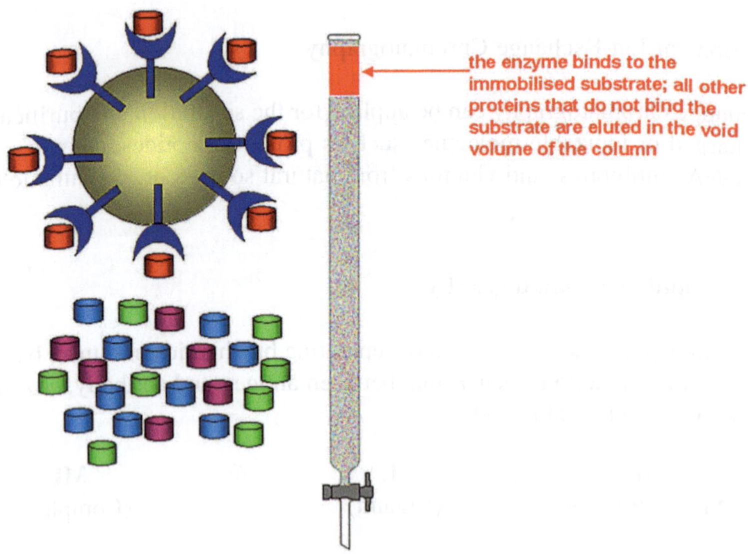

Fig. 3.69 Binding of ligand and macromolecular complex (matrix)

(stationary phase) in a variety of ways. This process involves three main steps: loading, separation, and elution (Fig. 3.68).

Most affinity purification strategies, especially those involving antibodies and other proteins, depend on covalent chemical conjugation of ligands to the solid support matrix (Fig. 3.69). Affinity ligands that have broad applicability are commercially available in a variety of ready-to-use, pre-immobilized forms. Examples include Protein A agarose resin for general antibody purification and streptavidin magnetic beads for purifications involving biotinylated molecules. This method finds its widest application in the field of protein purification and has also been useful in studying quantitative aspects of protein–ligand interaction [94, 95].

The most important components of affinity chromatography are described next.

Supporting Matrix

The ideal support material for use in affinity chromatography has a variety of physical and chemical characteristics that give it optimal behavior. In terms of its physical properties, it should have a high porosity to allow maximum access of a wide range of macromolecules to the immobilized ligand. It should be of uniform size and rigidity to allow good flow characteristics, and it must be mechanically and chemically stable under conditions used to immobilize the appropriate specific ligand. In terms of its chemical properties, it should have available a large number of groups that can be derivatized with the specific ligand, and it should not interact with proteins in general so that nonspecific adsorption effects are minimized.

Agarose beads have most of the desired features mentioned above. However, they have a susceptibility to contraction when a denaturizing solution is used for elution. Polyacrylamide beads lack porosity. This undesirable trait is heightened even further when they are attached to ligands. These beads provide mechanical rigidity and chemical inertness, in addition to providing very good flow rates. A high degree of nonspecific protein adsorption is the most serious drawback of polyacrylamide beads, which can be avoided to some extent by treatment with hexamethyldisilazane.

Although agarose derivatives approach the ideal, a common problem with the support material (one that is not restricted to agarose) is steric hindrance. Once the ligand has been immobilized on the support, simple steric hindrance can prevent the interaction of proteins having binding sites for the ligand with the immobilized ligand. To circumvent this problem, a "spacer arm" is usually included in the support material to allow the ligand to be extended away from the physical surface of the support and permit unhindered interaction of protein with the ligand. In principle, such a spacer arm readily overcomes problems of steric effects. In practice, its inclusion can lead to a major problem in affinity chromatography, that of nonspecific interaction of proteins with the support material. The most commonly used spacer arms consist of 6–10 carbon atoms. However, these have quite a hydrophobic nature that they impart to the immobilized ligand, which can lead to nonspecific binding.

Covalent Attachment of Ligand to the Activated Functional Groups

Coupling of an amino-containing ligand to CNBr-activated support is normally carried out by suspending the support and the ligand in a basic buffer solution (0.25 M $NaHCO_3$, pH 9.0). The suspension is stirred overnight in a cold room. During this time the ligand is covalently attached to the support medium. After the reaction is over, the matrix should be washed with a 0.1 M solution of pH 9.0 glycine buffer, which destroys any extra activated groups. The number of ligand groups bound is usually expressed in terms of capacity per milliliter of packed matrix rather than in terms of its dry weight.

Spacer Arm

To avoid steric repulsion between the ligands or activated groups of the matrix and the macromolecules of the matrix backbone, a spacer is introduced between the activated groups of the support and the ligand. This spacer is known as the "arm." The ligand then projects out from the macromolecule to prevent repulsion and allow unimpeded access of target protein. Examples of compounds used as spacer arms include hexamethylene, 3,3′-diamino propylamine, 1,6-diamino hexane, 6-amino-hexanoic acid, and 1,4-bis-(2,3-epoxypropoxy) butane. These spacer arms have two different functional groups: one to react with the functional groups of the matrix and the other to react with the ligand.

Activation of the Matrix Functional Groups

There are several different activated agarose gels that can be used to attach ligands. CNBr-agarose is easy to use for the attachment of amines, but does not have a long spacer between the gel beads and the bound ligand, meaning that protein binding can be subject to steric hindrance. Aminohexanoic acid-agarose (CH-agarose, for reaction with amines) and diaminohexane-agarose (AH-agarose, for reaction with carboxylic acids) have relatively long methylene chains that keep the ligand a significant distance from the gel beads. An alternative reagent for attaching amines is carbonyldiimadazole-agarose. Epoxy-activated agarose is used for alcohols.

The most common method of activation of polysaccharide supports (agarose) involves treatment with CNBr at alkaline pH (pH 11.0). Usually, 300 mg of powdered cyanogen bromide per milliliter of packed gel gives the maximum substitution. The reaction is exothermic and maintains the temperature constant at 20 °C at all times. To maintain the temperature, and the pH at 11, the mixture is continuously stirred and an electrode dipped into it at all times. The pH is maintained by the addition of 2 M NaOH. The activated suspension is then washed with about 20 times the gel volume of buffer (e.g., Tris, ammonium acetate, glycine) at a pH of 9.5–10. The reaction is usually completed within 10–15 min. Sodium bicarbonate and borate buffers are the usual choice.

Applications of Affinity Chromatography

- Affinity chromatography has been used to purify a wide range of enzyme and proteins, including immunoglobulins and DNA-related enzymes. Heparin-agarose is used to purify the heparin binding domain of platelet thrombospondin.
- Messenger RNAs are routinely isolated using poly(U)Sepharose 4B
- Clinical applications include immunoassay, therapeutic drug monitoring, and allergy testing.
- It has also been applied for concentration of dilute protein solutions and for storage of unstable protein in an immobilized form.

3.5 Formulation and Polishing

Microbial biomass processing has evolved from the use of biocatalyst activities, especially microorganisms, to obtain target products or services [95]. Industrial microbial processes have resulted in the control of safety, mixing, and environmental conditions and in the use of selected microorganisms to obtain higher productivity at lower cost. It could also help to solve the problem of increased international demand for natural products that are safe with respect to health and the environment. In fact, harvesting efficacy governs the marketability of a product by affecting potency and aiding in further processing during formulation development. Formulation is a crucial link between production and application and dictates economy, shelf life, ease of application, and enhanced field efficacy.

There are various environmental factors such as ultraviolet radiation, rain, pH, temperature, and foliage physiology that impede the efficacy of microbial product formulations for agricultural use. Various formulations have been developed (both solid and liquid) according to target application and feasibility to overcome these adverse environmental effects. Conventional formulations have given way to advanced versions such as microencapsulations and microgranules.

The formulation of low molecular weight products (solvents, organic acids) can be achieved by concentrating them through removal of most of the water. For certain small molecules (antibiotics, citric acid), formulation involves crystallization by adding salts. Proteins are highly susceptible to loss of biological activity; hence, their formulation requires special care. Certain stabilizing additives can be added to protein formulations to prolong their shelf life. Stabilizers include sugars (sucrose, lactose), salts (sodium chloride, ammonium sulfate), polymers (polyethylene glycol), and polyhydric alcohols (glycerol). Proteins can be formulated in the form of solutions, suspensions, or dry powders.

The improvements in techniques for biological control, and the increasingly widespread use of metabolites produced by microorganisms to obtain products that can be applied in various fields, are pushing scientists to provide and monitor economically viable supply chains for products of microbial origin. Generally, before a fermentation-based product can be incorporated into the supply chain it must be free of moisture to prevent deterioration and maintain stability. There are several processes that can be used for drying, but spray drying and freeze-drying are the most regularly used for biopharmaceuticals.

The process used for drying microbially derived biopharmaceuticals depends on the final product characteristics required, both in quality (active ingredients) and quantity (end biopharmaceuticals). For example, it is well known that there is a direct relation between the absorption of the active ingredient and its characteristic crystalline or polymorphic form. In some cases, the latter is related to the procedure and conditions of drying. The consequences of the process used to dry a biopharmaceutical product can vary according to the development phase (preclinical study versus downstream process).

The drying process also greatly influences the quality of the product regarding aspects such as the polymorphic form of the active ingredient, structural changes that could affect the activity of the substance, degradation products formed as a result of the drying conditions, and the presence of residual solvents undesirable or above the permitted limits. Consequently, it is important to know the influence of all the factors and to establish product specifications, as well as the nature and limits of residual solvents, in agreement with current regulations.

3.5.1 Spray Drying

The spray drying process is older than might commonly be imagined. Earliest descriptions date from 1860, with the first patented design recorded in 1872 (Fig. 3.70). The basic idea of spray drying is the production of highly dispersed powders from a fluid feed by evaporating the solvent. This is achieved by mixing a heated gas with an atomized (sprayed) fluid of high surface-to-mass ratio droplets, ideally of equal size, within a vessel (drying chamber). This causes the solvent to evaporate uniformly and quickly through direct contact [96–101].

Fig. 3.70 Spray drying process

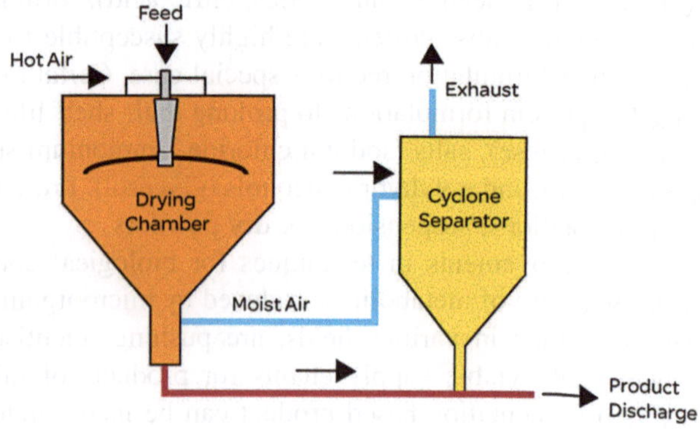

Spray drying can be used in a wide range of applications where the production of a free-flowing powder is required. This method of dehydration has become the most successful in the following areas:

- Bone and tooth amalgams
- Beverages
- Flavors, colorings, and plant extracts
- Milk and egg products
- Plastics, polymers, and resins
- Soaps and detergents
- Textiles
- Pharmaceuticals

Almost all other methods of drying, including use of ovens, freeze dryers, or rotary evaporators, produce a mass of material that requires further processing (e.g., grinding and filtering) and, therefore, produces particles of irregular size and shape. Spray drying, on the other hand, offers very flexible control over powder particle properties such as density, size, flow characteristics, and moisture content.

3.5.2 *Freeze-Drying*

Freeze-drying or lyophilization is the preferred method for drying and formulation of a wide range of products, including pharmaceuticals, foodstuffs, diagnostics, bacteria, and viruses. This is mainly because freeze-drying does not usually cause loss of biological activity of the desired product (Fig. 3.71). Freeze-drying is mainly used to remove the water from sensitive products, mostly of biological origin, without damaging them, so they can be preserved easily in a permanently storable

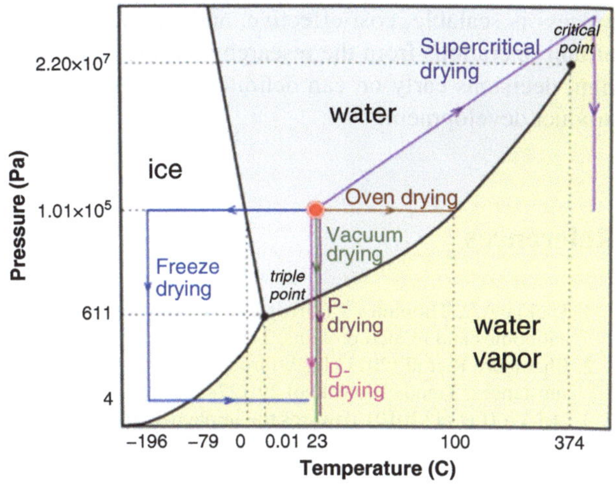

Fig. 3.71 In a typical phase diagram, the boundary between gas and liquid runs from the triple point to the critical point. Freeze-drying (blue arrow) brings the system around the triple point, avoiding the direct liquid-gas transition seen in ordinary drying time (green arrow). (Source: www.spscientific.com/freeze-drying-lyophilization-basics)

state and be reconstituted simply by adding water. Examples of freeze-dried products include antibiotics, bacteria, sera, vaccines, diagnostic medications, protein-containing products, biotechnological products, cells, tissues, and chemicals. The product to be dried is frozen under atmospheric pressure. Then, in an initial drying phase referred to as primary drying, the water (in the form of ice) is removed by sublimation. In the second phase, called secondary drying, water is removed by desorption. Freeze-drying is carried out under vacuum. The conditions under which the process takes place determine the quality of the freeze-dried product.

3.5.3 *Integration of Different Processes*

It is ideal to integrate fermentation and downstream processes to finally obtain the desired product. However, this has not been practicable for various reasons. Integration of certain stages in downstream processing for purification of product has met with some success. For instance, protein concentration by extraction into two-phase systems, combined with clarification and purification, can be done together. Microbial process development and integration of unit operations during the biopharmaceutical development cycle for a typical microbial or mammalian cell culture process is important. Most biopharmaceutical companies have their own in-house process development programs. In addition, several CMOs are involved in process development and the production of material for preclinical, phase I–III, or even GMP manufacturing.

Despite all this, there is often a lack of vision regarding process development, especially from small research companies, resulting in unscalable and highly expensive processes. Developing a scalable, robust, and regulatory-compliant process requires time, sophisticated equipment, skilled scientists and engineers, and, above all, money. Process development can only be justified if the developed process is scalable, cost-effective, and can be validated. Integration and selection of unit operations from the research phase is of prime importance, and making the right decisions early on can definitely save time and money in the later stages of product development.

References

1. Shukla AA, Thömmes J (2010) Recent advances in large-scale production of monoclonal antibodies and related proteins. Trends Biotechnol 28:253–261
2. Bhambure R et al (2011) High-throughput process development for biopharmaceutical drug substances. Trends Biotechnol 29:127–135
3. Del Va IJ et al (2010) Towards the implementation of quality by design to the production of therapeutic monoclonal antibodies with desired glycosylation patterns. Biotechnol Prog 26:1505–1527

4. Chon JH, Zarbis-Papastoitsis G (2011) Advances in the production and downstream processing of antibodies. New Biotechnol 28:458–463
5. Kelley B (2009) Industrialization of mAb production technology: the bioprocessing industry at a crossroads. MAbs 1:443–452
6. Strube J, Grote F, Ditz R (2012) Bioprocess design and production technology for the future. In: Subramanian G (ed) Biopharmaceutical production technology, vol 2, 1st edn. Wiley-VCH, Weinheim
7. Low D et al (2997) Future of antibody purification. J Chromatogr B 848:48–63
8. Singh RP, Heldman DR (1993) Introduction to food engineering, 2nd edn. Academic Press, Harcourt Brace & Company, San Diego, CA
9. Wesner J (2008) Reduction of host-cell DNA in mammalian cell culture using charged filters at protein-A capture for monoclonal antibodies. BPI, Korea. www.bioprocessintl.com
10. Zhou JX, Gottschalk U (2008) Orthogonal virus clearance applications in monoclonal antibody production. Wiley, Boca Raton, FL, pp 169–189
11. Zhou JX (2008) Viral clearance using disposable system in MAb commercial downstream process. Biotechnol Bioeng 100:488–496
12. Cheremisinoff NP et al (1993) Filtration equipment for wastewater treatment. Prentice-Hall, Englewood Cliffs
13. Halberthal J (2016) Nutsche filter. Engineering aspects in solid-liquid separation. www.solidliquid-separation.com
14. Koichi I, Dennis R (2016) Industrial gas filtration. Filtration: principles and practices. encyclopedia.che.engin.umich.edu
15. Loff Lawrence G (1990) Filter media, filter rating. In: Svarovsky L (ed) Solid-liquid separation, 3rd edn. Butterworth & Co., London
16. Maas Joseph H (1988) Gas-solid separations. In: Schweitzer PA (ed) Handbook of separation techniques for chemical engineers, 2nd edn. McGraw-Hill, New York
17. Osborne DG, Pierson HG (1999) Vacuum filtration. In: Svarovsky L (ed) Solid-liquid separation, 3rd edn. Butterworth & Co., London
18. Perry Robert H, Don WG (1997) Perry's chemical engineers' handbook, 7th edn. McGraw Hill, New York
19. Phelan M (2008) Focus on filtration. Chem Eng 8:53–54
20. Svarovsky L (1992) Pressure filtration. In: Svarovsky L (ed) Solid-liquid separation, 3rd edn. Butterworth & Co., London
21. Watson JHP (1990) High gradient magnetic separation. In: Svarovsky L (ed) Solid-liquid separation, 3rd edn. Butterworth & Co., London, pp 661–662
22. Zievers JF (1998) Batch filtration. In: Schweitzer PA (ed) Handbook of separation techniques for chemical engineers, 2nd edn. McGraw-Hill, New York
23. Hill P et al (2007) Cell harvesting. Wiley, Boca Raton, FL, pp 305–330
24. Yigzaw (2006) Exploitation of the adsorptive properties of depth filters for host-cell protein removal during monoclonal antibody purification. Biotechnol Progr 22:288–296
25. Susan R et al (2004) Bioanalytical chemistry, Ch. 13. In: Centrifugation methods. Wiley, Hoboken, NJ, pp 247–267
26. Basel RM et al (1983) Monitoring to use the present Method to distinguish between Live and microbial number in food by density centrifugation. Appl Environ Microbiol 45:1156–1159
27. Leung WW-F (1998) Industrial centrifugation technology. McGraw-Hill Education, New York. ISBN-13: 9780070371910
28. Ford TC, Graham JM (1991) An introduction to centrifugation. BIOS Scientific, Oxford. ISBN: 1 872748 40 6
29. Baldwin C, Robinson CW (1990) Disruption of *Saccharomyces cerevisiae* using enzymatic lysis combined with high-pressure homogenization. Biotechnol Tech 4:329–334
30. Kleinig AR, Middelberg APJ (1998) On the mechanism of microbial cell disruption in high-pressure homogenisation. Chem Eng Sci 53:891–898

31. Dunnill P (1982) Biosafety in the large-scale isolation of intracellular microbial enzymes. Chem Ind 22:877–879
32. Cameron P, Hambleton P, Melling J. et al. (1987) Assessing the microbiological integrity of biotechnology equipment. In: Verrall MS, Hudson MJ. (Eds.): Separations for biotechnology, 490–496. Chichester: Ellis Horwood.
33. Engler CR, Robinson CW (1979) New method of measuring cell-wall rupture. Biotechnol Bioeng 21:1861–1869
34. Middelberg APJ (1995) Process-scale disruption of microorganisms. Biotechnol Adv 13:491–551
35. Kleinig AR et al (1995) Influence of broth dilution on the disruption of *Escherichia coli*. Biotechnol Tech 9:759–762
36. Hetherington PJ et al (1971) Release of protein from baker's yeast (*Saccharomyces cerevisiae*) by disruption in an industrial homogeniser. Trans Instn Chem Engrs 49:142–148
37. Bailey JE, Ollis DF (1987) Biochemical engineering fundamentals, 2nd edn. McGraw-Hill, New York, NY
38. Naglak TJ et al (1990) Chemical permeabilization of cells for intracellular product release. In: Separation processes in biotechnology. Marcel Dekker, New York
39. Bulock J, Christiansen B (1989) Basic biotechnology. Academic Press, Cambridge, MA
40. Brookman J (1974) Mechanism of cell integration in high pressure homogenizer. Biotechnol Bioeng 16:371–383
41. Felix H (1982) Cell disruption. Anal Biochem 107:207–214
42. Felix H (1982) Permeabilized cells. Anal Biochem 120:234
43. Follows M et al (1971) Release of enzymes from Baker's yeast by disruption in an industrial homogenizer. Biotechnol Bioeng 13:549–560
44. Ayazi Shamlou P et al (1995) A physical model of high-pressure disruption of Bakers' yeast cells. Chem Eng Sci 13:83–1391
45. Wuytack EY et al (2002) Bacterial inactivation by high-pressure homogenisation and high hydrostatic pressure. Int J Food Microbiol 77:205–212
46. Stang M et al (2001) Emulsification in high-pressure homogenizers. Eng Life Sci 1:151–157
47. Floury J et al (2003) Effect of high pressure homogenisation on methylcellulose as food emulsifier. J Food Eng 58:227–238
48. Sanguansri P, Augustin MA (2006) Nanoscale materials development – a food industry perspective. Trends Food Sci Technol 17:547–556
49. Cortés-Muñoz M et al (2009) Characteristics of submicron emulsions prepared by ultra-high pressure homogenisation: effect of chilled or frozen storage. Food Hydrocolloids 23:640–654
50. Halim R et al (2012) Microalgal cell disruption for biofuel development. Appl Energy 91:116–121
51. Marie P et al (2002) Influence of major parameters in emulsification mechanisms using a high-pressure jet. J Food Eng 53:43–51
52. Cushen M et al (2011) Nanotechnologies in the food industry – recent developments, risks and regulation. Trends Food Sci Technol 24:30–46
53. Ji Y et al (2011) Effect of heat treatments and homogenisation pressure on the acid gelation properties of recombined whole milk. Food Chem 129:463–471
54. Masson LMP et al (2011) Effect of ultra-high pressure homogenization on viscosity and shear stress of fermented dairy beverage. LWT-Food Sci Technol 44:495–501
55. Ciron CIE et al (2012) Modifying the microstructure of low-fat yoghurt by microfluidisation of milk at different pressures to enhance rheological and sensory properties. Food Chem 130:510–519
56. Biasuttia M et al (2010) Rheological properties of model dairy emulsions as affected by high pressure homogenization. IFSET 11:580–586
57. Innocente N et al (2009) Effect of high-pressure homogenization on droplet size distribution and rheological properties of ice cream mixes. J Dairy Sci 92:1864–1875

58. Heffernan SP (2011) Efficiency of a range of homogenisation technologies in the emulsification and stabilization of cream liqueurs. Innov Food Sci Emerg Technol 12:628–634
59. Schultz S et al (2004) High-pressure homogenization as a process for emulsion formation. Chem Eng Technol 27:361–368
60. Naganagouda K, Mulimani VH (2008) Aqueous two-phase extraction (ATPE): an attractive and economically viable technology for downstream processing of Aspergillus oryzae [alpha]-galactosidase. Process Biochem 43:1293–1299
61. Walter H, Johansson G (1995) Aqueous two-phase systems. Methods in enzymology, vol 24. Academic Press, New York
62. Veide A et al (1983) A process for large scale isolation of b-galactosidase from E. coli in an aqueous two-phase system. Biotechnol Bioeng 25:1789–1800
63. Van Berlo M (1998) Poly(ethylene glycol)-salt aqueous two-phase systems with easily recyclable volatile salts. J Chromatogr B Biomed Sci Appl 711:61–68
64. Belter PA et al (1988) Bioseparations: downstream processing for biotechnology. Wiley, New York
65. Ladisch MR (2001) Bioseparations engineering: principles, practice and economics. Wiley, New York
66. Klein E (1991) Affinity membranes: their chemistry and performance in adsorptive separation processes. Wiley, New York
67. Azarkan ME et al (2003) Fractionation and purification of the enzymes stored in the latex of *Carica papaya*. J Chromatogr B 790:229–238
68. Benavides J, Rito-Palomares M (2008) Practical experiences from the development of aqueous two-phase processes for the recovery of high value biological products. J Chem Technol Biotechnol 83:133–142
69. Cleland JL et al (1992) Polyethylene glycol enhanced refolding of bovine carbonic anhydrase B. Reaction stoichiometry and refolding model. J Biol Chem 267:13327–13334
70. Nitsawang S et al (2006) Purification of papain from *Carica papaya* latex: aqueous two-phase extraction versus two-step salt precipitation. Enzyme Microbial Technol 39:1103–1107
71. Vernau J, Kula MR (1990) Extraction of protein from biological raw material using aqueous (polyethylene) glycol-citrate. Biotechnol Appl Biochem 12:397–404
72. Adam F et al (2012) "Solvent-free" ultrasound-assisted extraction of lipids from fresh microalgae cells: a green, clean and scalable process. Bioresource Technology 114:457–465
73. Azmir J et al (2013) Techniques for extraction of bioactive compounds from plant materials: a review. J Food Eng 117:426–436
74. Chavan Y, Singhal RS (2013) Ultrasound-assisted extraction (UAE) of bioactives from arecanut (Areca catechu L.) and optimization study using response surface methodology. Innov Food Sci Emerg Technol 17:106–113
75. Herrero M et al (2013) Compressed fluids for the extraction of bioactive compounds. Trends Anal Chem 43:67–83
76. Hidalgo D et al (2014) Integrated and sustainable system for multi-waste valorization. Environ Eng Manage J 13:2467–2475
77. Roseiro LB et al (2013) Supercritical, ultrasound and conventional extracts from carob (Ceratonia siliqua L.) biomass: effect on the phenolic profile and antiproliferative activity. Indus Crops Prod 47:132–138
78. Li Y et al (2013) Solvent-free microwave extraction of bioactive compounds. Trends Anal Chem 47:1–11
79. Mulder M (2012) Basic principles of membrane technology. Kluwer, Dordrecht
80. Jornitz MW (2006) Sterile filtration. Springer, Berlin. ISBN-10: 354028625X; ISBN-13: 9783540286257
81. Van Reis R, Zydney A (2007) Bioprocess membrane technology. J Mem Sci 297:16–50
82. Ripperger S, Schulz G (1986) Microporous membranes in biotechnical applications. Bioprocess Eng 1:43–49
83. Melin T, Rautenbach R (2007) Membranverfahren. Springer, Berlin. ISBN 3-540-00071-2

84. Hostettmann K et al (1998) Preparative chromatography techniques applications in natural product isolation, 2nd edn. Springer, Berlin, p 50
85. Ettre LS, Sakodynskii KI (1993) Discovery of chromatography II: Completion of the development of chromatography (1903–1910). Chromatographia 5–6:329–338
86. Still WC et al (1978) Rapid chromatographic technique for preparative separations with moderate resolution. J Org Chem 43(14):2923–2925
87. Ettre LS (1993) Nomenclature for chromatography (IUPAC Recommendations 1993). Pure Appl Chem (4):819–872
88. Harwood Laurence M, Moody Christopher J (1989) Experimental organic chemistry: principles and practice (Illustrated edn). Wiley-Blackwell, Oxford, pp 180–185
89. Anfinsen Christian B, Tileston EJ, Middlebrook RF (eds) (1976) Advances in protein chemistry. pp 6–7
90. Paper chromatography. www.chemguide.co.uk
91. Edwin H (2007) Vegetable tannins – lessons of a phytochemical lifetime. Phytochemistry 68:2713–2721
92. Laurence M, Moody Christopher J (1989) Experimental organic chemistry: principles and practice (Illustrated edn). Wiley-Blackwell, Oxford, pp 180–185
93. Fair JD, Kormos CM (2008) Flash column chromatograms estimated from thin-layer chromatography data. J Chromatogr A 1211:49–54
94. Harrison RG et al (2003) Bioseparations science and engineering. Oxford University Press, New York, NY
95. Hage DS (1999) Affinity chromatography: a review of clinical applications. Clin Chem 45:593–615
96. Lilly MD (1992) Biochemical engineering research: its contribution to the biological industries. Trans IchemE 70(C):3–7
97. Buckton G et al (2002) The effect of spray-drying feed temperature and subsequent crystallization conditions on the physical form of lactose. AAPS PharmSciTech 3:1–6
98. Gharsallaoui A et al (2007) Applications of spray-drying in microencapsulation of food ingredients: an overview. Food Res Int 40:1107–1121
99. Gonnissen Y et al (2008) Coprocessing via spray drying as a formulation platform to improve the compactability of various drugs. Eur J Pharm Biopharm 69:320–334
100. Hanrahan FP, Webb BH (1961) USDA develops foam-spray drying. Food Eng 33:37–38
101. Jorge MCP, Filipe G (2004) Spray drying technology for better API crystals. Process Dev:38–39. www.sp2.uk.com

Chapter 4
Microbial Products Supply Chain

4.1 Introduction

A supply chain is a suitable logistic network through which a manufactured product (s) is distributed to the end user without any change in quality of the product (Fig. 4.1). For biotechnology and biopharmaceutical industrial products, supply chain activities involve the transformation of natural resources, raw materials, and components into a finished product that is delivered to the end customer.

Products derived from microbial sources (Fig. 4.2), such as biopharmaceuticals, enzymes, nutraceuticals, food additives, food supplements, and anti-oxidants, are extremely susceptible to environmental stresses and strains, and therefore need efficient, effective logistics from point of origin to point of consumption so that they conform to customer requirements.

A supply chain in the biotechnology industry should have a well-connected network of individuals, organizations, resources, activities, and technologies involved in the manufacture and sale of the product or service. A supply chain starts with the delivery of formulated and polished bioproduct to the factory outlet, and continues with distribution of the product through a proper logistic system to the consumer. Supply chain management (SCM) oversees each touch point of a company's product or service, from initial creation to final sale. With so many places along the supply chain that can add value through efficiency or lose value through increased expenses, proper SCM can increase revenues, decrease costs, and affect company profits. To develop the efficacy of a supply chain, systematic management practices should be followed to make the product available at the consumer level with perfect satisfaction.

B. Kumara Behera, A. Varma, *Microbial Biomass Process Technologies and Management*, DOI 10.1007/978-3-319-53913-3_4

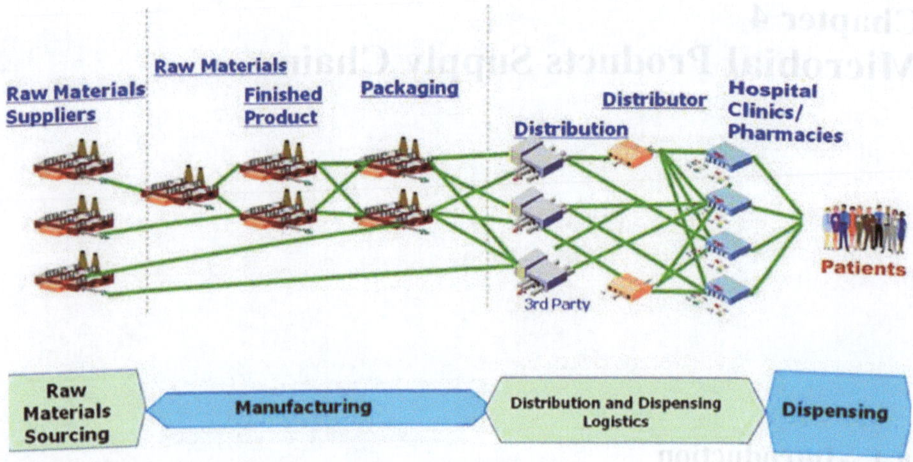

Fig. 4.1 Supply chain for bio-pharmaceuticals from production site to end users (Source: https://www.google.co.in/imgres)

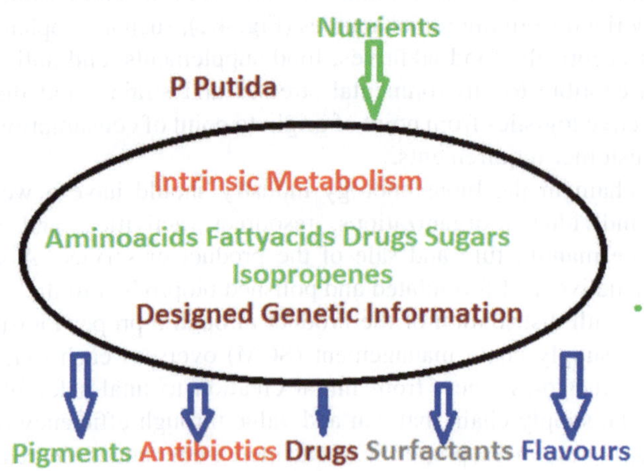

Fig. 4.2 Bacteria as source of sustainable biopharmaceuticals

4.2 Microbial Product Supply Chain Management

Biotechnology and biopharmaceutical industries have shown great interest in SCM since 1980, when top companies started realizing the benefit of SCM and encouraged collaborative relationships within and beyond their own organizations

[1]. Since then, with the passing of time and experience, the concept of SCM has changed. The big company promoters started realizing that SCM is an integrated philosophy for managing the total flow from the supply of raw materials to sale of product to the final customer [2–4]. SCM is a process by which products are manufactured and then distributed to end users [5]. In other words, SCM extends the concept of integrated management of an organization to all organizations involved in the process. Cooper and Ellram [6] suggested that the implementation of SCM has three major objectives: (1) reduce inventory investment in the chain; (2) increase customer service through increased stock availability and reduced order cycle time; and (3) help build competitive advantage to create customer value. Competition has now started between SCM groups for better service with a smaller budget [7]. Research supports the point of view that SCM provides a vision that focuses everyone in an organization on product, production, and quality improvements, and that these improvements are not only required by the market, but are also driven by the need for companies to survive [8].

The biotechnology industry is still in its early years and is characterized by an enormous amount of innovative potential and fast market growth. Within a short period of time, small businesses have become global players. However, these dynamics put special requirements on successfully managing the supply chain. One example is Miltenyi Biotec, a successful global manufacturer of reagents, instruments, and accessories for cell separation and analysis, headquartered in Bergisch Gladbach, Germany, with approximately 1400 employees. Miltenyi Biotec supplies research institutions and clinics around the world and is the world's leading provider of many products, thanks to its innovativeness. Because companies in the biotechnology industry are experiencing a rapid pace of development, they are left with little time to make adjustments to their supply chain processes.

In most developed countries, in the pharmaceutical, biopharmaceutical, and biotechnology industries, almost all large companies, have adopted advanced management methods in SCM to cope with changes in the healthcare environment so that they can manage business operations more efficiently [9]. SCM has become one of the most important infrastructures in the biopharmaceutical and biotechnology industries. Information systems play a significant role in the industry [10, 11] and multinational solution providers such as SAP, Oracle, and Info offer mature packages. As competition between biotechnology-based industry manufacturers intensifies, an urgent need arises to improve the efficiency of the whole supply chain and reduce cost. SCM is the strategy for optimizing the overall supply chain by sharing information between material suppliers, manufacturers, distributors, and retailers. In some industries such as the motor and semiconductor industries, the practice of SCM and programs such as Just in Time, continuous replenishment, quick response, and Kanban (an inventory control system) have been adopted and rely on the dissemination of scheduling, shipment, and manufacturing information to the supply chain [12].

A 2010 report from the World Economic Forum estimated that by 2020 the market for biofuels, biobased bulk chemicals and plastics, and bioprocessing enzymes would approach $95 billion. The market for technology to increase agricultural productivity is estimated to approach $15 billion. Producing and transporting biomass to biorefineries could generate an additional $120 billion in economic activity, and conversion of biomass to heat and power could be worth $65 billion [13]. According to a 2011 report from investment analysts at Clean Edge, the ethanol and biodiesel industries reached a combined wholesale value of $56.4 billion in 2010, representing more than 27.2 billion gallons of production, and will grow to $112.8 million by 2020 [14]. The World Economic Forum report notes that mandates for biofuel production around the world drive the market for biofuels, whereas economics and sustainability criteria drive the smaller market for renewable chemicals. At present, chemical companies have only sought to replace selected chemical intermediates already in their portfolio, rather than seek markets for novel chemical compounds that might best be suited to the inherent properties of biomass [13]. The strategy is limited to replacement of petrochemicals with biobased chemicals that offer identical functionality and performance.

The industrial enzymes market was valued at $4.2 billion in 2014 and is projected to grow at a compound annual growth rate (CAGR) of 7.0% from 2015 to 2020. The market for food and beverages is projected to reach a value of $2.0 billion by 2020. The food and beverage segment dominates the industrial enzymes market. The factors driving the industrial enzymes market are the increase in investments, research and development of industrial enzymes, increase in demand for consumer goods and biofuel, and the need for cost reduction and resource optimization in the production process. The market for industrial enzymes is projected to reach $6.2 billion by 2020, growing at a CAGR of 7.0% from 2015 to 2020 [15, 16].

With the increase in per capita income, expenditures on food products, detergents, and beverages has increased, resulting in an increase in the demand for industrial enzymes. These factors provide opportunities for emerging market players. Different regulatory structures and lack of transparency in protection laws are the major factors restraining the market. The global industrial enzymes market can be segmented on the basis of application (e.g., food and beverages, cleaning agents, animal feed). Industrial enzymes can also be segmented according to type. By type, the carbohydrase segment accounted for a share of 31.7% in 2014 and is projected to reach $2.0 billion.

The industrial enzymes market is a competitive market, with key market players undertaking strategic decisions on new product development, expansion, and investment to maintain their market positions. The leading players are involved in the research and development of various innovative enzyme products in attempts to attain a majority market share. The key players in this market are BASF SE (Germany), E.I. du Pont de Nemours and Company (USA), Associated British Foods plc (UK), Koninklijke DSM N.V (the Netherlands), and Novozymes A/S (Germany).

The Organization for Economic Co-operation and Development has projected that worldwide plastic consumption will grow from 250 million tons currently to about 1000 million tons by 2020. Five petroleum-derived polymers currently make up two-thirds of the plastics market: low-density polyethylene, high-density polyethylene, polypropylene, polyvinyl chloride, and polyethylene terephthalate [17]. Biobased analogs to these plastics and novel bioplastics are also being developed, either through fermentation of building blocks (such as ethanol or polylactic acid) or by direct metabolic production of new types of plastics (such as polyhydroxyalkanoate). These new products could eventually shift market consumption.

Currently, global bioplastics consumption is around a million tons, or 0.4% of total plastics consumption. The bioplastics industry expects to grow rapidly, reaching 3.45 million tons annually by 2020 [17]. The global market for algal biomass is poised for explosive growth in the next ten years (Fig. 4.3). Algae is attracting increased investment and interest from the biofuel, petroleum, and agribusiness industries (Table 4.1).

The Algae 2020 study found that the USA and Europe cannot grow enough corn, soy, or rapeseed to meet their biofuel targets. Fast growth in US and European biodiesel markets from 2007 to 2010 has led to increased biorefining capacity but a shortage of feedstock. Algae 2020 reports that long-term demand for biofuels in the USA, EU, and Asia will create new opportunities for algae and other nonfood feedstocks to meet ambitious targets for biodiesel and ethanol, and for advanced biorefineries to produce biofuels and biochemicals (Fig. 4.4).

Around the world, an urgent demand for alternative, sustainable fuels and feedstocks is growing. Compared with other feedstocks, algae can provide a high-

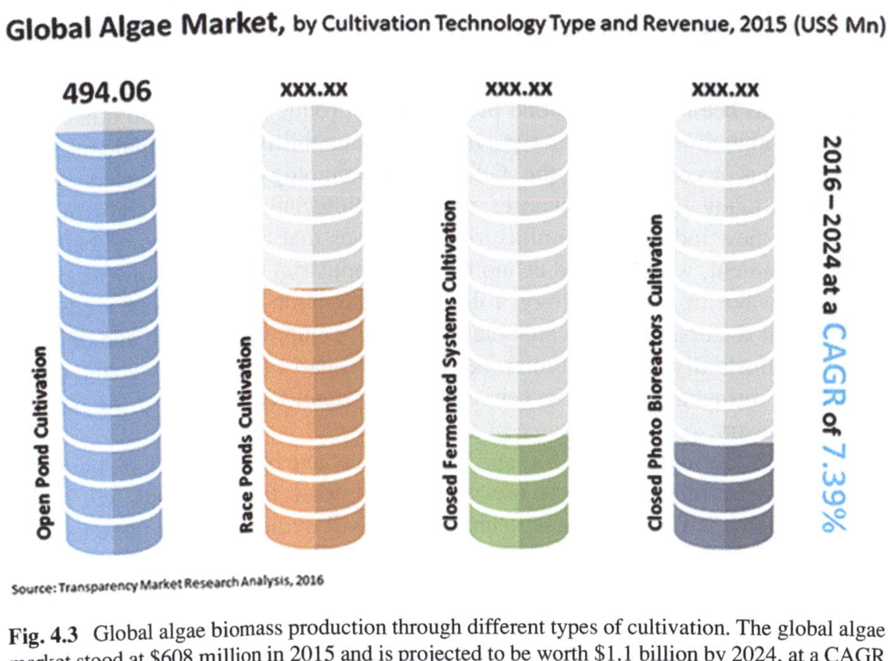

Fig. 4.3 Global algae biomass production through different types of cultivation. The global algae market stood at $608 million in 2015 and is projected to be worth $1.1 billion by 2024, at a CAGR of 7.39%. By volume, the market is poised to expand at 5.32% CAGR between 2016 and 2024 (Source: https://www.google.co.in/imgres)

Table 4.1 Production of biodiesel from various feedstocks (Source: Algae 2020 and Biodiesel 2020 studies, Emerging Markets Online)

Feedstock	US gallons per acre
Soy	40–50
Canola/Rapeseed	120–150
Jatropha	175–250
Palm	650
Algae	5–10,000

yield source of biodiesel, ethanol, and aviation fuel without compromising food supplies, rainforests, or arable land. Algae strains with high oil productivity can bring much-needed feedstock to biodiesel and biochemical refineries worldwide (Fig. 4.5). In view of the future demand for algal biodiesel all over the world, mass cultivation of algae for biofuel production is increasingly demanded (Table 4.2).

Working within complex environments, biopharmaceutical and microbial product supply chains and their supporting information technology (IT) infrastructure have tended to evolve into a series of separate processes serving a number of internal and external customers. With the increasing and changing demands of regulation and greatly increased pressure of costs, biotechnology companies are responding on a number of fronts by improving SCM process efficiency and effectiveness (Fig. 4.6).

4.2.1 *Feedstock Availability*

Mankind has been using microbial processes for producing foods and other goods since Neolithic times. However, the evolution of our ability to manipulate microbial genomes has revolutionized the field of biotechnology and produced a rapid increase in innovation. In most cases, industrial biotechnology tools allow industry to develop new industrial manufacturing processes that are cleaner and better for the environment, with reduced economic cost. Industrial biotechnology represents the third wave in biotechnology, following innovations in the health care and agricultural sectors. Finding a suitable feedstock resource for growth and development of industrially important microbes is the first part of the supply chain. Cheap sources of nitrogen, carbon, oxygen, etc. are the main feedstocks for the biotechnology industry.

Although raw materials such as sugar and starch have been the easy pickings that have been exploited so far, the use of lignocellulosic raw materials will be the breakthrough. Scaling up these technologies will require access to enormous amounts of biomass, at the right price over the long term and with the right logistics in place. This quest has recently brought a number of companies (mostly from the USA) to Malaysia for the 2011 Pacific Rim Summit on Industrial Biotechnology and Bioenergy, organized by the USA-based Biotechnology Industry Organization

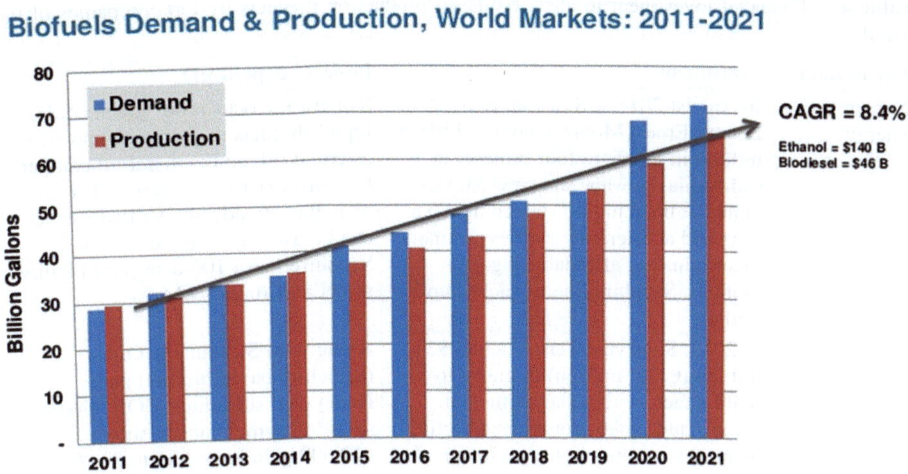

Fig. 4.4 The global market for biofuel: production and demand (Source: https://www.google.co.in/imgres)

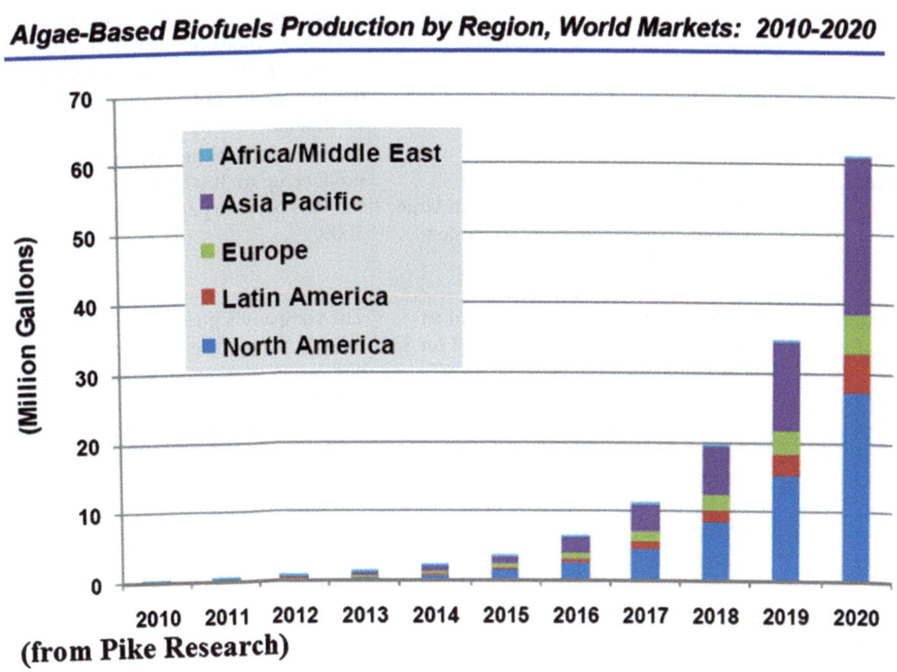

Fig. 4.5 Algae biodiesel forecast based on production milestones (Source: www.renewableenergyworld.com)

Table 4.2 Financial investment in algae biodiesel production projects by top companies of the world

Organization	Investment	Project scope/R&D
Sappire energy	In August 2013, US Secretary of Energy Ernest Moniz announced $16.5 million in grants to four projects in California, Hawaii, and New Mexico aimed at breaking down technical barriers and accelerating the development of sustainable, affordable algae biofuels. Sapphire Energy picked up $5 million	Sapphire's focus is on "green crude," a liquid that has the same composition as crude oil, and is therefore compatible with existing refineries. The company has already shown that its fuel can be used in cars and even jets. Sapphire has a 100-acre pilot facility near Las Cruces, N.M.
Solazyme	In 2012, Solazyme/TerraVia had $20 million (€17.7 million) in grants to build a factory to make biofuels in Illinois and broke ground on a biofuel plant in Brazil through a joint venture with Bunge	Along with Sustainable Oils (camelina-based biofuel) and Honeywell subsidiary UOP (biodiesel), Solazyme plans to supply 400,000 gallons of fuel to the Air Force and 190,000 gallons to the U.S. Navy, plus 1,500 gallons of jet fuel for the US Navy by 2010
Seambiotic	Seambiotic grows microalgal cultures in open ponds using flue gases such as carbon dioxide and nitrogen from a nearby coal plant as feedstocks	Its 1000-m^2 facility produces roughly 23,000 g of algae per day. Three tons of algal biomass would yield around 100–200 gallons of biofuel. It recently formed a partnership with NASA to optimize the growth rates of its microalgae
Solix biofuels	Solix uses specialized photobioreactors in which batches of microalgal cultures are grown in large, closed-growth chambers under controlled light and temperature conditions	A demonstration facility that could produce up to 3000 gallons of algal biofuels per acre per year by the end of 2009
Algenol biofuels	$850 million committed to build an algae farm that sells ethanol fuel for $3 per gallon. In India, Algenol and Reliance Industries have successfully deployed India's first Algenol algae production platform, the world's largest oil refinery with a nameplate capacity of 668,000 barrels per day at a 7500 acre complex in Gujarat state	The company's goal is to produce fuel directly from algae without killing or harvesting the creatures, allowing a shorter turnaround time to make fuel. The company claims its process can produce around 6000 gallons per acre per year

(BIO). It is interesting to note that the technology for converting biomass to chemicals is being developed in the USA and Europe, but the biomass is in Southeast Asia and in Latin America.

CO_2 is widely available from a variety of fossil sources (e.g., syngas production, natural gas sweetening, coal power production) and biobased sources (e.g., distilleries and anaerobic digestion). The use of CO_2 as a feedstock for chemicals and

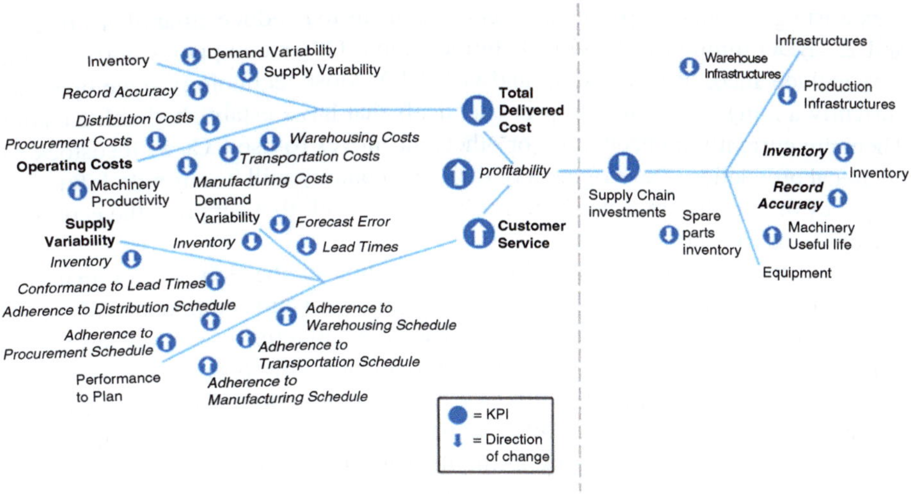

Fig. 4.6 Improving supply chain management; *KPI* key performance indicator (Source: https://www.google.co.in/imgres)

materials is attractive for two reasons: (1) CO_2 is the only raw material that Europe has in abundance, and (2) its use is completely outside the food chain. Utilization of CO_2 is a concept in which CO_2 is considered a valuable commodity rather than a pollutant. Currently, the largest example of direct use of CO_2 is for enhanced oil recovery (EOR). Other direct uses of CO_2 involve the production of carbonated drinks, inert gas for the packaging industry, use as a fire extinguisher, and as a cooling fluid (e.g., for fridges). CO_2 is also used as a feedstock for the production of urea, methanol, and acetylsalicylic acid (aspirin). There are many other methods under development, using both industrial biotechnology and chemical catalysts, by which CO_2 can be converted to chemicals and plastics. Some industrial biotechnology routes include:

- Microalgae technologies, where microalgae produce biomass and, within this biomass, chemicals from CO_2 and sunlight
- CO_2 fermentation, where CO_2 is fermented to a desired molecule using hydrogen as an energy carrier
- Advanced biotechnological processes, where the abilities of cyanobacteria (a branch of microalgae) to use CO_2 as a carbon source and sunlight for energy are merged with the metabolic pathways of known microorganisms
- Bioelectrical systems, where enzymes or microorganisms use CO_2 as a carbon source and electricity as an energy source for their synthesis
- Artificial photosynthesis, where CO_2 is converted to desired chemicals using a (bio)catalyst, with photocatalytic water splitting as an energy source

The first pre-commercial applications of waste gas fermentation to ethanol and 2,3-butanediol (2,3-BDO) have been developed. Lanzatech (a private company) uses waste gas from coal power and steel mill plants to produce ethanol, acetic acid, and 2,3-BDO through CO and CO_2 fermentation. Other companies active in this area include Ineos Bio, Coskata, and OPXBIO. Most technology developers are currently aiming to develop drop-in chemicals that have established value chains. There are currently no economic (or other) incentives to use CO_2 as feedstock for chemical production. Development of the CO_2 economy will be strongly influenced by policy drivers. There is need for a policy framework that supports the economic feasibility of CO_2 utilization.

One of the major tenets of SCM is coordination of the flow of raw materials (mainly cheap sources of nitrogen, carbon, oxygen, etc.) and components efficiently from various suppliers to biotechnology-based manufacturing companies for the purpose of maintaining the growth and development of microbes during megascale fermentation. The raw materials are used for the production of target metabolites (finished products) and thus to fulfil customer's expectations. Suppliers' capabilities are directly linked to consistency in the demand for raw materials. So, a perfect relation and coordination should be maintained so that consistency in the feedstock supply chain can be maintained [18].

Collaborative effort between focal company and supplier is the main ingredient of green supply chain management (GSCM), which facilitates supply-side environmentally and socially responsible activities. Collaboration in terms of the environment is an approach that helps firms to develop and support the environmental prowess of their supply partners [19]. According to Paulraj [20], environmental collaboration includes cooperating with suppliers to achieve environmental objectives and improve waste reduction initiatives, providing suppliers with design specifications that include environmental requirements for purchased items, encouraging suppliers to develop new source reduction strategies, and working with suppliers for cleaner production [21, 22].

4.2.2 Logistic System Operation

As defined by The American Council of Logistics Management, logistics is "the process of planning, implementing and controlling the efficient, cost-effective flow and storage of raw materials, in-process inventory, finished goods and related information from point of origin to point of consumption for the purpose of conforming to customers' requirements" (Fig. 4.7).

In microbial processes, microbial metabolites such as enzymes, biopolymers, antioxidants, antibiotics, food supplements, food additives, and biodetergents are highly susceptible materials that need a special type of logistic management for delivery to the consumer level through a chain of suppliers and retail dealers. Material required by the consumer must be delivered on time, not ahead of schedule nor behind schedule. Proper planning of the transportation modes, with availability of inventory, can ensure this. Most microbial products are perishable because of the effect of various environmental stresses and strains, mainly as a result of improper

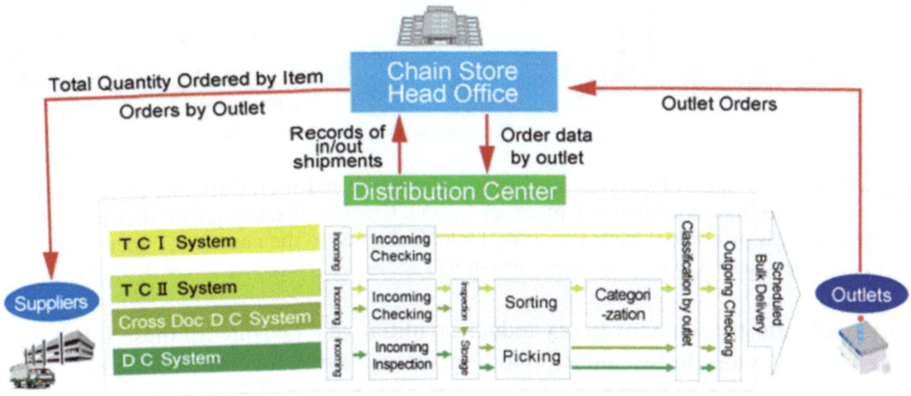

Fig. 4.7 Presentation of optimum logistics system corresponding to customer needs

packaging. This damage adds to the logistics cost. The use of proper logistic packaging, mechanized material handling equipment, etc. reduces this damage. Sometimes freight is a major source of cost in logistics. This can be reduced by following measures such as selecting the proper mode of transport, consolidation of freight, route planning, long-distance shipments, etc. A system of supply through logistics is an extremely dynamic process involving a series of steps: (1) order processing, (2) inventory planning and management, (3) warehousing, (4) transportation, and (5) packing. These steps are discussed in the rest of this section.

4.2.2.1 Order Processing

Order processing is related to three main components of the supply chain: manufacturers, distributors, and retailers. The retailers place orders; manufacturers and distributors work in collaboration with each another to fulfil these orders (Fig. 4.8).

4.2.2.2 Inventory Planning and Management

Inventory management is the supervision of non-capitalized assets (inventory) and stock items. As a component of SCM, inventory management supervises the flow of goods from manufacturers to warehouses and from these facilities to the point of sale. A key function of inventory management is to keep a detailed record of each new or returned product as it enters or leaves a warehouse or point of sale.

SCM and logistics encompass all the activities involved in getting products to consumers, including planning, storing, moving, and accounting for inventory. Inventory availability is the most important aspect of customer service, and the cost of inventory is one of the most important entries on a company's balance sheet. Recognition of the balance sheet implications of inventory in SCM has launched a

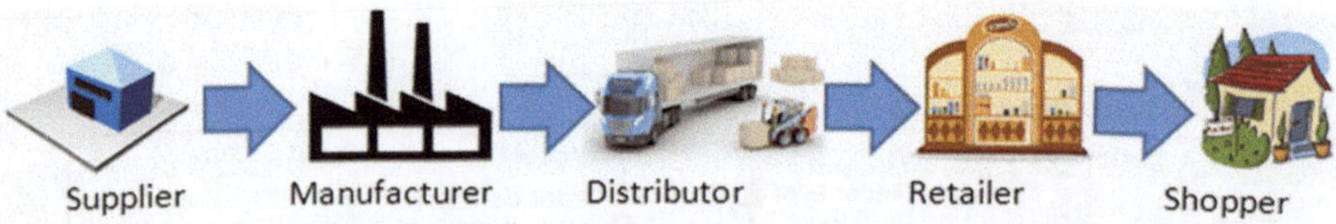

Fig. 4.8 The movement of goods from manufacturing site to end-user: Customer places an order. Order is received by manufacturer. Order is processed. Credit is checked (credit department) and verified. Order is picked and loaded on truck. Order is shipped to the customer. Order is received by customer and added to the customer's inventory

variety of industry-wide inventory reduction initiatives. Despite all these initiatives to reduce inventory in the supply chain, inventory levels for most companies have remained the same or increased.

4.2.2.3 Warehousing

Warehousing management in the biopharmaceutical industry and biotechnology industry has its own set of unique rules that have to be adhered to in order to stock, pick, and ship product in a way that conforms to the many regulations that the market sector demands.

Storage areas should be designed or adapted to ensure good storage conditions. In particular, they should be clean and dry and maintained within acceptable temperature limits. Where special storage conditions are required (e.g., temperature, relative humidity), these should be provided, checked, monitored, and recorded. Materials and pharmaceutical products should be stored off the floor and suitably spaced to permit cleaning and inspection. Pallets should be kept in a good state of cleanliness and repair (Fig. 4.9). Storage areas should be clean and free from accumulated waste and vermin. A written sanitation program should be available, indicating the frequency of cleaning and the methods to be used to clean the premises and storage areas. There should also be a written program for pest control. The pest-control agents used should be safe, and there should be no risk of contamination of the materials and pharmaceutical products. There should be appropriate procedures for the cleanup of spillage to ensure complete removal of any risk of contamination.

Materials and pharmaceutical products should be stored in conditions that assure that their quality is maintained, and stock should be appropriately rotated. The "first expired/first out" (FEFO) principle should be followed. Recorded temperature monitoring data should be available for review. The equipment used for monitoring should be checked at suitable predetermined intervals and the results of such checks should be recorded and retained. All monitoring records should be kept for at least the shelf-life of the stored material or product plus 1 year, or as required by national legislation. Temperature mapping should show uniformity of the temperature across the storage facility. It is recommended that temperature monitors be located in areas that are most likely to show fluctuations.

Fig. 4.9 Photo of a temperature-controlled warehouse showing serialized package material bearing bar codes

In recognition of patient safety, and to reduce errors, it has been a regulatory requirement in the USA since 2006 that all pharmaceutical products sold to hospitals must now bear a bar code on the smallest unit of use (i.e., the size dispensed to the patient). Bar-coding of medicines to the single unit also assists in the management of medicine alerts and recalls. The European Association of Hospital Pharmacists (EAHP) has therefore called on decision makers, politicians, and national administrations to introduce national and European regulations for implementation of bar-coding of medicines to the single dose administered in hospitals. While storing biopharmaceutical and microbially derived product in a temperature-controlled warehouse, the packaging should be well serialized with proper bar codes (Fig. 4.9).

4.2.2.4 Transportation

Transportation refers to the movement of product from one location to another as it makes its way from the beginning of a supply chain to the customer. This requires a new wider look at the transportation supply chain (Fig. 4.10), including SCM, logistics, and procurement. SCM can be divided into three main areas: purchasing, manufacturing, and transport. From end to end, this includes decisions about which input materials to use, production quantities, inventory levels, distribution network configuration, and transportation of both input materials and finished products.

Logistics management is the component of SCM that focuses on how and when to get raw materials, intermediate products, and finished goods from their respective origins to their destinations. Today, international trade is commonplace and an increasing market share in emerging markets is highly desirable. It is therefore safe to say that goods are rarely consumed where they are produced and that transportation services are the essential connection between all the elements of the supply chain. Effective, cost-efficient logistics management can be a real point of competitive differentiation. But how does a company achieve this? To practice effective, cost-efficient logistics management, an organization must lay the foundation for a responsive, economic transportation network. With this in place, an organization is able to implement major strategic changes to reduce costs and increase customer service levels with very little disruption to the overall supply chain flow.

Materials and biopharmaceutical and biological products should be transported in such a way that their integrity is not impaired and that storage conditions are maintained. Special care should be exercised when using dry ice in cold chains. In addition to observing safety precautions, it must be ensured that the materials or products do not come in into contact with dry ice, as this can adversely affect quality product (e.g., by freezing). Where appropriate, the use of devices to monitor conditions such as temperature during transportation is recommended. Monitoring records should be available for review.

4.2.2.5 Packaging

In the USA, a biological product is defined as "a virus, therapeutic serum, toxin, antitoxin, vaccine, blood, blood component or derivative, allergenic product,

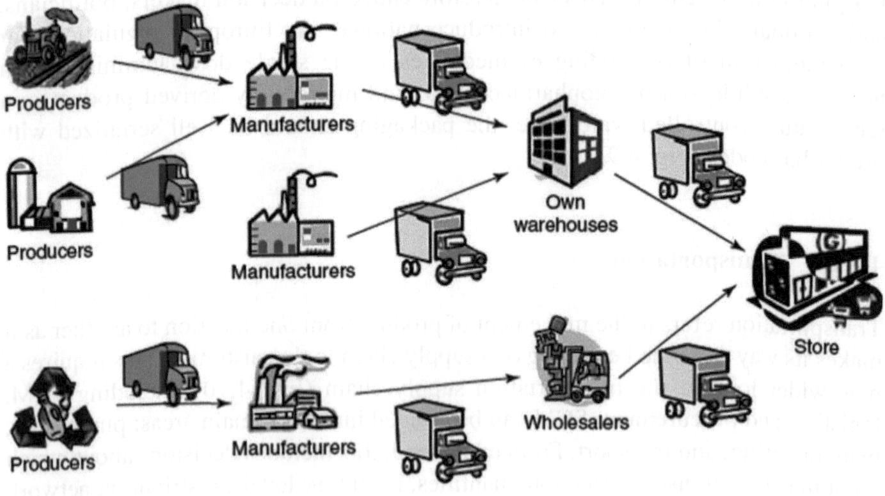

Fig. 4.10 Role of the transport system in supply chain management

protein (except any chemically synthesized polypeptide), or analogous product, or arsphenamine, or derivative of arsphenamine (or any other trivalent organic arsenic compound), applicable to the prevention, treatment or cure of a disease or condition of human beings" (Public Health Services Act 42 U.S.C. § 262(i)).The manufacturing process for a biological product usually differs from the process for drugs. The manufacture of biological medicinal products involves certain specific considerations arising from the nature of the products and processes. Persons responsible for production and quality control should have an adequate background in the relevant scientific disciplines, such as bacteriology, biology, biometry, chemistry, medicine, pharmacy, pharmacology, virology, immunology, and veterinary medicine. The degree of environmental control of particulate and microbial contamination of the production premises should be adapted to the product and the production steps.

For a sustainable supply chain of biopharmaceuticals and biotechnology-based finished goods, a suitable packaging system is essential [23–25]. As stated earlier, biopharmaceuticals and microbial products are extremely susceptible to external stresses [26, 27], so safe, secure, efficient, and effective handling of these products during transport, distribution, storage, retailing, consumption, recovery, and reuse or disposal (with related information) must be ensured.

For this purpose, integrated packaging design helps to increase the efficacy of supply chains. This practice is known "packaging logistics." Another managerial approach related to efficiency and sustainability is the implementation of "meta-standards" (ISO 9000 and ISO 14000). Both packaging logistics and meta-standards can contribute positively to supply chain efficiency and sustainability.

As packaging is one of the key and global elements that support an efficient and sustainable supply chain [28–30], the process of designing, implementing, and controlling integrated packaging not only helps in ensuring the sustainability of the supply chain but also adds profit for company. In this context, packaging is one of the key elements that can support the implementation of efficient and sustainability-oriented strategies. Beyond the traditional and basic view of packaging as a means of protecting products [26, 31], over the last few years new design requirements have been added, on the one hand, to improve the differentiation capacity of the product (commercial function) and, on the other hand, to improve the efficiency of the product at a logistic and production level (logistic function).

4.2.3 *Flow of Manufacturing Costs*

The flow of manufacturing costs refers to the manner in which costs move through a firm. Typically, the flow of costs is relevant to a manufacturing environment where accountants must quantify what costs can be attributed to raw materials, work in process, finished goods inventory, and goods sold. Flow of costs does not only apply to inventory, but also to factors in other processes to which a cost is attached, such as labor and overheads (Fig. 4.11).

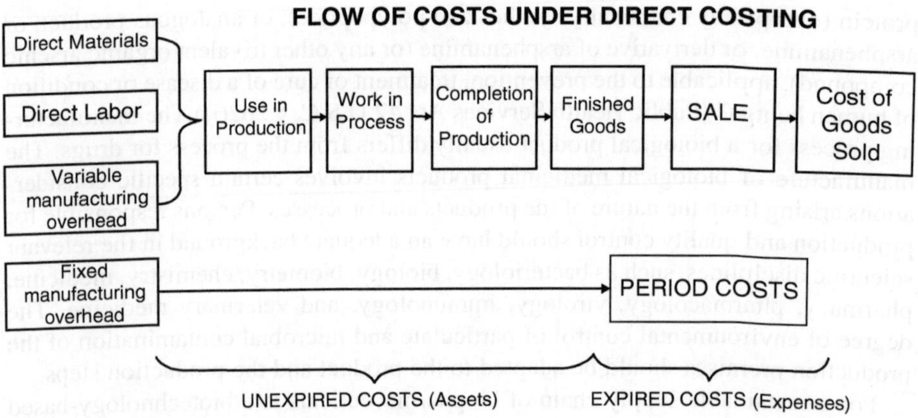

Fig. 4.11 Methods of accounting for the flow of costs

4.2.3.1 Breaking Down the Flow of Costs

There are several methods of accounting for the flow of costs. These include LIFO (last in, first out), FIFO (first in, first out), specific identification, and weighted-average cost.

The world of accounting is full of inventory management conventions used to help keep track of the cost and value of inventory. For this, accountants have developed two systems of inventory management, FIFO and LIFO, which are calculations used by companies to monitor inventory goods.

LIFO is an asset management and valuation method that assumes assets produced or acquired last are the ones used, sold, or disposed of first; LIFO assumes that an entity sells, uses, or disposes of its newest inventory first. If an asset is sold for less than it is acquired for, then the difference is considered a capital loss. If an asset is sold for more than it is acquired for, the difference is considered a capital gain. Using the LIFO method to evaluate and manage inventory can be tax advantageous, but it may also increase tax liability.

LIFO liquidation happens when a company uses the LIFO method of inventory costing, and then liquidates its older LIFO inventory. A LIFO liquidate occurs if current sales are higher than its purchase of inventory, which, as a result, forces the company to liquidate any inventory not sold in a previous period. This liquidation causes a distortion in a company's net operating income, because the lower-cost inventory is recognized on its income statement.

The LIFO method is used by companies during periods of rising prices, or periods of rising inflation, when the cost to purchase inventory increases over time. This is used because the higher priced inventory is matched to current revenue, increasing COGS and decreasing earnings before tax, which lowers a company's tax burden. However, if a company faces a LIFO liquidation event when sales outpace current inventory, the business is forced to use older and older inventory at increasingly lower prices. This results in a higher tax liability, which is the opposite of the desired effect. The tax advantage of LIFO accounting, then, quickly turns into a disadvantage due to the fact that the older, lower costs of inventory are matched with current revenues.

The LIFO method is used by companies during periods of rising prices, or periods of rising inflation, when the cost to purchase inventory increases over time. This is used because the higher priced inventory is matched to current revenue, increasing COGS and decreasing earnings before tax, which lowers a company's tax burden. However, if a company faces a LIFO liquidation event when sales outpace current inventory, the business is forced to use older and older inventory at increasingly lower prices. This results in a higher tax liability, which is the opposite of the desired effect. The tax advantage of LIFO accounting, then, quickly turns into a disadvantage due to the fact that the older, lower costs of inventory are matched with current revenues.

FIFO is an asset-management and valuation method in which the assets produced or acquired first are sold, used, or disposed of first. The method may be used by an individual or a corporation. For taxation purposes, FIFO assumes that the assets remaining in inventory are matched to the assets that were most recently purchased or produced. FIFO is used for cost-flow assumption purposes. Because items being manufactured progress to a later development stage and finished inventory items are sold, the costs associated with that product must be recognized as an expense. The dollar value of total inventory decreases as this occurs because inventory has been removed from the company's ownership. The costs associated with the inventory may be calculated in numerous ways, one being the FIFO method.

US GAPP financial reporting standards require that companies that use the LIFO method report the difference between that method and FIFO in a line item (called the LIFO reserve). This allows analysts to compare firms using different cost-flow assumptions. The LIFO reserve is an accounting term that measures the difference between FIFO and LIFO cost of inventory for book-keeping purposes. The LIFO reserve bridges the gap between FIFO and LIFO costs when a company uses the FIFO method to track its inventory but reports under the LIFO method in the preparation of its financial statement. In periods of rising prices, constant increases in costs can create a credit balance in the LIFO reserve, which results in reduced inventory costs when reported on the balance sheet. LIFO reserve comes about because most businesses use the FIFO, or standard cost method, for internal use and the LIFO method for external reporting, as is the case with tax preparation. This is advantageous in periods of rising prices because it reduces a company's tax burden when it reports using the LIFO method. For example, when using the LIFO method for inventory accounting in periods of rising prices, the cost of reported inventory is higher than with the FIFO method, which increases a company's COGS and decreases its pre-tax earnings. Then, for internal purposes, as is the case with investor reporting, the same company can use the FIFO method of inventory accounting, which reports lower costs and higher margins.

4.2.4 Tracelink and Serialization of Biological Products

Top multinational biopharmaceutical, pharmaceutical and biotechnology companies implement traceability and serialization capabilities to meet regulatory requirements, quality, and recall purposes [32]. This requires a community that can share that information effectively and collaborate with regulatory agencies at national, state, and local levels.

After implementation of traceability and serialization practice, the US pharmaceutical industry has achieved more than 50-fold reduction in pharmaceutical theft over the last 4 years. In this connection, a theft incident occurred in an Eli Lily warehouse in 2010 ($76 million of drugs stolen) that was most alarming for pharmaceutical and biotechnology-based industries [33].

The primary objective of traceability is typically not theft control, but rather quality control and enabling recall capabilities. All biological products should be clearly identified by labels. The labels used must remain permanently attached to the containers under all storage conditions, and an area of the container should be left uncovered to allow inspection of the contents. If the final container is not suitable for labeling (e.g., a capillary tube), it should be in a labeled package. The information given on the container label and on the package label should be approved by the national or international control authority. The label on the container should show:

- Name of the drug product
- List of active ingredients and the amount of each present
- Batch or final lot number assigned by the manufacturer
- Expiration date
- Recommended storage conditions
- Directions for use, and any warning and precautions that may be necessary
- Name and address of the manufacturer or the company

The label on the package should show at least the nature and amount of any preservative or additive in the product. The leaflet in the package should provide instructions for use of the product, and mention any contraindications or potential adverse reactions.

4.2.5 Expiration Date Determination

US regulations (CFR 114.13) state the following: Unless otherwise provided for in a Standard Requirement of filed Outline of Production, the expiration date for each product shall be computed from the date of the initiation of the potency test. Prior to licensure, stability of each fraction shall be determined by methods acceptable to

Animal and Plant Health Inspection Service. Expiration dates based on this stability data shall be confirmed as follows:

- Products consisting of viable organisms. Each serial shall be tested for potency at release and at the approximate expiration date until a statistically valid stability record has been established.
- Nonviable biological products. Each serial presented in support of licensure shall be tested for potency at release and at or after the dating requested.
- Subsequent changes in the dating period for a product may be granted, based on statistically valid data submitted to support a revision of the Outline of Production.

4.2.6 Impact of New Technology Development on the Supply Chain

High customer demand for healthcare bioproducts, food additives, biopharmaceuticals, etc. has forced multinational groups to accelerate manufacturing processes and SCM to fulfil the requirements of the end user, timely and without much financial burden. To achieve this, the interlinking of IT with SCM is unavoidable in bringing sustainability to SCM [33]. In reality, IT can create a link with the supply chain and help the consumer keep in close touch with the manufacturing process and quality control information. In addition, the alignment of IT goals and objectives with strategic SCM can increase efficiency, productivity, and profitability [34].

Integration of SC with IT can achieve immense benefits in product development, procurement, manufacturing, physical distribution, customer relationship management, and performance measurement [35]. This sort of practice supports the promoter by providing the means to link technology and people, and to align the technology with the capabilities of the organization and its trading partners [36]. SCM enables trading partners to coordinate their processes through information sharing to facilitate supplier–customer interactions and minimize transaction costs.

A competitive advantage is the extent to which an organization has the competency to create a defensible position over its competitors as a result of crucial management decisions, which differentiates itself from its rivals. Although empirical research has indicated cost, quality, delivery, and flexibility to be important competitive capabilities [33], time and innovation have recently been identified as the next sources of competitive advantage [37]. This stresses the importance of gaining sustainable competitive advantage from IT. Moreover, in 2008, Ketchen et al. determined supply chain information systems as one of the key areas where best-value supply chains differ from traditional supply chains [33]. Competitive advantage throughout the supply chain can be built through the development of IT systems for SCM that support and speed up all business activities and improve decision making and productivity. This is accomplished through the exploitation of IT for internal and external integration of business processes.

By promoting IT, a company can integrate internal business functions and keep track of company activities in a timely manner. This can help the promoter to fulfil customer requirements and catalyze the quality control process [34]. SCM systems aided by IT can improve logistics management, transportation management, strategic planning, warehousing, inventory, manufacturing, supplier management, and customer management [38].

Enterprise resource planning (ERP) systems are included as part of the broader SCM software. ERP systems are employed to integrate business processes by organizing, codifying, and standardizing business processes and data [39]. Functionally, ERP helps in systematic data management and in transferring data within the internal system, preventing the expense of transportation of data from one department to another. ERP reports can be used to forecast production and make decisions. Another key supply chain process is customer relationship management (CRM), which is the management of relationships between the organization and its customers.

A crucial issue for internal integration of business processes is the integration of ERP with CRM. Many ERP vendors now provide an ERP–CRM integration package, as many enterprises have expressed an interest in integrating a new CRM system with their legacy ERP system. Moreover, the establishment of ERP and CRM systems should be a primary concern for any enterprise that is interested in embarking on e-business [40]. Generally, a company shares crucial information about demand prediction, actual orders, and inventory levels in an efficient manner [41], while protecting proprietary data [33]. The adoption of an interorganizational information system (IOS) for SCM has been proposed as a way to enable external integration [36, 42, 43]. There are some technologies that, overall, are considered of high relevance for future supply chains, such as cloud solutions and mobile devices.

Even though security concerns are always a consideration, cloud computing has become an integral part of today's SCM solutions portfolio and remains of interest. The main advantages of cloud computing in SCM are in finding suitable options for company investment and addressing functional requirements at various stages of development. It is applicable to a number of business areas in global logistics. Mobile computing and mobile devices have gained greater importance in recent years as a result of their ability to efficiently support supply chain processes "on the go." Technical advancement has enabled significant development in their scope of capabilities, but their most valuable feature continues to be the ability to integrate into ERP and operational systems to gather and share data in a variety of forms and formats.

Comprehensive, sophisticated SCM systems can meet the requirements of supply chains in the twenty-first century, that is, adaptability, collaborative ability, transparency, compliance, and speed in all processes. Streamlining processes and cutting costs in the long run are all vital for future business growth. By choosing the right software solution, supply chain managers can future-proof their business and gain a competitive advantage.

4.3 Green Supply Chain Management

Green supply chain management (GSCM) is eco-friendly performance of the supply chain, in which attention is paid to conservation of ecosystem balance while distributing product from manufacture site to consumer level without any change in the quality of product. In this process, the traditional supply chain process (Fig. 4.12) is integrated with environmental conservation so that waste generated during any supply chain can be adequately recycled in order to induce sustainability in SCM practice and bring profit to the manufacturer.

Microbial processes for the production of biopharmaceuticals and health care products such as nutraceuticals, food supplements, food additives, and antioxidants may threaten the environment in terms of carbon monoxide emissions, discarded packaging materials, scrapped toxic materials, traffic congestion, and other forms of industrial pollution [44]. GSCM is an eco-friendly practice in which ecosystem conservation and safety issues are integrated with the traditional supply chain to minimize or eliminate wastage (including hazardous chemicals, emissions, energy, and solid waste) during supply chain processes such as product design, material resourcing and selection, manufacture, delivery of final product, and end-of-life management of the product [45, 46].

Hence, the impact of GSCM practice on the environment is to ensure minimum harm to the sustainability performance of a supply chain. During the beginning of the 1990s, competition between big multinationals increased in order to impress the consumers on product purity from the aesthetic and social responsibility points of view, without adding any environmental cleanness issues [47]. Over time, GSCM has gained acceptance and adoption of this technique could convince manufacturers of the reality of the market demand [48].

Having these practices in mind, firms have developed environmental management strategies in response to the dynamic nature of environmental requirements and their impacts on supply chain operations [49]. A fruitful SCM is an integrated process for the supply of goods to the end user by means of a wide network consisting of all parties (e.g., supplier, manufacturer, distributor, wholesaler, retailer, and customer.) involved in the manufacture of current good manufacturing practice (cGMP)-certified product and its delivery to the customer, with the assurance of traceability during transportation, and involves both upstream and downstream processes [50].

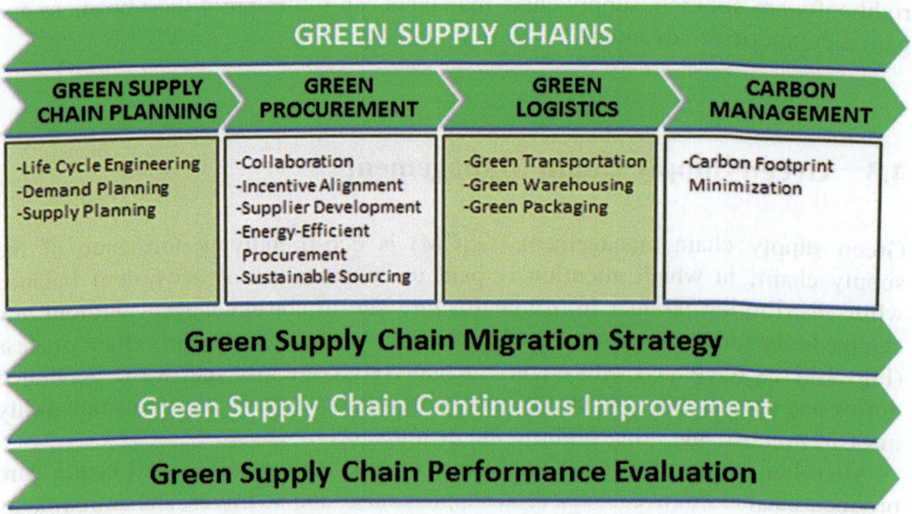

Fig. 4.12 Green supply chain management, showing four phases of the process

4.4 Quality Control

Quality control is the pillar of success for biotechnology and biopharmaceutical industries. In this age of cut-throat competition and large-scale production, only manufacturers who can supply better quality goods and render service to consumers on a need basis can survive. Because of an increase in quality awareness, it has been mandatory to incorporate a blueprint for quality control (QC) processes while establishing an industrial undertaking. Proper QC ensures the most effective utilization of available resources and a reduction in production costs.

The word "quality control" comprises two words, "quality" and "control." It would be appropriate to explain these two words separately to understand clearly the meaning of quality control. According to Dr. W.K. Spriegel, "The quality of a product may be defined as the sum of a number of related characteristics such as shape, dimension, composition, strength, workmanship, adjustment, finish and colour." In the words of John D. McIlellan, "Quality is the degree to which a product conforms to specifications and workmanship standards." It is clear from these definitions that quality refers to various characteristics of a product and their excellence. Quality is a relative term and is never absolute, depending upon the use of the product and circumstances under which it is used.

4.4.1 Objectives of Quality Control

The following are the important objectives of QC:

- Establish the desired quality standards that are acceptable to the customers

- Discover flaws or variations in the raw materials and the manufacturing processes in order to ensure smooth and uninterrupted production
- Evaluate the methods and processes of production and suggest further improvements in their functioning
- Study and determine the extent of quality deviation in a product during the manufacturing process
- Analyze in detail the causes responsible for such deviation
- Undertake such steps as are helpful in achieving the desired product quality

Successful businesses inevitably place great emphasis on managing QC and take carefully planned steps to ensure that the products and services offered to their customers are consistent and reliable and truly meet their customers' needs. Multinational corporations have entire departments of highly trained specialists to design and implement their quality assurance programs. For a custom countertop fabricator, as for other small businesses that make unique handcrafted products, QC is also essential. Such companies cannot rely on a staff of trained experts or the standardization inherent in mass production to take care of QC. One or a handful of managers must handle it themselves, along with every other management function. That is part of being a small business. However, it is all too often true that manufacturers fail to institute their own carefully planned QC procedures.

It is said that when the Japanese business that later became Sony Corporation was founded, the co-founder Mr. Ibuka established the company philosophy by stating, "If it were possible to establish conditions where persons could become united with a firm spirit of teamwork and exercise to their hearts' desire their technological capacity, then such an organization could bring untold pleasure and untold benefits." For decades, Japanese businesses have pioneered management techniques intended to continuously improve quality. American and European companies have spent years catching up. To maintain or enhance the quality of their offerings, manufacturers use two techniques, QC management and quality assurance (QA). These two practices ensure that the end product or service meets the quality requirements and standards defined for that product or service.

4.4.2 *Quality Control Management*

Many people are confused between QC and QA. As already discussed, organizations can define their own internal quality standards, processes, and procedures. An organization develops these over time and then relevant stakeholders are required to adhere to them. The process of making sure that the stakeholders adhere to the defined standards and procedures is called QC. In QC, a verification process takes place. Certain activities and products are verified against a defined set of rules or standards. Every organization that practices QC needs to have a quality manual. The quality manual outlines the quality focus and the objectives of the organization.

The quality manual gives quality guidance to different departments and functions. Therefore, everyone in the organization needs to be aware of his or her responsibilities, as mentioned in the quality manual.

There are many methods followed by organizations to achieve and maintain the required level of quality. Some organizations believe in the concepts of total quality management (TQM) and others believe in the use of internal and external standards. The standards usually define the processes and procedures for organizational activities and assist in maintaining the quality in every aspect of organizational functioning. When it comes to standards for quality, there are many. The International Organization for Standardization (ISO) is one of the most prominent bodies for defining quality standards for different industries. Therefore, many organizations try to adhere to the quality requirements of ISO. In addition, there are many other standards that are specific to various industries. As an example, SEI-CMMi is a standard followed in the field of software development. Because standards have become a symbol for product and service quality, customers are now keen to buy a product or service from a certified manufacturer or service provider. Therefore, complying with standards such as ISO has become a necessity when it comes to attracting the customers.

The importance of properly established and managed QC and QA systems, with their integral well-written standards of procedure (SOPs) and other documents for the achievement of company business objectives, cannot be ignored. They serve as a passport to success by assisting the company to achieve high-quality processes, procedures, systems, and people, with eventual high-quality products and services. These all lead to enhancement of the following:

- Customer satisfaction and, therefore, customer loyalty, repeat business and referral
- Timely registration of drugs by eliminating waste and the need for rework
- Operational results such as revenue, profitability, market share, and export opportunities
- Alignment of processes with achievement of better results
- Understanding and motivation of employees toward the company quality policy and business objectives, as well as participation in continual quality improvement initiatives
- Confidence of interested parties in the effectiveness and efficiency of the company, as demonstrated by the financial and social gains from company performance and reputation

4.4.3 Quality Control Assurance

QA is a broad practice used for assuring the quality of products or services. There are many differences between QC and QA. In QA, a constant effort is made to enhance the quality practices in the organization.

The QA department must operate independently of the operational units and must regularly perform quality review activities (self-inspection audits, internal audits) to ensure compliance of operational units with company quality standards, good working practices, and local, national, regional, and international legal, ethical, and regulatory requirements. Therefore, continuous improvements are expected in quality functions in the company. For this, there is a dedicated QA team. Sometimes, in larger organizations, a "process" team is also allocated to enhance the processes and procedures. The QA team of an organization has many responsibilities. The first and foremost responsibility is to define a process for achieving and improving quality. Some organizations develop their own process and others adopt a standard process such as ISO or CMMi. Processes such as CMMi allow organizations to define their own internal processes and adhere by them. The QA function of an organization uses a number of tools for enhancing the quality practices. These tools vary from simple techniques to sophisticated software systems. QA professionals should also go through formal industrial training and become certified. This is especially applicable for QA functions in software development houses. The QA teams of organizations constantly work to enhance the existing quality of products and services by optimizing existing production processes and introducing new processes. Two important practices in QA, stability testing and regulatory approval, are described next in detail.

4.4.3.1 Stability Testing

Biotechnological and biopharmaceutical products are extremely susceptible to environmental stress and strain. A stability test assesses the change in quality of such a product in response to temperature, humidity, light and various physical factors during storage and transportation to the destination. Standard protocols for testing are well defined on the basis of the effects of climatic conditions in the three regions of the EU, Japan, and the USA.

On the basis of global climatic data, one can easily calculate the current climatic condition of a specific region. The principle has been established that stability information generated in any one of the three regions of the EU, Japan, and the USA would be mutually acceptable to the other two regions, provided the information is consistent with FDA guidance and the labeling is in accord with national and regional requirements.

FDA guidance documents do not establish legally enforceable responsibilities. Instead, they describe the FDA's current thinking on a topic and should be viewed only as recommendations, unless specific regulatory or statutory requirements are cited. The use of the word "should" in FDA guidances means that something is suggested or recommended, but not required.

Stress testing can be helpful in understanding the degradation behavior of biomolecules, which can ultimately be helpful in studying the biological conservation of biotechnological and biological products during distribution to the consumer. In practice, stress testing is likely to be carried out on a single batch of the drug substance. Testing should include the effect of temperature (in 10 °C increments, or more for accelerated testing), humidity (e.g., 75% relative humidity or greater), and, where appropriate, the effects of oxidation and photolysis on the drug substance. The testing should also evaluate the susceptibility of the drug to hydrolysis across a wide range of pH values when in solution or suspension. Photostability testing should be an integral part of stress testing. The standard conditions for photostability testing are described in ICH *Q1B Photostability testing of new drug substances and products*.

Sample analysis for stability study should be carried out with a minimum of three batches. The capacity of each batch should not be less than the size of pilot level production followed by the same synthetic pathway as practiced in turnkey-level production. The overall quality of the batches of drug substance placed in formal stability studies should be representative of the quality of the material to be made on a production scale. The stability studies should be conducted on the drug substance packaged in a container closure system that is the same as or simulates the packaging proposed for storage and distribution.

In stability testing, more attention should be given to biological samples, which are susceptible to change during storage and transport, in order to keep vigilance on the efficacy and quality of the biological product. The testing should cover, as appropriate, the physical, chemical, biological, and microbiological attributes. Validated stability-indicating analytical procedures should be applied. Whether and to what extent replication should be performed should depend on the results of validation studies.

For long-term studies, frequency of testing should be sufficient to establish the stability profile of the drug substance. For drug substances with a proposed re-test period of at least 12 months, the frequency of testing during long-term storage should normally be every 3 months over the first year, every 6 months over the second year, and annually thereafter through the proposed re-test period. Well-defined protocols are available for storage of biological/biotechnological/pharmaceutical products under variable physical conditions [51–58].

In some biotechnological/biological products, potency is dependent upon the conjugation of the active substance(s) to a second moiety or binding to an adjuvant. Dissociation of the active substance(s) from the carrier used in conjugates or adjuvants should be examined in real-time/real-temperature studies (including conditions encountered during shipment).

The following product characteristics, although not specifically relating to biotechnological/biological products, should be monitored and reported for the medicinal product in its final container:

- Visual appearance of the product (color and opacity for solutions/suspensions; color, texture, and dissolution time for powders), visible particulates in solutions or after the reconstitution of powders or lyophilized cakes, pH, and moisture level of powders and lyophilized products.
- Sterility testing or alternatives (e.g., container/closure integrity testing) should be performed at least initially and at the end of the proposed shelf life.

- Additives (e.g., stabilizers, preservatives) or excipients may degrade during the dating period of the medicinal product. If there is any indication during preliminary stability studies that reaction or degradation of such materials adversely affect the quality of the medicinal product, these items may also need to be monitored during the stability program.
- The container/closure has the potential to adversely affect the product and should be carefully evaluated.

4.4.3.2 Existing Legal Basis for Approval of Biologics

Two US statutes apply to the regulation of biological products, the Federal Food, Drug, and Cosmetic Act (21 U.S.C. §301 et seq.) (FFDCA), and the Public Health Services Act (42 U.S.C. §262) (PHSA). The US regulations charge the FDA with the protection of public health, in part by ensuring that human drugs and biological products are safe and effective. The FDA administers the FFDCA and PHSA (among other statues). The FFDCA applies to all drugs and medical devices, and the PHSA applies to "biological products." Marketing approval under the FFDCA is by means of a New Drug Application (NDA) whereas approval under the PHSA is by means of a Biologics License Application (BLA). Both drugs and biologics are subject to Investigational New Drug Application (INDA) regulations. Preclinical research on new compounds is carried out in a laboratory, using a wide variety of techniques. Promising candidates are then studied in animals and, subsequently, various clinical studies in humans are carried out following strict guidelines. There are three initial stages of clinical testing:

- Phase I: A small number of healthy volunteers is given the compound mainly to determine that the drug is safe for human use
- Phase II: A small number of patients is given the medicine to assess its efficacy and safety and to ensure that there are no unacceptable side effects
- Phase III: A large number of patients, usually thousands, take the medicine under supervision over a defined period of time, with the results used to establish efficacy

If the results show the drug to be efficacious and safe, the data are presented to the FDA. The FDA reviews the data and, if the data are acceptable, a marketing authorization is issued. Alternatively, the FDA may request additional studies or reject the application. Following the grant of marketing authorization, the drug product is studied in large numbers of patients in hospitals and clinics to further assess its clinical effectiveness. This stage is called phase IV or post-marketing study. Safety Assessment of Marketed Medicines (SAMM) studies help identify any unforeseen side effects. In order to be marketed, a biologic requires only proper labeling and an approved BLA that indicates the product has been determined safe, pure, and potent, and that the manufacturing facilities meet the requirements to ensure safety, purity, and potency.

Although biologics have traditionally been subject to much more scrutiny in manufacturing than drugs, those differences are being eroded. Biologics have been approved under the FFDCA and PHSA; thus, both NDA and BLA applications have been submitted for biologics. The exceptions are glucagon and follistim, which were approved under §505(b)(2), and insulin, which was approved under its own statute for a time. The default approval pathway for biologics is now a BLA, unless the product is a hormone, in which case §505(b) is used.

4.5 Quality Control and Supply Chain Integration

The past two decades have witnessed the positive impact of total quality management (TQM) on the success of a biotechnological or biopharmaceutical business [59, 60]. TQM plays a vital role in the development of management practices [61, 62]. It has been noticed that TQM is immensely helpful in upgrading a business, while fulfilling the complete requirements of customers [63]. In addition, TQM increases sustainable competitiveness in business at national and global levels [64] and, above all, enhances organizational performance through continuous improvement in an organization's activities [65–67]. Hence, it is strongly recommended that quality principles and procedures are incorporated into SCM. Thus, quality management is often related to a model of organizational change [68], the implementation of which largely relies on the organization's ability to adapt itself to these principles.

Quality is an important factor when it comes to any product or service. With the current strong market competition, quality has become the market differentiator for almost all products and services. So, the practice of QC is an integral part of SCM for the production and distribution of product at the consumer level. It helps in developing confidence in the customer on product acceptability. Without QC, waste becomes prevalent beyond a tolerable amount.

For biotechnology and biopharmaceutical industries, feedstock for bacterial growth and development is abundant. How to use it depends on the requirements of the industry. In order to maintain the quality of product, vendors and the materials they provide are often audited by supply chain staff members to ensure that raw materials meet specifications. By controlling the quality of production inputs, supply chain managers are protecting the integrity of their company's operations.

Genetically engineered microbes are in use to produce value-added biochemicals. QC helps in screening those genetically modified microbes against the possibility of possessing/producing toxic compounds harmful for human use. The US Department of Transportation prescribes important rules for the transport of hazardous substances. Non-compliance can lead to penalties or fines, which makes QC imperative. The more efficiently and effectively toxic materials are handled in the supply chain, the better for all internal and external stakeholders.

4.6 Future Prospects for Biotechnological Products

In recent decades selling biotechnological and biological products has been a tough challenge. Considering global economic uncertainties, increasing healthcare costs, and expiring patents, the industry appears to be in a state of turmoil. At the same time, markets are diversifying and new fields of growth are opening up. Rapid development of emerging markets, progress in drug research, rise in production of generics, availability of high-potency drugs, and innovations in manufacturing processes are modifying the global biopharmaceutical and biological landscape.

Hence, the biopharmaceutical industries have been facing intensive hurdles as the sector needs to review its economic model. On the other hand, the biotechnology sector seems to be the most promising, but in reality is not reaching the expected results regarding putting drugs on the global market. Knowing this, it is likely that the biotechnology industry will face the same challenges as those of the pharmaceutical industry.

Pharmaceutical companies are also aware of the value of innovative research and new product development. These firms are keen to understand the potential of biotechnology companies to achieve innovation. This strategy also helps biotechnology companies to develop new technologies because of the financial resources that pharmaceutical companies can bring. It has been noticed that the total value of merger and acquisition (M&A) involving American or European biotechnology companies in 2013 reached €42.5 billion [69]. Interestingly, in the same year biotechnology–biotechnology collaborations increased by 68% to reach €9.47 billion.

Generally, biotechnology faces four major problems while launching new products, both in local and global markets:

- R&D productivity is not remarkable due to the number of failures in clinical trials
- High R&D expenses due to total failure in clinical trial
- Expiry of patents, triggering the appearance of biosimilars
- Merger & acquisition is necessary to strengthen the pipeline

In spite of heavy market competition, some top players in biotechnology companies want to acquire the art of pharmaceutical value chain management to achieve success in their product development. Because of financial constraints and a changing economic model, some biotechnology companies have decided to adopt a new model called "Fully Integrated Pharma Companies" or FIPCO. One way to achieve this FIPCO model is via an M&A strategy to help companies survive in the long term. Today, of 4000 biotechnology companies, there are only around 60 that have adopted the model. For instance, we can mention Amgen, Celgene, Gilead, and Shire [70–73].

The quality of new drugs (NME/BLA) approved by the FDA in 2013 increased by 43%, compared with 2012, with nine of the top ten forecast to reach blockbuster status (>$1 billion sales in the USA) by 5 years post-launch. This exceeds the record-breaking year of 2012 and marks a step-change in innovation and output for the industry. In total, the class of 2013 is expected to add over $24.4 billion to total US prescription drug sales in 2018.

In December 2013, Gilead Sciences received FDA approval for its new hepatitis C drug, Sovaldi, and recorded an unprecedented $2.1 billion of sales in its first full quarter on the US market. Current full-year sales for 2014 are forecast to be an equally impressive $7.6 billion in the USA alone. Priced at $1000 per pill, or $84,000 per treatment, some in the industry are questioning the basis of Gilead's pricing. Proponents, on the other hand, argue that the price per pill is irrelevant and economic benefit is a more rational basis for pricing discussions. Certainly, compared with the estimated cost of over $500,000 for a liver transplant in the USA to treat the more severe long-term complications of the disease, $84,000 does not seem so disproportionate. The pricing debate will intensify as the industry continues to shift away from primary care and new technology waves offer innovative products that potentially take some costs out of healthcare. Governments and payers will need to be more flexible in budget planning to accommodate these revolutionary new products, but do not expect easy discussions on products offering marginal benefits over generic products that are the current standard treatment of care [74].

The global market for microbes and microbial products is projected to approach $154.7 billion and $306 billion in 2015 and 2020, respectively, reflecting a five-year (2015–2020) CAGR of 14.6%. As a segment, the microbial products market is expected to nearly double from $151 billion in 2015 to $300.1 billion in 2020, indicating a 5-year CAGR of 14.7%. The global market for microbes as a segment should gain $290 billion over the forecast period, leading to $5.9 billion in 2020 with a 5-year CAGR of 9.8% [74].

Healthcare, the largest end-user market for microbes and microbial products, should total $187.8 billion in 2020, up from $111.5 billion in 2015. Energy was the second-largest end-user sector at $25.2 billion in 2014. The energy sector is expected to decline to $22.6 billion in 2015 as a result of falling ethanol prices. However, this market should rebound to almost $74.4 billion by 2020, courtesy of a five-year CAGR of 26.6%, the highest of all sectors. Manufacturing is the third-largest sector of the market, with sales of nearly $15.7 billion in 2014, $18.2 billion in 2015, and more than $40 billion in 2020. Agricultural and environmental applications of microbes and microbial products, although large in absolute terms, account for relatively small shares of the overall market [74]. Technologies under development for producing biodiesel from biomass include using bacteria, fungi. and algae. Although algal production of biodiesel is drawing the greatest interest, judging by the number of companies that are pursuing it, bacterial biodiesel appears to be the closest to commercialization [74]. "At present, rapeseed accounts for the bulk of global biodiesel production, with sunflower oil providing most of the remainder and small contributions from other vegetable oils," says BCC research analyst Andrew McWilliams. "Because plant oils are often derived from seeds, biodiesel currently has a low yield relative to the amount of land, water, and farming energy required. Direct biodiesel production from biomass feedstock using engineered microorganisms would have both a potentially high yield and lower input requirements" [74].

For 30 years now, the biotechnology sector has brought many innovations to industry processes as well as to the final products. This has been made possible because of the discovery of new action mechanisms, better knowledge and understanding of diseases, and a desire to improve existing techniques. Scientific and industrial environments are becoming more and more specialized and diverse and use biotechnology as a tool in their processes to a greater or lesser extent. This diversity has, in turn, brought about the need for a system to classify biotechnology uses based on common features or final purpose. As a result, biotechnological applications have been divided into five main groups, identified by a color system.

Red biotechnology refers to the use of organisms for the improvement of medical processes. It includes the design of organisms for manufacture of pharmaceutical products such as antibiotics and vaccines, the engineering of genetic cures through genomic manipulation, and its use in forensics through DNA profiling. Some relevant examples of red biotechnology are cell therapy and regenerative medicine, gene therapy, and medicines based on biological molecules such as therapeutic antibodies.

White biotechnology or industrial biotechnology refers to the use of living cells and/or their enzymes to create industrial products that are more easily degradable, require less energy, create less waste during production, and sometimes perform better than products created using traditional methods. There are many examples of white biotechnology, such as the use of microorganisms in production of chemicals, the design and production of new materials for daily use (plastics, textiles, etc.), and the development of new sustainable energy sources such as biofuels.

Grey biotechnology or environmental biotechnology deals primarily with waste water, waste, and exhaust air problems. The most varied of analytical and modeling methods are employed for this purpose. Risk assessment of target objects can be carried out using experimentally generated data combined with mathematical models. For pollutant removal or bioremediation, grey biotechnology uses microorganisms and plants to isolate and dispose of different substances (e.g., heavy metals and hydrocarbons), with the added possibility of subsequently making use of these substances or by-products.

Green biotechnology refers to the application of biological techniques to plants with the aim of improving the nutritional quality, quantity, and production economics. Examples include the use of genetic modification for the production of disease-resistant plants, UV-resistant plants, or plants that have superior qualities. Other examples include production of biofuels (e.g., ethanol, methane) from crops such as corn or marine algae grown at land-based production facilities.

Three main objectives are pursued using this technology. First, it is expected to obtain plant varieties resistant to pests and diseases (e.g., maize varieties resistant to corn stalk borer). Second, use of transgenic plants is aimed at developing varieties with improved nutritional properties (e.g., higher content of vitamins). Third, transgenesis in plants is also studied as a means to develop plant varieties able to act as biofactories and produce substances of medical, biomedical, or industrial interest in quantities appropriate for isolation and purification.

Blue biotechnology is based on the exploitation of sea resources to create products and applications of industrial interest. Taking into account that the sea presents the greatest biodiversity, there is potentially a huge range of sectors to benefit from the use of this kind of biotechnology. Many products and applications from blue biotechnology are still the objects of study and research, although some are already used on a daily basis. Using raw materials from the sea is the most widespread blue biotechnology in many different sectors. These materials, mostly hydrocolloids and gels, are already widely used in food, health treatment, cosmetics, etc. Medicine and research are other major beneficiaries of developments in blue biotechnology. Some marker molecules from marine organisms are now commonly used in research. Enzymatically active molecules useful in diagnostics and research have also been isolated from marine organisms. Some biomaterials and pharmacological or regeneratively active agents are being produced or investigated for their use in these sectors. Sectors such as agriculture and cosmetics are analyzing the potential of blue biotechnology for future development.

Indeed, more and more new biological entities (NBEs) have been launched on the market since 2000. For instance, in 2011, of the 20 most sold drugs, eight were from the biotechnology industry, representing almost 50% of the approvals given by the regulatory agencies. These drugs are focused on niche markets, such as rare diseases. Besides, we know that one bioblockbuster is sufficient to make a company profitable. However, because of the expiry of patents in the biotechnology industry, biosimilars have started to appear. Today, there is strong competition between biotechnology companies concerning production of biosimilars. An analogy can be made with generics and the pharmaceutical industry. Ultimately, we can observe a more complex competition:

- Bioblockbusters versus blockbusters
- Bioblockbusters versus biosimilars
- Biosimilars versus blockbusters

The first ever biosimilar launched was Remicade, and biosimilars MabThera and Herceptin should appear in the next 5 years. As competition in the biotechnological and biological products market continues to increase, QA and QC management need to re-evaluate the role of their organizations within the corporation. QC and QA units must downplay the traditional "policeman" role and assume a more constructive role (Fig. 4.13). The key to this new role is confirming that databases meet corporate and regulatory standards and evaluating the sense/nonsense of regulatory submissions for management. Like economies and financial markets, as supply chains have become more global and interconnected, they have also increased their exposure to shocks and disruptions. Supply chain speed only exacerbates the problem. Even minor missteps and miscalculations can have major consequences as their impacts spread like viruses throughout complex supply chain networks. Looking at the competitive market and short life of biotechnological and biological products, executives are under pressure as compliance mandates, suppliers, and information flows multiply and supply chains become more complex, costly, and vulnerable. Executives are finding it increasingly difficult to respond to these challenges, especially with conventional supply chain strategies and designs.

Businesses and supply chains have become substantially more global over the last decade. Between 1995 and 2007, the number of transnational companies more than doubled, from 38,000 to 79,000, and foreign subsidiaries nearly tripled, from 265,000 to 790,000. In addition to spreading geographically, supply chains now involve more companies. Nearly 80% of executives say they expect the number of collaborative relationships with third parties to increase. Furthermore, an ever broader range of activities are being outsourced. Between 2007 and 2010, R&D outsourcing is forecast to increase by 65% and engineering services and product-design projects by more than 80%. Supply chains must also contend with rapidly expanding and contracting product portfolios. In the consumer products industry, for example, product introductions increased by 17% in 2006, which is more than double the 2005 rate. Confronted with such daunting complexity, supply chain executives have described five major challenges in bringing sustainability to supply chain management (Fig. 4.14). All are critically important, and must be addressed simultaneously. Together, they comprise what we call the "chief supply chain officer" agenda.

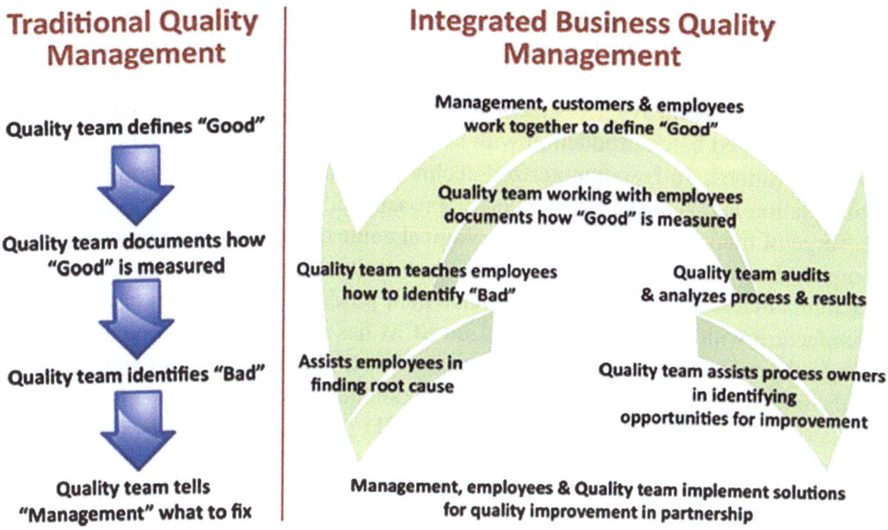

Fig. 4.13 Diagram showing how integrated quality management has taken the place of traditional quality management

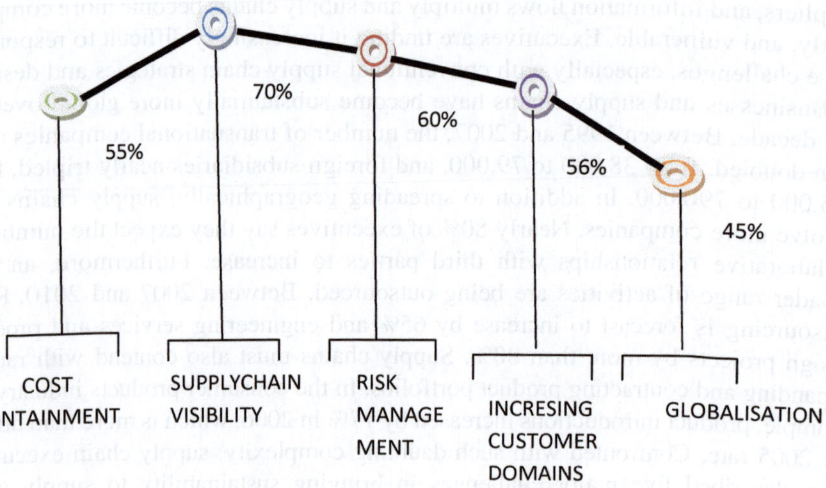

Fig. 4.14 Supply chain leaders wrestle with five major challenges

4.7 Continuous Manufacturing Process Technology

Biotechnological, biopharmaceutical, and biological industries are among the most advanced and sophisticated as a result of their major investment in research and development, understanding of the complex metabolism of microbes and complex human biology, and heavy investment with delay in results. Earlier, the manufacturing process involved batch production with some traditional touches. Thanks to production innovations, fully computerized technology, and an understanding of gene manipulation and sequencing, the new approach of continuous manufacturing is on the verge of transforming the biotechnological value chain. It has an impact on every company based on biotechnology, from giant multinationals to the third-party manufacturers that small start-ups hire to make their products. The process of continuous manufacture with completely modernized SCM has increased the scope of biotechnological product markets at the international level. The biotechnology industry now needs to be less expensive, more ecofriendly, and able to provide a timely supply of goods on a quality base. Digital fabrication (3D printing of drugs) is also gaining traction as a viable technology for making small batches of medicines that are generally too costly and impractical to produce.

Global spending on biopharmaceuticals, food additives, and industry-based biological products has been soaring, and several top players in these markets are keen in bring costs under control. The USA is on the process of understanding how to deal with the changes brought about by the Affordable Care Act. Around the world, reforms have unleashed numerous disruptive innovations, and growth in emerging markets is creating a large pool of prospective consumers. Biotechnology and biopharmaceutical industries are desperately trying to overcome the problem of expiring patents for blockbuster drugs, which have given billions of dollars in revenue. Biotechnology and biopharmaceutical industries are in constant touch with the latest developments in manufacturing processes and adopt new paradigms to control costs. These industries are under constraint to deliver a variety of products to a wide range of markets. This is where continuous manufacturing comes in.

Previously, biotechnology and biopharmaceutical industries produced products in a batch process rather than in an assembly-line fashion, as cars are. In batch production, the processes are carried out in sequence manner and consist of the blending of powder ingredients, formation of pellets, compression into tablets, and coating, which sometimes take place at different plants. Additives are mixed separately in a large vat. After the completion of formulation and polishing of drugs, the separate process of packaging is followed. The operation is time-consuming, asset-intensive, and expensive. The risk of contamination is always present because batches of partially finished medicines must be moved from place to place.

With continuous operation and ongoing inline analysis, process parameters are more closely monitored and maintained at optimal values, leading to more consistent processes and product quality, which in turn can reduce waste generation, product losses, and downtime. Because the process runs for long periods of time, smaller reactors can provide large quantities of product. A smaller footprint often equates to lower capital expenditure and can result in reduced consumption of energy, water, and raw materials, which means lower operating expenses. Greater automation and reduced human intervention are other important benefits, particularly for the production of highly potent active pharmaceutical ingredients (APIs) and formulated drugs. The smaller environmental footprint of continuous processing is also becoming increasingly important as the pharmaceutical industry moves toward implementation of green chemistry principles and considers values such as process mass intensity (PMI) and environmental factors. There are also unique benefits for small-molecule production using flow chemistry. Perhaps the most important is the ability to carry out hazardous reactions that would not be possible in a batch process. Because only small quantities of reagents, intermediates, and products are present at any given time in a flow reactor, exposure to toxic or energetic substances is minimized, and process parameters such as temperature and pressure can be readily controlled.

With the growing demand for biosimilars and an ever-increasing pressure for reduction in manufacturing costs, the biotechnology industry seems to exhibit a flourishing interest in the development of continuous biomanufacturing systems. Although the bulk of the research efforts so far have been focused on making individual unit operations continuous, a few studies have explored the possibility of a continuous end-to-end biomanufacturing process [75]. Novartis, a pioneer in such efforts, has partnered with the MIT Center for Continuous Manufacturing, and is investing $65 million in a joint 10-year research project. The two parties have already concluded that continuous manufacturing will benefit patients, healthcare providers, and the pharmaceutical industry. This project has demonstrated, for example, that continuous manufacturing can accelerate the introduction of new drugs through efficient production. It also tends to minimize waste, energy consumption, and raw material use, and enhance companies' flexibility in responding to market needs [76].

Bristol-Myers Squibb uses flow chemistry for an oxidative rearrangement process to minimize the potential for thermal runaway. The process has been implemented at the pilot plant and commercial scales for the production of the investigational liver cancer drug brivanib alaninate [77]. GlaxoSmithKline, meanwhile, employ a continuous flow strategy for the synthesis of multikilogram quantities of bromomethyltrifluoroborate, a key reagent for a Suzuki–Miyaura coupling reaction, because of the highly exothermic nature of the reaction [78].

Contract manufacturing organizations (CMOs) have also developed continuous processes to overcome the hazards associated with batch reactions. For instance, the ring opening of a highly strained (and thus energetic) cyclopropane ring was performed by AMRI Inc. in a continuous-flow reactor to increase the safety of a process step in the production of taxadienone on a decagram scale [79]. It is interesting to note that safety is the most important metric for respondents of the Nice Insight annual survey when considering a technological innovation that would influence CMO selection. Perfusion technology also offers a particular advantage for biopharmaceutical production. In perfusion mode, the medium is fed at a constant rate, and spent cells and product are removed continually. As a result, the product is exposed to the culture conditions for a minimum amount of time. Proteins that are highly sensitive and easily degraded under conventional production conditions can therefore be manufactured in a continuous culture process.

Despite its clear advantages and real potential, the adoption of continuous processing is proceeding relatively slowly. The main reason for this is the need for further technology development. In the biopharmaceutical industry, perfusion technology has been available for many years, but practical continuous solutions for downstream processing have only recently begun to reach the market. The situation is similar for small-molecule API production. Although some downstream processes, such as distillation and extraction, are well suited to continuous operation, others (particularly handling of solids) are more challenging. The existing infrastructure presents another major block to widespread adoption of continuous processing. Companies are more likely to use existing equipment than to invest in new continuous processing systems, unless there is a distinct advantage in doing so. Thus, at present, continuous/flow chemistry is implemented on a case-by-case basis. The lack of clear regulatory guidelines also acts as barrier to adoption of continuous manufacturing. Harmonized guidelines that are accepted by regulatory bodies around the world are needed to ensure that the monitoring of product outputs from continuous processes is conducted in a consistent manner. Additional approval of continuous processes and experience in auditing of continuous process steps are required before this situation can be adequately addressed.

Other issues facing manufacturers and CMOs considering implementation of continuous processes are resistance to change, the need to integrate activities across different groups, and the aversion to investment in new technologies at early process development stages [80]. A lack of people with expertise in continuous flow technology is also hindering its adoption. There is both a real dearth of experienced operators and a continuing perception that only people with extensive experience and specific technical know-how can successfully achieve the scale-up and commercialization of continuous processes. The situation will not be resolved until universities incorporate continuous manufacturing for both small- and large-molecule drugs into their curricula and more continuous processes are successfully implemented [81].

Although the adoption of continuous processing remains limited in both pharmaceutical and biopharmaceutical industries, most large companies are investigating the potential benefits [82]. In the Nice Insight survey, 62% of respondents indicated that they have learned of new technological innovations for the biopharmaceutical industry in the past year that would benefit their company. Many CMOs have also recognized the need to have flow/continuous manufacturing capability, and leading companies are developing the necessary expertise. The FDA is also very supportive of continuous manufacturing because it can lead to real increases in product quality consistency. Continuous processes are also generally becoming better characterized and understood, which can only have a positive impact on the final product. The FDA has spoken positively about continuous manufacturing since 2004 and has been more vocal recently [83]. In addition, the proposed 21st Century Cures Act, which seeks to speed up drug development and commercialization, requires the FDA to support the development and implementation of continuous manufacturing for drugs and biologics [84].

As well as the development of new continuous crystallization techniques and equipment, advances are being made in microreactor technology for the continuous manufacture of small-molecule APIs. Collaborative efforts between pharmaceutical companies and equipment manufacturers have resulted in microreactor technology suitable for use in pilot and small commercial-scale production, which is important for overcoming scale-up concerns. CMOs including DPx Fine Chemicals and Lonza Custom Manufacturing have leveraged these developments to install commercial-scale continuous production capabilities. Equally important is the expansion of flow chemistry for API synthesis to systems other than those involving liquid reagents and products. Flow chemistry is also now being applied to reactions that were not typically used in the pharmaceutical industry because they are not easy to control or scale up beyond the bench top. Two examples include photochemical and electrochemical reactions [85]. For biopharmaceutical manufacturing, key advances in upstream continuous production equipment have focused on improving reliability. Upstream technologies that will have a future impact include alternating tangential flow systems, continuous centrifuges, acoustic resonance devices, and cell settlers. Downstream, some real progress has been made with respect to simulated moving bed chromatography, continuous filtration systems, and flow-through absorbers.

One area where significant further development will be crucial is process analytical technology. In-line analytical monitoring that provides real-time data is crucial for effective implementation of integrated continuous processes. Not only is the information necessary for maintaining optimum operating conditions at each step, it is also necessary to ensure that changes in one step do not affect other integrated downstream processes. Many trials are being carried out on downstream processes in continuous form, and a variety of technology is available to interlink downstream processes to save time, cost, and increase the quality of production. Various regulatory bodies such as the FDA and the European Medicines Agency are in the process of bringing changes to manufacturing processes for better production in the shortest possible time [86, 87]. For example, the concept of "batch" has been changed to highlight the quality of product rather than describe a manufacturing process [87].

References

1. Lummus RR, Vokurka RJ (1999) Defining supply chain management: a historical perspective and practical guidelines. Ind Manag Data Syst 99:11–17
2. Tyndall G et al (1998) Supercharging supply chains: new ways to increase value through global operational excellence. Wiley, New York
3. Ellram LM, Cooper MC (1990) Supply chain management, partnerships, and the shipper – third party relationship. Int J Logist Manag 2:1–10
4. Houlihan JB (1988) International supply chains: a new approach. Manag Decis 26:13–19
5. Agus A (2011) The structural influence of supply chain management on product quality and business performance. Int J Trade Econ Finan 2:269–275
6. Cooper MC, Ellram LM (1993) Characteristics of supply chain management and the implications for purchasing and logistics strategy. Int J Logist Manag 4:13–24
7. Robinson JR, Malhotra MK (2005) Defining the concept of supply chain quality management and its relevance to academic and industrial practice. Int J Prod Econ 96:315–337
8. Agus A (2011) Supply Chain Management, production quality and business performance, 2001 International Conference on Sociality and Economics Development IPEDR, vol 10. IACSIT Press, Singapore
9. Beamon BM (2003) Supply chain design and analysis: models and methods. Int J Prod Econ 55:281–294
10. Pulakos ED (2009) Performance management: a new approach for driving business results. Wiley, Blackwell, UK
11. Alfaro JJ et al (2002) Performance measurement for e-business enterprises. Int J Bus Perform Manag 4(2–4):296–315
12. Nigel Slack et al (2009) Operations and process management. Prentice Hall, pp 209–215
13. King D et al (2010) Inderwildi OR, Williams A. The future of industrial biorefineries world economic forum
14. Ron P et al (2011) Clean Energy Trends (2011) Clean Edge. www.fondazionesvilupposostenibile.org
15. Microbial Products: Technologies, Applications and Global Markets BCC Research, April 2011. BIO086A. www.bccresearch.com
16. Enzymes in Industrial Applications: Global Markets BCC Research, Jan. 2011. BIO030F. www.bccresearch.com
17. Kircher M (2010) Discussion paper – session trends in technology and applications, OECD Workshop on "Outlook on Industrial Biotechnology." OECD Working Party on Biotechnology, Vienna, January 13–15
18. Sarkis JA (2003) Strategic decision making framework for green supply chain management. J Clean Prod 11:397–409
19. Klassen RD, Vachon S (2003) Collaboration and evaluation in the supply chain: the impact of plant-level environmental investment. Prod Oper Manag 12:336–352

20. Paulraj A (2011) Understanding the relationships between internal resources and capabilities, sustainable supply management and organizational sustainability. J Supply Chain Manag 47:19–37
21. Thoo AC et al (2014) The moderating effect of enviropreneurship on green supply chain management practices and sustainability performance. Adv Mater Res 869–870:773–776
22. Porter ME, Kramer MR (2006) Strategy and society: the link between competitive advantage and corporate social responsibility. Harv Bus Rev 84:78–92
23. Kye D et al (2013) The perceived impact of packaging logistics on the efficiency of freight transportation (EOT). Int J Phys Distrib Logist Manag 43:707–720
24. Grönman K et al (2013) Framework for sustainable food packaging design. Packag Technol Sci 26:187–200
25. García-Arca J (2015) Drivers and stages in "Packaging Logistics": an analysis in the food sector. In: Cortès P (ed) Enhancing synergies in a collaborative environment. Springer, Switzerland, pp 51–59
26. Williams H et al (2008) A life cycle perspective on environmental effects of customer focused packaging development. J Clean Prod 16:853–859
27. Molina-Besch K, Pålsson H (2014) Packaging for eco-efficient supply chains: why logistics should get involved in the packaging development process. Sustainable logistics. Transp Sustain 6:137–163
28. Azzi A et al (2012) Packaging design: general framework and research agenda. Packag Technol Sci 25:435–456
29. García-Arca J et al (2014) Packaging logistics: promoting sustainable efficiency in supply chains. Int J Phys Distrib Logist Manag 44:325–346
30. García-Arca J, Prado-Prado JC (2008) Packaging design model from a supply chain approach. Supply Chain Manag Int J 13:375–380
31. Verghese K et al (2015) Packaging's role in minimizing food loss and waste across the supply chain. Packag Technol Sci 28:603–620
32. Bill McBeath (2012) How the pharmaceutical industry has dramatically reduced thefts. Chainlink Research, www.chainlinkresearch.com
33. Ketchen DJ et al (2008) Best value supply chains: a key competitive weapon for the 21st century. Bus Horiz 51:235–243
34. Chandra C, Grabis J (2007) Supply chain configuration – concepts, solutions and applications. Springer, New York
35. Olson LD (2012) Supply chain information technology. In: Nahmias S (ed) The supply and operations management collection. Business Expert Press, New York
36. Shaik MN, Abdul-Kader W (2013) Interorganizational information systems adoption in supply chains: a context specific framework. Int J Inf Syst Supply Chain Manag 6:24–40
37. Nelson M (2001) Sustainable competitive advantage from information technology: limitations of the value chain. In: Papp R (ed) Strategic information technology: opportunities for competitive advantage. Idea Group Publishing, Hershey, PA, pp 40–55. doi:10.4018/978-1-878289-87-2.ch002
38. Turek B (2013) Information systems in supply chain integration and management. Retrieved from http://www.ehow.com/info_8337099_information-supply-chain-integration-management.html
39. Norris G et al (2000) E-business and ERP: transforming the enterprise. Wiley, Canada
40. Yanjing J (2009) Integration of ERP and CRM in E-commerce environment. Proceedings of the International Conference on Management and Service Science, pp 1–9
41. Marakas MG (2003) Decision support systems in the 21st century. Indiana University, NewJersey, Prentice Hall
42. Premkumar GP (2000) Interorganization systems and supply chain management: an information processing perspective. Inf Syst Manag 17:56–69

43. Grover V, Saeed KA (2007) The impact of product, market, and relationship characteristics on interorganizational system integration in manufacturer supplier dyads. J Manag Inf Syst 23:85–216
44. Wisner JD et al (2012) Supply chain management: a balanced approach, 3rd edn. South-Western Cengage Learning, Canada
45. Rao P (2006) Greening of suppliers/in-bound logistics – In the South East Asian Context. Greening Supply Chain 189–204
46. Srivastava SK (2007) Green supply-chain management: a state-of-the-art literature review. Int J Manag Rev 9:53–80
47. Diabat A, Govindan K (2011) An analysis of the drivers affecting the implementation of green supply chain management. Resour Conserv Recycl 55:659–667
48. Luthra S et al (2014) Green supply chain management. Implementation and performance – a literature review and some issues. J Adv Manag Res 11:20–46
49. Beamon BM (1999) Designing the green supply chain. Logist Inf Manag 12:332–342
50. Mentzer JT (2001) Supply chain management. Sage, United States of America
51. Guidance for Industry Q1A(R2) (2003) Stability testing of new drug substances and products, US Department of Health and Human Services Food and Drug Administration Center for Drug Evaluation and Research (CDER) Center for Biologics Evaluation and Research (CBER), ICH
52. ICH Q1B (1996) Photostability testing of new drug substances and products. International Conference on Horminisation of Technical Requirements for Registration of Pharmaceuticals for Human Use. Center for Biologics Evaluation and Research (CBER), ICH. www.ich.org/fileadmin/Public_Web_Site/ICH_Products/.../Q1B/.../Q1B_Guideline.pd
53. ICH Q1C Stability Testing for New Dosage Forms, Center for Biologics Evaluation and Research (CBER), www.ich.org/fileadmin/Public_Web.../ICH.../Guidelines/.../Q1C/.../Q1C
54. ICH Q3A Impurities in New Drug Substances, Center for Biologics Evaluation and Research (CBER), www.ich.org/fileadmin/Public_Web_Site/ICH.../Q3A.../Q3A_R2
55. ICH Q3B Impurities in New Drug Products, Center for Biologics Evaluation and Research (CBER), www.pharma.gally.ch/ich/q3b028295en.pdf
56. ICH Q5C Quality of Biotechnological Products: Stability Testing of Biotechnological/Biological, Center for Biologics Evaluation and Research (CBER), Products www.ich.org/products/.../quality/quality.../stability-testing-of-biotechnologicalbiologi
57. ICH Q6A Specifications: Test Procedures and Acceptance Criteria for New Drug Substances and New Drug Products: Chemical Substances, Center for Biologics Evaluation and Research (CBER), www.ich.org/fileadmin/Public_Web.../ICH_Products/Guidelines/.../Q6A/.../Q6Astep4
58. ICH Q6B Specifications: Test Procedures and Acceptance Criteria for New Drug Substances and New Drug Products: Biotechnological/Biological Products, Center for Biologics Evaluation and Research (CBER), www.ich.org/fileadmin/Public_Web_Site/ICH_Products/.../Q6B/.../Q6B_Guideline.pd
59. Arumugam V et al (2009) Self-assessment of TQM practices: a case analysis. TQM J 21:46–58
60. Mohanty RP, Behera AK (1996) TQM in the service sector. Work Study 45:13–17
61. Prajogo DI, Sohal SA (2003) The relationship between TQM practices, quality performance, and innovation performance: an empirical examination. Int J Qual Reliab Manag 20:901–918
62. Hoang DT et al (2006) The impact of total quality management on innovation: findings from a developing country. Int J Qual Reliab Manag 23:1092–1117
63. Oakland JS (1993) Total quality management, 2nd edn. Butterworth-Heinemann, Oxford
64. Terziovski M (2006) Quality management practices and their relationship with customer satisfaction and productivity improvement. Manag Res News 29:414–424
65. Claver-Cortés E et al (2008) TQM, managerial factors and performance in the Spanish hotel industry. Ind Manag Data Syst 108:228–244
66. Teh PL (2009) Does total quality management reduce employees' role conflict? Ind Manag Data Syst 109:1118–1136

67. Blackiston GH (1996) A barometer of trends in quality management. Natl Product Rev 16:15–23
68. Boronat P, Canard F (1995) Management par la qualite: a totale et changement organisationnel. Les NouvellesForms Organisationnelles, Paris, Economica
69. Ernst & Young (2014) Beyond borders Biotechnology Industry Report, tinyurl.com/EY-biotech
70. Sandner P, Ziegelbauer K (2008) Drug Discov Today 13:457–463
71. Amir-Aslani A et al (2013) Spectra Biologie 199:24–25
72. Czerepak E, Ryser S (2008) Nat Rev Drug Discov 7:197–198
73. Buisson R et al (2014) Biofutur 350:36–40
74. *1*- www.bccresearch.com (2014) Microbial Products: Technologies, Applications and Global Markets (BIO086C)
75. 2016 Nice Insight Contract Research (2016) Preclinical and Clinical Survey (CRO Outsourcing survey) and Nice Insight; 2 Contract Development & Manufacturing Survey (CDMO Outsourcing survey)
76. Pellegatti L, Sedelmeier J (2015) Synthesis of Vildagliptin utilizing continuous flow and batch technologies. Org Process Res Dev 19:551–554
77. Thomas L La Porte et al (2014) Reaping the benefits of a continuous process Org Process Res Dev I 18:1492–1502
78. Broom T et al (2015) Reaping the benefits of a continuous process. Org Process Res Dev doi: 10.1021/op400090a
79. Krasutsky SG et al (2015) Reaping the benefits of a continuous process. Org Process Res Dev 19:284–289
80. Challener CA (2014) Potential benefits drive interest in continuous manufacturing. Pharm Technol, www.pharmtech.com/bioprocessing-and-sterile-manufacturing
81. Challener CA (2015) Lack of expertise hinders adoption of continuous API synthesis. Pharm Technol 39:30–33
82. Poechlauer P et al (2013) Reaping the benefits of a continuous process. Org Process Res Dev 17(12):1472–1478
83. Rockoff JD (2015) Drug making breaks away from its old ways. Wall Street J, www.wsj.com/articles/drug-making-breaks-away-from-its-old-ways-1423444049
84. Energy & Commerce Committee, United States House of Representatives. Full Committee Vote on the 21st Century Cures http://energycommerce.house.gov/markup/full-committee-vote-21st-century-cures-act, accessed June 2, 2015
85. Kamarck ME (2006) Building biomanufacturing capacity – the chapter and verse. Nat Biotechnol 24:503–505
86. Rathore AS et al (2015) Continuous processing for the production of biopharmaceuticals. Prep Biochem Biotechnol 45:836–849
87. CFR Title 21, Part 210.3 (Government Printing Office, Washington, DC)

Chapter 5
Material-Balance Calculation of Fermentation Processes

5.1 Introduction

About the time of the American Revolution, the people of England began to use machines to make cloth, and steam engines to run the machines [1]. A little later, they invented locomotives. Productivity began a spectacular climb. The first successful atmospheric steam engine was made in England by Thomas Savery in 1697 and then by Thomas Newcomen in 1712 [2–6]. This unique discovery sprouted the idea of thermodynamics. The word is derived from Greek terms representing heat (*therme*) and power (*dynamis*). The basic philosophy behind this was to obtain power (energy per unit time) from a given amount of energy stored in some fuel (which was typically turned into heat). Basically, thermodynamics is a branch of science dealing with the properties of matter [7–17]. Thermodynamic principles are useful in setting up material balances. Material balance is mainly based on the law of conservation of mass, which is applied to atoms, molecular species, and total mass, The universal principle (Lavoisier principle) that matter cannot be created or destroyed (unless there is a nuclear reaction) holds in biochemical systems. In any closed system, the total mass of every element (carbon, nitrogen, oxygen, hydrogen, phosphorus, etc.) is constant over time. Material balance templates have been used in fermentation to understand the nature and the quantities of biochemical species consumed and produced by analyzing the stoichiometry of biochemical reactions taking place in a close system (fermenter). This also provides instant information on transformation yields during the fermentation processes [18–23].

Such information is helpful in successfully monitoring a process so that maximum yield of a target metabolite can be achieved. It is also useful for understanding the bioenergetic aspect of the process supporting microbial growth, as well as the quantity of biomass produced. Material balance technology has become an integral part of fermentation technology. Using this method, many complex situations are simplified by looking at the movement of mass and equating what comes out to what goes in. In addition, one can understand the CO_2 concentration in the off-gas;

B. Kumara Behera, A. Varma, *Microbial Biomass Process Technologies and Management*, DOI 10.1007/978-3-319-53913-3_5

the fraction of substrate consumed, not converted, or converted to product; amounts of reactants needed to produce a certain amount of product(s); and oxygen requirements for a fermentation process.

Computerized technology, in association with advanced biometry and biostatistics, has successfully enabled maximum accuracy in quantitative analysis of microbial growth and development. On the other hand, the application of basic mass balance methods has catalyzed the process of gaining instant information to monitor the fermentation process.

5.2 History

In 1957, Hans Kreb and Hans Kornberg published the book *Energy Transformations in Living Matter* [24], explaining how biochemical reactions also obey the basic principles of thermodynamics. In addition, the appendix contained the first-ever published thermodynamic tables, written by Kenneth Burton on equilibrium constants and the Gibbs free energy of formation for chemical species, which enabled calculation of biochemical reactions that had not yet occurred. With this understanding, the biophysicist Ilya Prigogine developed nonequilibrium thermodynamics to explain how biological organisms can develop from disorder. He called organisms “dissipative systems” and explained how they are formed and maintained by the dissipative processes that exchange matter and energy between the system and its environment, and disappear if that exchange ceases. It may be said that they live in symbiosis with their environment.

About 2×10^{23} J of solar energy per year [25] is captured through photosynthesis by green plants. This amount is equivalent to about 4% of the total energy that reaches Earth. The energy transformations in biological communities surrounding hydrothermal vents are exceptions; they oxidize sulfur and obtain their energy via chemosynthesis rather than photosynthesis.

Based on the principles of chemical thermodynamics (first law of thermodynamics, second law of thermodynamics, Gibbs free energy, statistical thermodynamics, reaction kinetics), the stoichiometry of biochemical processes such as ATP hydrolysis, protein stability, DNA binding, membrane diffusion, enzyme kinetics [26], and other such essential energy-controlled pathways can be well expressed. In terms of thermodynamics, the amount of energy capable of doing work during a chemical reaction is measured quantitatively by the change in the Gibbs free energy. The physical biologist Alfred Lotka attempted to unify the change in the Gibbs free energy with evolutionary theory.

Material balance plays an important role in understanding metabolic economy. On the basis of material balance, the concept of stoichiometric coefficients for the overall reaction representing a biological system was introduced and can provide information about yield factors, substrate consumption, and product formation rates [27–39]. Despite the excellent performance and suitability of this approach for in-line bioprocess monitoring and control, its application to the deduction of cellular behavior is limited because the model does not contain any intracellular mechanisms and is therefore unable to provide essential information at the cellular level.

The mass balance of fermentation has become more and more recognized as a valuable tool for analytical data validation, detection of measurement errors, detection of unnoticed products, and estimation of variables for which no direct analytical methods are available [18], and for improvement of the accuracy and reliability of fermentation parameter estimations [35, 39]. The principles of mass balancing theory were developed in the late 1970s by Minkevich, Erickson, Roels, and others [18, 33, 34, 36, 40]. The theory uses the formalism of linear algebra to express the relationships between measurable (macroscopic) flows. Practical applications are mostly in the field of numerical procedures for solving a system for unknown flows or to calculate maximum likelihood estimators in the case of overdetermined systems. Thus, it is now recognized that biological reaction stoichiometry is mandatory for analysis of biological processes.

5.3 Application of the Basic Principles of Thermodynamics in Fermentation

5.3.1 Systems and Processes

In thermodynamics, a system consists of any matter identified for investigation (Fig. 5.1). The system is set apart from the surroundings, which are the remainder of the universe, by a system boundary. The system boundary may be real and tangible, such as the walls of a beaker or fermenter, or imaginary. If the boundary does not allow mass to pass from the system to surroundings and vice versa, the system is a closed system with constant mass. Conversely, a system able to exchange mass with its surroundings is an open system.

A process causes changes in the system or its surroundings. Fermentation processes can be categorized into four types: batch, semibatch, fed-batch, and continuous.

5.3.1.1 Batch Process

A batch fermentation system is a closed system. At time $t = 0$, the sterilized nutrient solution in the fermenter is inoculated with microorganisms and incubation is allowed to proceed at a suitable temperature and in a gaseous environment for a suitable period of time. Over the course of the entire fermentation, nothing is added except oxygen (in the case of aerobic microorganisms), antifoam agent, and acid or base to control pH. The composition of the medium, the biomass concentration, and the metabolite concentration generally change constantly as a result of cell

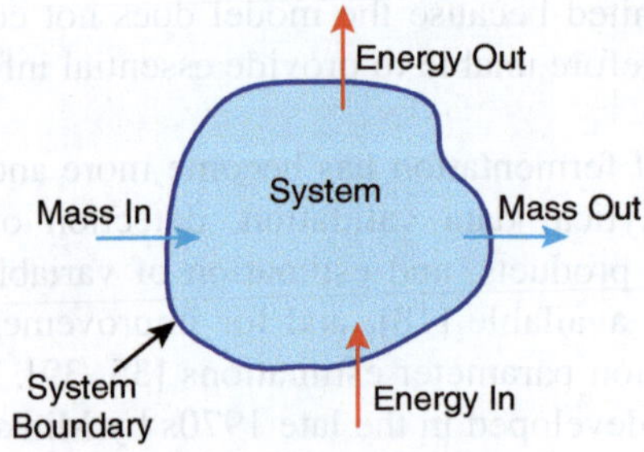

Fig. 5.1 A schematic presentation of thermodynamic system

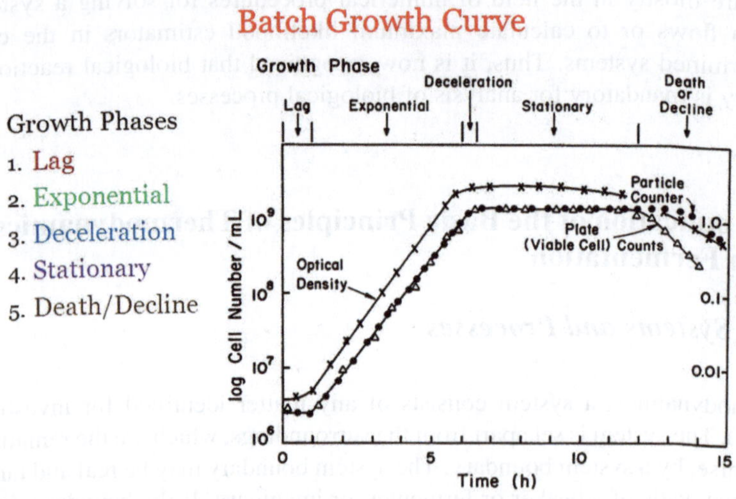

Fig. 5.2 Bacterial growth curve in a batch process showing different stages of growth (Source: https://www.google.co.in/imgres)

metabolism. After inoculation of a sterile nutrient solution with microorganisms and cultivation under physiological conditions, six typical phases of growth are observed: lag, acceleration, exponential, deceleration, stationary, and decline and death (Fig. 5.2).

Growth is a result of consumption of nutrients. The initial lag phase is a time of no apparent growth, but biochemical analyses do show metabolic turnover, indicating that cells are in the process of adapting to the environmental conditions and that new growth will eventually begin. Then a transient acceleration phase occurs as the microbes begin to grow (not marked in Fig. 5.2), which is quickly followed by an exponential phase. In the exponential phase, microbial growth proceeds at the highest possible rate for that organism with excess nutrients, ideal environmental parameters, and absence of growth inhibitors. However, in batch cultivations, exponential growth is of limited duration and, as nutrient conditions change, the growth rate decreases and enters the deceleration phase. This is followed by the stationary phase, when overall growth can no longer be maintained as a result of nutrient exhaustion. The final phase of the cycle is the death phase when growth has ceased. Most biotechnological batch processes are stopped before this stage because of decreasing metabolism and cell lysis. Typical microbial cultures in the laboratory (in a flask) are batch cultures. In batch cultivation, the bacteria are inoculated into a bioreactor (always stirred tank bioreactors). Then, under certain conditions (temperature, pH, aeration, etc.), the bacteria go through all the growth phases. Finally, the fermentation is stopped and the product is collected. After cleaning and sterilizing the fermenter, it is ready for another batch. All materials are added to the system at the start of the process; the system is then closed and products are removed only when the process is complete.

5.3.1.2 Semibatch or Semicontinuous Process

A semibatch reactor is a variation of a batch reactor in which one reactant is added intermittently or continuously to other reactants contained in the vessel as the reaction proceeds (Fig. 5.3). The reaction may be single-phase or multiphase. As in a batch reactor, the operation is inherently in an unsteady state and is usually characterized by a cycle of operation, although in a more complex manner [1].

In a homogeneous liquid-phase reaction of the type $A + B \rightarrow$ Product, reactant A is initially charged to the vessel, and reactant B is added at a prescribed rate as the reaction proceeds (Fig. 5.3). For a liquid phase reaction in which a gaseous product is formed, the gas is removed as the reaction proceeds.

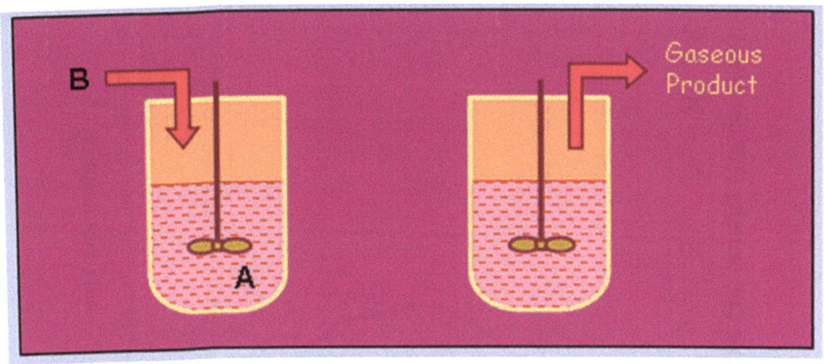

Fig. 5.3 Mode of operation of a semibatch reactor (Source: https://www.google.co.in/imgres?)

5.3.1.3 Fed-Batch Process

A fed-batch process allows input of material to the system, but not output. The fed-batch method is characterized by the addition of small concentrations at the beginning of the fermentation and continued addition of these substances in small doses during the fermentation process. Despite the apparent similarity between the fed-batch reactor model and the continuous culture model, they are very different. Whereas the chemostat process (continuous culture) for biomass accumulation is composed of a growth and removal process, the fed-batch procedure is composed of a growth and dilution process (Fig. 5.4).

The concept of steady state cannot be easily applied to a fed-batch reactor. It is significantly more difficult to maintain a specific growth rate in a fed-batch system than in continuous culture. Because cells are not removed during the fermentation, fed-batch cultures are suitable for the production of compounds produced during very slow or zero growth. Unlike a continuous culture, the feed does not need to contain all the nutrients needed to sustain growth. The feed may contain only a nitrogen source or a metabolic precursor. Contamination and/or mutation does not have the same dramatic effect on a fed-batch fermenter.

A fed-batch fermenter can be operated in a variety of ways; for example, the reactor can be operated in the sequence, batch → fed-batch → batch. The feed can also be manipulated to maximize product formation. During fermentation, the feed composition and feed flow rate can be adjusted to match the physiological state of the cells. Fed-batch reactors can maintain low nutrient and substrate concentrations. They are therefore thus suitable for the production of product or cells when the substrate is inhibitory by allowing the maintenance of non-inhibitory low levels of substrate. These reactors are very useful for the production of vinegar and amylase.

Fed-batch fermentations are also useful when the product or biomass yield is highest at low substrate concentrations, as in the case of mammalian cell systems for recombinant protein, baker's yeast products, and antibiotic production. Another

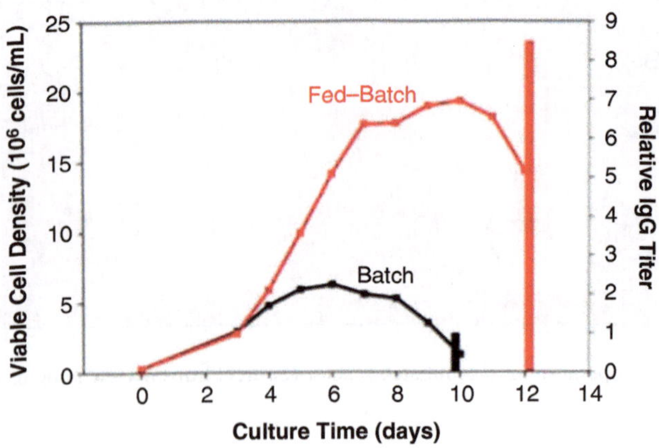

Fig. 5.4 Bacterial growth curve in a fed-batch process, showing no distinguishable steady state (Source: https://www.google.co.in/imgres)

suitable application is when product formation is dependent on a specific nutrient composition (e.g., specific carbon to nitrogen ratio).

5.3.1.4 Continuous Process

A continuous process allows matter to flow in and out of the system. In continuous fermentation, an open system is set up. Sterile nutrient solution is added to the bioreactor continuously and an equivalent amount of converted nutrient solution with microorganisms is simultaneously taken out of the system (Fig. 5.5). A homogenously mixed bioreactor can be either a chemostat or turbidostat. In a chemostat, in the steady state, cell growth is controlled by adjusting the concentration of one substrate. In the turbidostat, cell growth is kept constant by using turbidity to monitor the biomass concentration and adjusting the rate of feed of nutrient solution as appropriate. In the chemostat, a constant chemical environment is maintained, whereas in a turbidostat constant cell concentration is maintained.

In a chemostat, the growth chamber is connected to a reservoir of sterile medium. Once growth is initiated, fresh medium is continuously supplied from the reservoir. The volume of fluid in the growth chamber is maintained at a constant level by some sort of overflow drain. Fresh medium is allowed to enter the growth chamber at a rate that limits the growth of the bacteria. The rate of addition of fresh medium determines the rate of growth because the fresh medium always contains a limiting amount of an essential nutrient. Thus, using a chemostat relieves the insufficiency of nutrients, accumulation of toxic substances, and accumulation of excess cells in the culture (which are the parameters that initiate the stationary phase of the growth cycle). In continuous cultivation, the fresh medium flows into the fermenter continuously and some of the medium in the reactor is withdrawn at the same flow rate as the inlet flow.

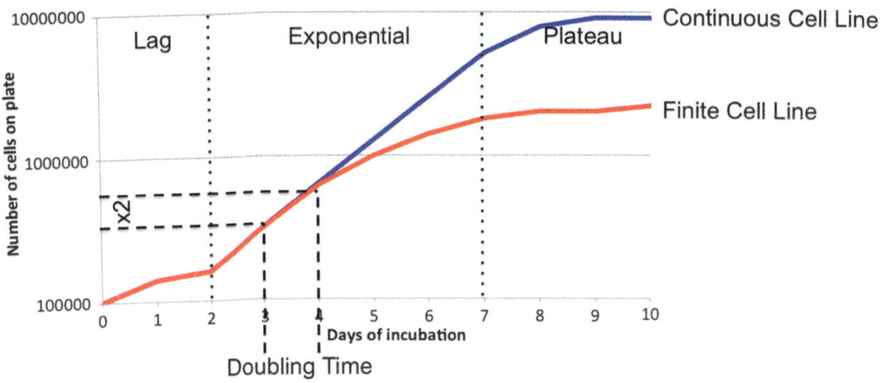

Fig. 5.5 Bacterial growth curve in a continuous process, showing different stages of growth (Source: https://www.google.co.in/imgres)

5.3.2 *Steady State and Equilibrium*

Microorganisms can be grown in a physiological steady state under constant environmental conditions. In this steady state, growth occurs at a constant specific growth rate and all culture parameters remain constant (culture volume, dissolved oxygen concentration, nutrient and product concentrations, pH, cell density, etc.). Thus, if we monitor any variable of a steady-state system, its value is unchanging with time. Batch, fed-batch, and semibatch processes cannot operate under steady-state conditions. The mass of these system either increases or decreases with time. Even though the total mass is constant, changes occurring inside the system cause the system properties to vary with time. Continuous processes may be either steady state or transient. It is usual to run continuous processes as close to steady state as possible; however, unsteady-state conditions exist during start-up and for some time after any change in operating conditions.

5.3.3 *Law of Conservation of Mass*

Mass is conserved in ordinary chemical and physical processes. Consider a system (Fig. 5.6) operating as a continuous process with input and output streams containing glucose. The mass flow rate of glucose into the system is M_{i} kg h^{-1}; the mass flow rate out is M_{o} kg h^{-1}. If M_{i} and M_{o} are different there are four possible explanations:

1. Measurements of M_{i} and M_{o} are wrong
2. The system has a leak, allowing glucose to enter or escape undetected
3. Glucose is consumed or generated by a chemical reaction within the system
4. Glucose accumulates within the system

If we assume that the measurements are correct and there are no leaks, the difference between M_{i} and M_{o} must be due to consumption, generation by reaction, and/or accumulation. A mass balance for the system can be written in a general way to account for these possibilities:

Fig. 5.6 Flow sheet for mass flow of glucose

$$\left\{\begin{array}{l}\text{Mass in through}\\ \text{system boundary}\end{array}\right\} - \left\{\begin{array}{l}\text{Mass out}\\ \text{through system}\\ \text{boundary}\end{array}\right\} + \left\{\begin{array}{l}\text{Mass generated}\\ \text{within system}\end{array}\right\}$$
$$- \left\{\begin{array}{l}\text{Mass consumed}\\ \text{within system}\end{array}\right\} = \left\{\begin{array}{l}\text{Mass}\\ \text{assumulated}\\ \text{within system}\end{array}\right\} \quad (5.1)$$

The accumulation term in the above equation can be either positive or negative; negative accumulation represents depletion of pre-existing reserves. Equation 5.1 is known as the general mass-balance equation. The mass referred to in the equation can be total mass, mass of a particular molecular or atomic species, or biomass.

Example of use of the general mass-balance equation

A 10 gallon fish tank contains 2% salt.

How much dry salt must be added to bring the salt concentration to 3.5% by weight?

Solution:

(1) Given and find statements

Given:

10 gallon fish tank
Initial concentration: 2% salt
Final concentration: 3.5% salt

Find:

Amount of salt to be added

Instructor comment

Do not simply restate or repeat the problem. Strip it down to its essential pieces

(2) Draw a picture

Instructor comment

Pictures are very important, and it is hard to create a good schematic on the first try. Drawing different versions of the schematic is part of the problem-solving process. Here we show the picture in two stages. Only show the final schematic in your solution.

Start by identifying the high level process. There are two inputs, the initial saltwater and the dry salt. There is one output, the final saltwater mixture.

We will add more detail to this schematic as we move through the problem-solving process (Fig. 5.7a).

(3) Assign variables to the unknown

There are two unknowns:

X = mass of dry salt to be added
Y = the total mass of the final mixture

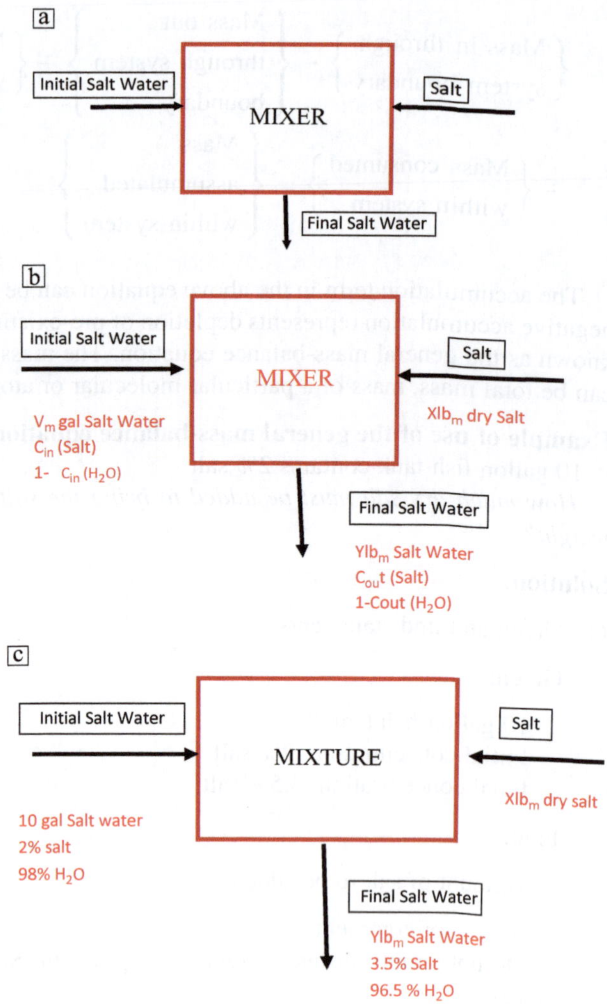

Fig. 5.7 Flow sheet for oxygen enrichment and humidification of air (**a**) Schematic to identify the process. Start by identifying the high level process. There are two inputs: the initial salt water and the dry salt. There is one output: the final salt water mixture. (**b**) Process involvement in second stage showing concentration of salt changes from initial to final state. Here, Cin and Cout to designate the initial and final salt concentrations. The unknowns X and Y, and the known parameters Vin, Cin and Cout are added to the diagram. (**c**) Final process parameters indicating quantity of salt water and substrate (salt) added to the system and final output

At the beginning of a typical engineering problem, the unknowns may not be obvious. Sometimes what appears to be an unknown is just a parameter that determines the outcome, but will be a constant for the given problem.

For example, in the current problem, the mass of the water is not given as an unknown. We could list it as an unknown and then compute the mass of water from the concentration of salt in the initial mixture. However, to avoid making this sample problem too cluttered, we will simply note that the mass of water is a known parameter and let it show up in the analysis/solution of the problem.

There are several known parameters:

V_{in} = volume of water in the initial mixture
C_{in} = concentration of salt in the initial mixture (a known input)

C_{out} = concentration of salt in the final mixture (the desired output)

C is a fractional concentration. The mass percent (or weight percent) is $100 \times C$

(4) Quantify the water in the initial and final mixtures

The mass of the mixture is the sum of the masses of water and salt, given by $m_{H2O} + m_{NaCl}$.

The concentrations of water and salt are:

$$C_{H_2O} = \frac{m_{H_2O}}{m_{NaCl} + m_{H_2O}} \qquad C_{NaCl} = \frac{m_{NaCl}}{m_{NaCl} + m_{H_2O}}$$

Rearrange to obtain the masses of water and salt:

$$m_{H_2O} = C_{H_2O}(m_{NaCl} + m_{H_2O}) \quad m_{NaCl} = C_{NaCl}\,(m_{NaCl} + m_{H_2O})$$

So

$$m_{H_2O} + m_{NaCl} = C_{H_2O}(\,m_{NaCl} + m_{H_2O}) + C_{NaCl}(m_{NaCl} + m_{H_2O})$$

Divide through by $m_{H2O}+m_{NaCl}$:

$$1 = C_{H_2O} + C_{NaCl}$$

This definition allows us to treat C_{NaCl} as the only unknown concentration.

(5) Update the picture

The preceding analysis involves very basic definitions that allow us to identify the inputs and outputs. We know that the concentration of salt changes from initial to final state. We use C_{in} and C_{out} to designate the initial and final salt concentrations.

The unknowns X and Y, and the known parameters V_{in}, C_{in}, and C_{out} are added to the diagram (Fig. 5.7b)

Alternatively, we could use numerical values for the parameters, as in the version of the system diagram (Fig. 5.7a–c). Note that X and Y are still variables (Fig. 5.7c).

(6) Is this a batch or a rate process?

This is a batch process. We have a starting state with a known volume and concentration. We add salt to obtain an end state. The material is not flowing continuously through the system. There are no species created or destroyed (i.e., there are no chemical reactions).

(7) Write the mass balance for the system and each component

The overall mass of the salt and saltwater mixture is unchanged:

$$m_n = m_{out} \tag{1}$$

The initial mass consists of two ingredients: saltwater with a known concentration of salt and an unknown mass (X) of salt. Thus, the left -hand side of Eq. (1) is:

$$m_{\text{in}} = \left(m_{\text{H}_2\text{O}} + m_{\text{Nacl}}\right)_{\text{initial}} + X \tag{2}$$

To calculate the mass of 10 gal of the initial saltwater mixture, assume that the saltwater has the same density as fresh water (1 g/ml, which is 8.33 lb_m/gal).

$$\left(m_{\text{H}_2\text{O}} + m_{\text{NaCl}}\right)_{\text{initial}} = \rho_{\text{H}_2\text{O}} V_{\text{in}} = 83.3\,\text{lb}_\text{m} \tag{3}$$

So Equation (2) becomes:

$$m_n = 83.3\,\text{lb}_\text{m} + X \tag{4}$$

From the schematic and the definition of the unknowns, the overall mass balance is:

$$83.3\ \text{lb}_\text{m} + X = Y \tag{5}$$

Mass balance for salt
The total mass of salt before and after the mixing is unchanged:

$$m_{\text{in}} = m_{\text{out}}$$

For the salt alone this is:

$$C_{\text{in}} m_{\text{in}} + X = C_{\text{out}} Y$$

Or, substituting the known concentrations:

$$(0.02)(83.3\ \text{lb}_\text{m}) + X = 0.035\ Y \tag{6}$$

Mass balance for water
The total mass of water before and after the mixing is unchanged:

$$m_{\text{in}} = m_{\text{out}}$$

For the water alone this is:

$$(1 - C_{\text{in}}) m_{\text{in}} + X = (1 - C_{\text{out}}) Y$$

Or, substituting the known concentrations:

$$(0.98)(83.3\ \text{lb}_\text{m}) = 0.965 Y \tag{7}$$

5.3.4 Types of Material Balance

The general mass-balance equation (Eq. 5.1) is applicable in fermentation processes. For continuous processes at particular instant of time, the amounts of mass entering and leaving the system are specified using flow rates; for example, molasses enters the system at a rate of 25 kg h^{-1} and, at the same instant in time, fermentation broth leaves at a rate of 10 kg h^{-1}. The two quantities can be used directly in Eq. (5.1) as the input and output terms. A mass balance based on rates is called a differential balance.

For batch and semibatch processes, information is collected over a period of time; for example, 100 kg substrate is added to the reactor and, after 3 days of incubation, 45 kg of product is recovered. Each term of the mass-balance equation in this case is a quantity of mass, not a rate. A mass balance based on mass is known as an integral balance.

5.3.5 Simplification of the General Mass-Balance Equation

The general mass-balance equation can be simplified under certain conditions. If a continuous process is in a steady-state condition, then accumulation will be zero:

$$\text{Mass in} + \text{Mass generated} = \text{Mass out} + \text{Mass consumed} \qquad (5.2)$$

This general steady-state mass-balance equation is also applied over the entire duration of batch and fed-batch processes. "Mass out" in this case is the total mass harvested from the system so that at the end of the process there is no accumulation. If reaction does not occur in the system, or if the mass balance is applied to a substance that is neither a reactant nor product of reaction, the generation and consumption terms in Eqs. 5.1 and 5.2 are zero. At steady state, for balances on total mass or atomic species or when reaction does not occur, Eq. 5.2 can be further simplified to:

$$\text{Mass in} = \text{Mass out} \qquad (5.3)$$

5.3.6 Procedure for Material-Balance Calculation

Before proceeding to the material balance in a fermentation process, one should be clear about several aspects of stoichiometry (Table 5.1).

Table 5.1 Some important terminology used in stoichiometry expressions

Name	Unit
Mole ratio	*X* moles A/*Y* moles B
Formula weight	*Z* grams A/mole A
Normal gas volume	24.4 L gas/mole gas

What is a mole ratio? Consider the balanced chemical reaction for the fermentation of glucose (solid) to give carbon dioxide (gas) and ethanol:

$$C_6H_{12}O_6(s) \rightarrow 2CO_2(g) + 2C_2H_5OH$$

The number 2 in front of CO_2 and C_2H_5OH is called the stoichiometric coefficient. The mole ratio is simply the ratio of two stoichiometric coefficients from the same balanced equation. The mole ratios for this equation are:

2 moles CO_2/1 mole $C_6H_{12}O_6$
2 moles C_2H_5OH/1 mole $C_6H_{12}O_6$
2 moles C_2H_5OH/2 moles CO_2

Which ratio is needed depends on the question we wish to answer.

What is the stoichiometric coefficient of glucose, $C_6H_{12}O_6$, in the above equation?

If a number is not explicitly given, it is implicitly understood to have a value of 1.

Four simple problems and their solutions are given below as examples of material balance calculations. But first, some important tips are given for working with stoichiometry problems:

- Always state "moles A," "moles B," and never "moles" alone
- Only cancel "moles A" with "moles A" and not with "moles B"
- Only use 24.4 L mol^{-1} for gases

By paying attention to the following points one can successfully work out stoichiometry problems while calculating material balance:

1. Presenting the data in a simplified manner is the first step in successful solution of a problem related to material balance in a fermentation process. A simple schematic diagram highlighting the process of entering or leaving the system should be well depicted and give all given quantitative information on variables such as masses, mass flow rates, and mass compositions. If information about particular streams is given using volume or molar quantities, then mass flow rates and compositions should be calculated before labeling the flow sheet.
2. The process flow quantities in a system should be expressed using consistent units, as practiced at international level.
3. Select a basis for the calculation and state it clearly. Focus on a specific quantity of material entering or leaving the system. For continuous processes at steady state we usually base the calculation on the amount of material entering or

leaving the system within a specified period of time. For batch or semibatch processes, it is convenient to use either the total amount of material fed to the system or the amount withdrawn at the end.

4. State all assumptions applied to the problem. To solve, it is necessary to apply some "engineering" judgments. Real-life situations are complex, and there will be times when one or more assumptions are required before one can proceed with calculations. The details omitted can be assumed, provided the assumptions are reasonable. Making assumptions when an assumption is permissible and what constitutes a reasonable assumption is one of the marks of a skilled engineer. It is important to state all assumptions exactly.

   Differential mass balances on continuous processes are performed under steady state conditions; we can assume that mass flow rate and compositions do not change with time and the accumulation term of Eq. 5.1 is zero. Another assumption is that the system under investigation does not leak.
5. Identify which components of the system, if any, are involved in the reaction. This is necessary for determining which mass-balance equation (Eq. 5.2) or (Eq. 5.3), is appropriate. The simpler Eq. 5.3 can be applied to molecular species that are neither reactants nor products of reaction.

Example 1
How many moles of glucose are needed to produce 25 mol ethanol?

Solution
As stated above, the mole ratio is 2 mol C_2H_5OH/1 mol $C_6H_{12}O_6$, meaning that 2 mol ethanol are produced for every mole of glucose.

Moles glucose needed:

$$25 \text{ mol ethanol}(1 \text{ mol glucose}/2 \text{ mol ethanol}) = 12.5 \text{ mol glucose}$$

So, to make 25 mol ethanol, 12.5 mol glucose are needed. But how to measure out 12.5 mol glucose?

The conversion factor for moles to grams is called the formula weight. It can be obtained from the atomic weight for each atom in the formula and adding them all up. From the Periodic Table, information on the following atomic weights can be noted:

- Carbon: 12.011 g mol^{-1}
- Hydrogen: 1.008 g mol^{-1}
- Oxygen: 16.000 g mol^{-1}

The number of digits may vary from one table to another depending on the precision used. For our purposes we can round off to integer values. This gives the formula weight of glucose as:

$$6 \times 12 + 12 \times 1 + 6 \times 16 = 180 \text{ g glucose/mole glucose}$$

For carbon dioxide and ethanol, we have:

$$1 \times 12 + 2 \times 16 = 44\,\mathrm{g}\ CO_2/\mathrm{mole}\ CO_2$$

$$2 \times 12 + 6 \times 1 + 1 \times 16 = 46\ \mathrm{g\ ethanol/mole\ ethanol}$$

Example 2
How many grams of glucose are needed to produce 1000 g ethanol?

Solution
Grams glucose needed:

$$\begin{aligned}&1000\ \mathrm{g\ ethanol}\ (1\ \mathrm{mol\ ethanol}/46\ \mathrm{g\ ethanol})(1\ \mathrm{mol\ glucose}/2\ \mathrm{mol\ ethanol})\\&\times(180\ \mathrm{g\ glucose}/1\ \mathrm{mol\ glucose})\\&= 1000 \times (1 \times 1 \times 180)/(46 \times 2 \times 1)\ \mathrm{g\ glucose}\\&= 1956\ \mathrm{g}\ \ \mathrm{glucose}\end{aligned}$$

Example 3
What volume of carbon dioxide is produced when 100 g glucose is fermented to ethanol?

Solution
According to the ideal gas equation, under normal conditions (25 °C and normal atmospheric pressure) 1 mol of gas occupies approximately 24.4 L, no matter what the identity of the gas is. The volume mainly depends on temperature and pressure.

Liters carbon dioxide produced:

$$\begin{aligned}&100\,\mathrm{g\,glucose}\,(1\,\mathrm{mol\,glucose}/180\,\mathrm{g\,glucose})(2\,\mathrm{mol\,carbon\,dioxide}/1\,\mathrm{mol\,glucose})\\&\times(24.4\,\mathrm{L\,carbon\,dioxide}/1\,\mathrm{mol\,carbon\,dioxide})\\&=100(1\times2\times24.4)/(180\times1\times1)\\&=27\,\mathrm{L\,carbon\,dioxide}\end{aligned}$$

Example 4: Setting up a flow sheet
Material balances can be based on total mass, mass of dry solids, or mass of particular components (e.g., protein). Skim milk is prepared by the removal of some of the fat from whole milk. This skim milk contains 90.5% water, 3.5% protein, 5.1% carbohydrate, 0.1% fat, and 0.8% ash.

If the original milk contained 4.5% fat, calculate its composition assuming that only fat was removed to make the skim milk and that there were no losses in processing.

Solution:
100 kg of skim milk contains 0.1 kg of fat. Let the mass of fat that was removed to make skim milk be x kg.

Total original fat = $(x + 0.1)$ kg
Total original mass = $(100 + x)$ kg

The original fat content was 4.5%, so:

$(x + 0.1)/(100 + x) = 0.045$

where

$x + 0.1 = 0.045(100 + x)$

Therefore

$x = 4.6$ kg and total original mass is 104.6 kg.

Calculation of the composition of the whole milk gives fat = 4.5%, water = 90.5/104.6 = 86.5%, protein = 3.5/104.6 = 3.3%, carbohydrate = 5.1/104.6 = 4.9%, and ash = 0.8%

Material-Balance Worked Examples

Example 5: Batch mixing
A solution of common salt in water is prepared by adding 20 kg of salt to 100 kg of water, giving a liquid of density 1323 kg/m^3.

Calculate the concentration of salt in this solution as (a) weight fraction, (b) weight/volume fraction, (c) mole fraction, and (d) molar concentration.

Solution:

(a) Weight fraction

20/(100 + 20) = 0.167 or 16.7% (w/w)

(b) Weight/volume

A density of 1323 kg/m^3 means that 1 m^3 of solution weighs 1323 kg.
1323 kg of salt solution contains (20 × 1323 kg of salt)/(100 + 20) = 220.5 kg salt/m^3. In other words, 1 m^3 solution contains 220.5 kg salt.
Weight/volume fraction = 220.5/1000 = 0.2205 or 22.1% (w/v)

(c) Mole fraction

Moles of water in 100 kg water = 100/18 = 5.56
Moles of salt in 20 kg salt = 20/58.5 = 0.34
Mole fraction of salt = 0.34/(5.56 + 0.34) = 0.058

(d) The molar concentration is 220.5/58.5 = 3.77 mol/m^3

Note that the mole fraction can be approximated by the value of moles of salt/moles of water because of the excess number of moles of water; that is, the mole fraction is close to 0.34/5.56 = 0.061. As the solution becomes more dilute, this approximation improves and, for dilute solutions, the mole fraction of solute closely approximates the moles of solute/moles of solvent.

In solid–liquid mixtures, all these methods can be used. In solid mixtures, concentrations are normally expressed as simple weight fractions (w/w). For gases, concentrations are mostly measured in weight concentrations per unit volume (w/v), or as partial pressures. These can be related through the gas laws, using the form:

$$pV = nRT$$

where p is the pressure, V the volume, n the number of moles, T the absolute temperature, and R the gas constant, which is equal to 0.08206 m^3 atm/mol K.

The molar concentration of a gas is give by $n/V = p/RT$.

The weight concentration is then nM/V, where M is the molecular weight of the gas.

The SI unit of pressure is pascal (Pa), which is defined as one newton per square meter (N/m^2). This is an inconvenient size for many purposes, so standard atmospheres (atm) are often used as units of pressure. The conversion is 1 atm = 1.013 × 10^5 Pa or, very nearly, 1 atm = 100 kPa.

Example 6: Continuous fermentation

Butyric acid is an organic acid produced via fermentation and is used in perfumes, as a food additive, and as an intermediate in alternative fuels.

Clostridium tyrobutyricum can produce several acid species, including lactate, acetate, and butyrate. All of these liquid products, as well as the gaseous by-products CO_2 and H_2, are possible end products of C. *tyrobutyrium* fermentation under various culture conditions.

Develop a suitable model to (a) quantify butyric acid, (b)understand the effect of different media on butyric acid production, and (c) compare the growth rates of batch and continous systems.

Solution:

To understand batch and continuous fermentation systems, a mass balance accounting for the specific growth rate, μ, and cell retention time, σ, was derived. For batch fermentation, the specific growth rate can be approximated as $\frac{\Delta \times X}{\Delta t}/X$ and the productivity of butyric acid as $\frac{\Delta C_b}{\Delta t}$. For continuous fermentation, as shown as Fig. 5.8, a material balance for cell mass yields:

$$FX_0 + V\mu X_1 - F_1X_1 - F_2X_2 = V - \frac{dX_1}{dt}$$

$$F = F_1 + F_2$$

$$F_2 = RF$$

In this model (Fig. 5.8), F is the feed flow rate, F_1 is the flow rate of the purge, and F_2 is the flow rate of the permeate from the UF unit. Parameter X indicates the

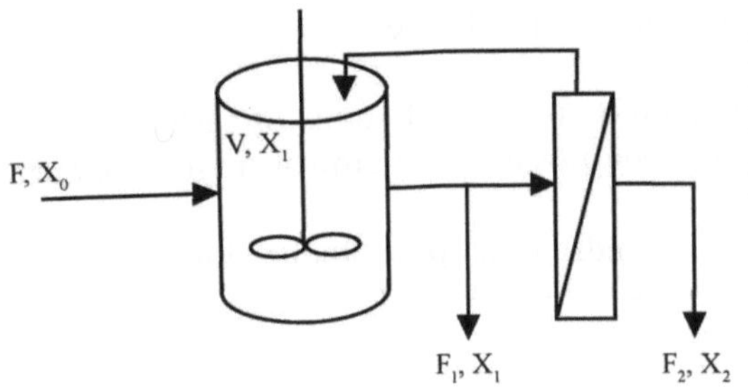

F: feed flow
F_1: flowrate of reactor purge
F_2: flowrate of ultrafiltration permeate
X_0: cell concentration in feed
X_1: cell concentration in reactor
X_2: cell concentration in permeate
V: reactor volume

Fig. 5.8 Schematic diagram for mass balance of continuous fermentation with cell recycle by C. *tyrobutyricum*

cell mass present in the same positions. When the continuous fermentation reaches steady state, $\frac{dX_1}{dt} = 0$ and the specific growth rate is given by:

$$\mu = (C_1 - R)\frac{F}{V} = (1 - R)D$$

The cell retention time for continuous fermentation is defined as:

$$\sigma = \frac{1}{D} = 1/\mu$$

The cell retention time for continuous fermentation with cell recycle is defined as:

$$\sigma = \frac{V}{F_1} = \frac{V}{(1 - R)F} = 1/\mu$$

These equations allow comparison of the growth rate of batch and continuous systems. This model is not media-independent. The medium used affects both the cell growth rate and the quantity of butyric acid produced. Different media produce different effects. By adapting this model, we can specifically find the weight percentage (g/g) of butyric acid produced compared with the total organic acids produced.

5.3.7 *Stoichiometry of Growth and Production*

For mass balances with reaction, the stoichiometry of conversion must be known before the mass balance can be solved. Under growth conditions, cells are a product of reaction and must be represented in the reaction equation. Metabolic stoichiometry has many applications in bioprocessing in mass and energy balances. It can be used to compare theoretical and actual product yields.

5.3.7.1 Growth Stoichiometry and Elemental Balance

Cell growth obeys the law of conservation of matter. All atoms of carbon, hydrogen, oxygen, nitrogen and other elements consumed during growth are incorporated into new cells or excreted as products.

Confining our attention to those compounds taken up or produced in significant quantity, if the only extracellular products formed are CO_2 and H_2O and we can write the following equation for aerobic cell growth:

$$C_wH_xO_yN_z + aO_2 + bH_gO_hN_i \rightarrow cCH_\alpha O_\beta N_\delta + dCO_2 + eH_2O \tag{5.4}$$

As illustrated in Fig. 5.9, the equation represents a macroscopic view of metabolism. It ignores the detailed structure of the system and considers only those components that have net interchange with the environment.

Compounds such as vitamins and minerals taken up during metabolism could be included; however, because these growth factors are generally consumed in small quantities we assume here that their contribution to the stoichiometry and energetics of the reaction can be neglected. Other substrates and products can easily be added if appropriate.

Bacteria tend to have slightly higher nitrogen contents (11–14%) than fungi (6.3–9.0%). Table 5.2 shows the elemental composition of *Escherichia coli*. For a particular species, cell composition also depends on culture conditions and substrate utilized, hence the different entries in Table 5.3 for the same organism. $CH_{1.8}O_{0.5}N_{0.2}$ can be used as a general formula when composition analysis is not available. The average molecular weight of biomass based on C, H, O, N content is 24.6. Residual ash (5–10%) is often added to account for those elements not included in the formula.

Equation 5.4 is not complete unless the stoichiometric coefficients a, b, c, d, and e are known. Once a formula for biomass is obtained, these coefficients can be evaluated using normal procedures for balancing equations (i.e., elemental balances and solutions of simultaneous equations).

$$\text{C balance}: \quad \omega = c + d \tag{5.5}$$

$$\text{H balance}: \quad x = bg + = c\alpha + 2\,e \tag{5.6}$$

$$\text{O balance}: \quad y + 2a + bh = c\beta + 2d + e \tag{5.7}$$

$$\text{N balance}: \quad z + bi = c\delta \tag{5.8}$$

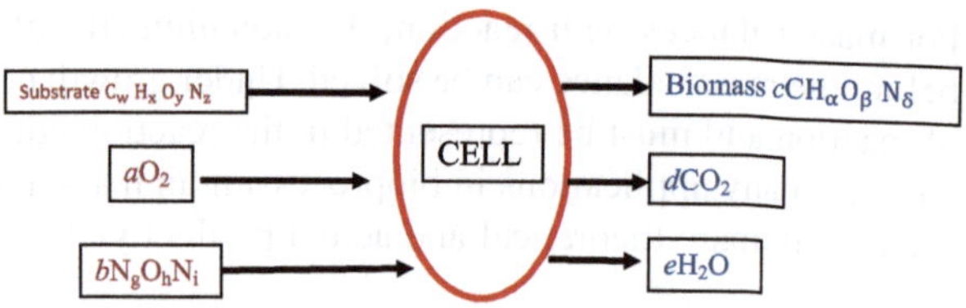

Fig. 5.9 Conversion of substrate, oxygen, and nitrogen for cell growth

Table 5.2 Elemental composition of *Escherichia coli* bacteria (From [41])

C	50
O	20
N	14
H	8
P	3
S	1
K	1
Na	1
Ca	0.5
Mg	0.5
Cl	0.5
Fe	0.2
Others	0.3

Table 5.3 Elemental composition and degree of reduction for selected organisms [42]

Organism	Elemental formula	Degree of reduction related to NH_3
Escherichia coli	$CH_{1.77}O_{0.49}N_{0.24}$	4.07
Klebsiella aerogens	$CH_{1.75}O_{0.43}N_{0.22}$	4.23
Klebsiella aerogens	$CH_{1.73}O_{0.43}N_{0.22}$	4.15
Klebsiella aerogens	$CH_{1.75}O_{0.47}N_{0.17}$	4.30
Klebsiella aerogens	$CH_{1.73}O_{0.43}N_{0.24}$	4.15
Pseudomonas C_{12} B	$CH_{2.00}O_{0.52}N_{0.23}$	4.27
Aerobacter rerogenes	$CH_{1.83}O_{0.55}N_{0.25}$	3.98
Paracoccus denitrificance	$CH_{1.81}O_{0.51}N_{0.20}$	4.19
Paracoccus denitrificance	$CH_{1.51}O_{0.46}N_{0.19}$	3.96
Saccharomyces cerevisiae	$CH_{1.64}O_{0.52}N_{0.16}$	4.12
Saccharomyces cerevisiae	$CH_{1.83}O_{0.56}N_{0.17}$	4.20
Saccharomyces cerevisiae	$CH_{1.81}O_{0.51}N_{0.17}$	4.28
Candida utilis	$CH_{1.83}O_{0.54}N_{0.10}$	4.45
Candida utilis	$CH_{1,81}O_{0.56}N_{0.20}$	4.15
Candida utilis	$CH_{1.83}O_{0.46}N_{0.19}$	4.34
Candida utilis	$CH_{1.82}O_{0.56}N_{0.20}$	4.15
Average	$CH_{1.79}O_{0.50}N_{0.20}$	4.19

Notice that we have five unknown coefficients (a, b, c, d, and e) but only four balance equations. This means that additional information is required before the equations can be solved. Usually this information is obtained from experiments. A useful measurable parameter is the respiratory quotient (RQ):

$$\text{Respiratory quotient} = \frac{\text{moles } C_2 \text{ produced}}{\text{moles } O_2 \text{ consumed}} = \frac{d}{a} \tag{5.9}$$

We must solve this set of simultaneous equations. A solution can be achieved in many different ways. Usually it is a good idea to express each variable as a function of only one other variable. Variable b is already written simply as a function of c in (4); let us try expressing the other variables solely in terms of c. From (1):

$$D = 16 - c \quad (6)$$

From (5):

$$\alpha = \frac{d}{0.43} = 2.326d \quad (7)$$

Combining (6) and (7) gives an expression for α in terms of c only:

$$\alpha = 2.326\,(16 - c)$$

$$\alpha = 37.22 - 2.326c$$

Substituting (4) into (2) gives:

$$34 + 3(0.20c) = 1.66\mathrm{c} + 2e$$
$$34 = 1.06c + 2e$$
$$e = 17 - 0.53c$$

Substituting (8), (6) and (9) into (3):

$$2(37.22 - 2.326c) = 0.27c + 2(16 - c) + (17 - 0.53c)$$
$$25.44 = 2.39c$$
$$c = 10.64$$

Using this result for c in (8), (4), (6) and (9) gives:

$$a = 12.48$$

$$b = 2.13$$

$$d = 5.37$$

$$e = 11.36$$

Check that these coefficient values satisfy (1)–(5).
The complete reaction is therefore:

$$C_{16}H_{34} + 12.5O_2 + 2.13NH_3 \rightarrow 10.6CH_{1.66}O_{0.27}N_{0.20} + 5.37CO_2 + 11.4H_2O$$

Although elemental balances are useful, the presence of water in Eq. (5.4) causes some problems in practical application. Because water is usually present in great excess and changes in water concentration are inconvenient to measure or experimentally verify, H and O balances can present difficulties. Instead, a useful principle is conservation of reducing power or available electrons, which can be applied to determine quantitative relationships between substrates and products. An electron balance shows how available electrons from the substrate are distributed in reaction.

5.3.7.2 Electron Balance

Available electrons are the electrons available for transfer to oxygen on combustion of a substance to CO_2, H_2O, and nitrogen-containing compounds. In organic material, it is calculated from the valence of the various elements: 4 for C, 1 for H, −2 for O, 5 for P, and 6 for S. For N, it depends on the reference state: −3 if ammonia is the reference, 0 for molecular nitrogen N_2, and 5 for nitrate. The reference state for cell growth is usually chosen to be the same as the nitrogen source in the medium. It is assumed for convenience that ammonia is used as nitrogen source; this can easily be changed if other nitrogen sources are employed. The degree of reduction, γ is the number of equivalents of available electrons in the quantity of material containing 1 g of atomic carbon. Therefore, for substrate $C_wH_xO_yN_z$, the number of available electrons is $(4w + x - 2y - 3z)$.

The degree of reduction for a substrate, γ_s, is therefore $(4w + x - 2y - 3z)/w$. The degree of reduction for CO_2, H_2O, and NH_3 is zero. Electrons available for transfer to oxygen are conserved during metabolism. In a balanced growth equation, the number of available electrons is conserved by virtue of the fact that the amounts of each chemical element are conserved.

Applying this principle to Eq. (5.4) with ammonia as nitrogen source, the available-electron balance is:

$$\omega\gamma_s - 4\alpha = c\gamma_B \tag{5.10}$$

Where γ_s and γ_B are the degrees of reduction of substrate and biomass, respectively.

Note that the available-electron balance is not independent of the complete set of elemental balances. If the stoichiometric equation is balanced in terms of each element, including H and O, the electron balance is implicitly satisfied.

5.3.7.3 Biomass Yield

Typically, Eq. (5.10) is used for carbon and nitrogen balances, and Eqs. (5.5) and (5.8) and a measured value of RQ for evaluation of stoichiometric coefficients. During cell growth there is a linear relationship between the amount of biomass produced and the amount of substrate consumed. This can be expressed quantitatively as biomass yield, Y_{xs}.

Factors influencing biomass yield (Y) are medium composition, nature of the carbon and nitrogen sources, pH, and temperature. Biomass yield is greater in aerobic than in anaerobic cultures. The choice of electron acceptor (e.g., O_2, nitrate, or sulfate) can also have a significant effect. When Y_{xs} is constant throughout growth, its experimentally determined value can be used to determine the stoichiometric coefficient c in Eq. (5.4). Equation (5.11) expressed in terms of the stoichiometric Eq. (5.4) is as given in the Eq. 5.12

$$Y_{xs} = \frac{c\ (\text{MW cell})}{(\text{MW substrate})} \tag{5.12}$$

where MW is molecular weight and "MW cell" means the biomass formula-weight plus any residual ash.

Before applying measured values of Y_{xs} and Eq. (5.12) to evaluate c, we must be sure that the experimental culture system is well represented by the stoichiometric equation. (i.e., we must be sure that substrate is not used to synthesize extracellular products other than CO_2 and H_2O).

One complication with real cultures is that some fraction of substrate consumed is always used for maintenance activities such as maintenance of membrane potential and internal pH, turnover of cellular components and cell motility. These metabolic functions require substrate but do not necessarily produce cell biomass, CO_2, and H_2O in the way described by Eq. (5.4). It is important to account for maintenance when experimental information is used to complete stoichiometric equations. We assume that available values for biomass yield reflect substrate consumption for growth only.

5.3.7.4 Product Stoichiometry

Consider formation of an extracellular product $C_jH_kO_lN_m$ during growth. Equation (5.4) can be extended to include product synthesis as follows:

$$\begin{aligned} C_wH_xO_yN_z + aO_2 + bH_gO_hN_i \rightarrow\ & cCH_\alpha O_\beta N_\delta + dCO_2 + eH_2O \\ & + fC_jH_kO_lN_m \end{aligned} \tag{5.13}$$

where f is the stoichiometric coefficient for product.

Product synthesis introduces one extra unknown stoichiometric coefficient into the equation; thus, an additional relationship between coefficients is required. This is usually provided as another experimentally determined yield coefficient, the product yield from substrate:

$$Y_{ps} = \frac{g\ \text{product formed}}{g\ \text{substrate consumed}} = \frac{f(\text{MW product})}{(\text{MW substrate})}. \tag{5.14}$$

As mentioned above with regard to biomass yields, we must be sure that the experimental system used to measure Y_{ps} conforms to Eq. (5.13). Equation (5.13) does not hold if product formation is not directly linked with growth; accordingly it cannot be applied for secondary-metabolite production or for biotransformations such as steroid hydroxylation. In these cases, independent reaction equations must be used to describe growth and product synthesis.

5.3.8 *Theoretical Oxygen Demand*

Oxygen is often the limiting substrate in aerobic fermentations and is represented by the stoichiometric coefficient *a* in Eqs. (5.4) and (5.13), which are repeated here for convenience:

$$C_wH_xO_yN_z + aO_2 + bH_gO_hN_i \rightarrow cCH_\alpha O_\beta N_\delta + dCO_2 + eH_2O \quad (5.4)$$

$$C_wH_xO_yN_z + aO_2 + bH_gO_hN_i \rightarrow cCH_\alpha O_\beta N_\delta + dCO_2 + eH_2O + fC_jH_kO_iN_m \quad (5.13)$$

Oxygen requirement is related directly to the electrons available for transfer to oxygen. Oxygen demand can be derived from an appropriate electron balance. When product synthesis occurs as represented by Eq. (5.13), the electron balance is:

$$\omega\gamma_s - 4\alpha = c\gamma_B + fj\gamma_p \quad (5.15)$$

where γ_p is the degree of reduction of the product. Rearranging gives:

$$\alpha = 1/4(\omega\gamma_s - c\gamma_B - fj\gamma_p). \quad (5.16)$$

Equation (5.16) means that if we know which organism (Y_B), substrate (w and Y_s), and product (j and Y_P,) are involved in cell culture, and the yields of biomass (c) and product (f), we can quickly calculate the oxygen demand. We can also determine *a* by solving for all the stoichiometric coefficients of Eq. (5.13). Equation (5.16) allows more rapid evaluation and does not require that the quantities of NH_3, CO_2, and H_2O involved in the reaction be known.

5.3.9 *Maximum Possible Yield*

From Eq. (5.15), the fractional allocation of available electrons in the substrate can be written as:

$$1 = \frac{4\alpha}{\omega\gamma_s} + \frac{c\,\gamma_B}{\omega\gamma_s} + \frac{fj\gamma_p}{\omega\gamma_s} \quad (5.17)$$

where $\frac{4\alpha}{\omega\gamma_s}$ is the fraction of available electrons transferred from substrate to oxygen, $\frac{c\gamma_B}{w\gamma_s}$ is the fraction of available electrons transferred to biomass, and $\frac{fj\gamma_p}{\omega\gamma_s}$ is the fraction of available electrons transferred to product.

This relationship can be used to obtain upper bounds for the yields of biomass and product from substrate. Let us define ζ_B as the fraction of available electrons in the substrate transferred to biomass:

$$\zeta_B = \frac{c\gamma_B}{\omega\gamma_s} \tag{5.18}$$

In the absence of product formation, if all available electrons were used for biomass synthesis, ζ_B would equal unity. Under these conditions, the maximum value of the stoichiometric coefficient c is:

$$c_{max} = \frac{\omega\gamma_s}{\gamma_B} \tag{5.19}$$

The coefficient c_{max} can be converted to a biomass yield with mass units using Eq. (5.12).

If we do not know the stoichiometry of growth, we can quickly calculate an upper limit for biomass yield from the molecular formulae for substrate and product. If the composition of the cells is unknown, Y_B can be taken as 5.2, corresponding to the average biomass formula $CH_{1.8}O_{0.5}N_{0.2}$.

Maximum biomass yields for several substrates are listed in Table 5.4. Maximum biomass yield can be expressed in terms of mass ($Y_{xs,max}$) or as the number of C atoms in the biomass per substrate C atom consumed ($C_{max/w}$). These quantities are sometimes known as thermodynamic maximum biomass yields. Table 5.4 shows that substrates with high energy content, indicated by high Y_s values, give high maximum biomass yields. The maximum possible product yield in the absence of biomass synthesis can be determined from Eq. (5.17):

$$f_{max} = \frac{\omega\gamma_s}{j\gamma_p} \tag{5.20}$$

Equation (5.20) allows us to quickly calculate an upper limit for product yield from the molecular formulae for substrate and product.

Example 7: Product yield and oxygen demand

The chemical reaction equation for respiration of glucose is:

Candida utilis cells convert glucose to CO_2 and H_2O during growth. The cell composition is $CH_{1.84}O_{0.55}N_{0.2}$ plus 5% ash. Yield of biomass from substrate is 0.5 g g^{-1}. Ammonia is used as nitrogen source.

(a) *What is the oxygen demand with growth compared to that without?*

(b) *C. utilisis also able to grow with ethanol as substrate, producing cells of the same composition as above. On a mass basis, how does the maximum possible biomass yield from ethanol compare with the maximum possible yield from glucose?*

Table 5.4 Thermodynamic maximum biomass yields (adapted from [42])

Substrate	Formula	Y_s	Thermodynamic maximum yield corresponding to $\zeta_B = 1$	
			Carbon yield $C_{max/w}$	Mass yield $Y_{xs.max}$
Alkane				
Methane	CH_4	8.0	1.9	2.9
Hexane (*n*)	C_6H_{14}	6.3	1.5	2.6
Hexadecane (*n*)	$C_{16}H_{34}$	6.1	1.5	2.5
Alcohols				
Methanol	CH_4O	6.0	1.4	1.1
Ethanol	C_2H_6O	6.0	1.4	1.5
Ethylene glycol	$C_2H_6O_2$	5.0	12	0.9
Glycerol	$C_3H_6O_3$	4.7	1.1	0.9
Carbohydrate				
Formaldehyde	CH_2O	4.0	0.95	6.8
Glucose	$C_6H_{12}O_6$	4.0	0.95	6.8
Sucrose	$C_{12}H_{22}O_{11}$	4.0	0.95	0.8
Starch	$(C_6H_{10}O_3)_x$	4.0	0.95	0.9
Organic acid				
Formic acid	CH_2O_2	2.0	0.5	0.3
Acetic acid	$C_2H_4O_2$	4.0	0.95	0.8
Propionic acid	$C_3H_6O_2$	4.7	1.1	1.1
Lactic acid	$C_3H_6O_3$	4.0	0.95	0.8
Fumaric acid	$C_4H_4O_4$	3.0	0.7	0.6
Oxalic acid	$C_2H_2O_4$	1.0	0.24	0.1

Solution

The molecular weight of glucose is 180 and that of ethanol 46. The total molecular weight of biomass is (25.44 + ash). Because ash accounts for 5% of the total weight, 95% of the total is 25.44, therefore:

$$\text{MW biomass} = 25.44/0.95 = 26.78$$

From Table 5.4, Y_s for glucose is 4.0 and Y_s for ethanol is 6.0, giving:
$Y_\beta = (4 \times 1 + 1 \times 1.84 - 2 \times 0.55 - 3 \times 0.2) = 4.14.$
For glucose $w = 6$ and for ethanol $w = 2$, therefore:
$Y_{xs} = 0.5 \text{ g g}^{-1}.$
Converting this mass yield to a molar yield:

$$Y_{xs} = \frac{0.5 \text{ g biomass}}{\text{g glucose}} \times \left|\frac{180 \text{ g glucose}}{1 \text{ gmol glucose}}\right| - \left|\frac{1 \text{ gmol biomass}}{16.78 \text{ g biomass}}\right|$$

$$Y_{xs} = 3.36 \frac{\text{gmol biomass}}{\text{gmol glucose}} = c.$$

Oxygen demand is given by Eq. (5.16). In the absence of product formation: $a = 1/4\,[6(4.00) - 3.36(4.14)] = 2.52$.

Therefore, to answer question (a), the oxygen demand for glucose respiration with growth is 2.5 gmol O_2 per gmol glucose consumed. By comparison with the chemical reaction equation for respiration, this is only about 42% of that required in the absence of growth.

To answer question (b), the maximum possible biomass yield is given by Eq. (5.19). Using the data above for glucose and converting this to a mass basis:

$$C_{max} = \frac{6(4.00)}{4.14} = 5.80$$

$$Y_{xs,max} = \frac{5.80\ \text{g biomass}}{\text{gmol glucose}} \times \left|\frac{1\ \text{gmol glucose}}{180\ \text{g glucose}}\right| - \left|\frac{26.78\ \text{g biomass}}{1\ \text{gmol biomass}}\right|$$

$$Y_{xs,max} = 0.86 \times \frac{\text{g biomass}}{\text{g glucose}}$$

For ethanol:

$$C_{max} = \frac{2(6.00)}{4.14} = 2.90$$

$$Y_{xs.max} = \frac{2.90\ \text{gmol biomass}}{\text{gmol ethanol}} \times \left|\frac{1\ \text{gmol ethanol}}{46\ \text{g ethanl}}\right| - \left|\frac{26.78\ \text{g biomass}}{1\ \text{gmol biomass}}\right|$$

$$Y_{xs,max} = 1.69 \times \frac{\text{g biomass}}{\text{g ethanol}}.$$

Therefore, on a mass basis, the maximum possible amount of biomass produced per gram ethanol consumed is roughly twice that per gram glucose consumed. This result is consistent with the data in Table 5.4.

5.3.10 Material Balance With Recycle, By-Pass, and Purge Systems

The material balances for simple single-unit processes have been demonstrated. Steady-state systems incorporating recycle, by-pass, and purge streams are common in bioprocesses [43]. Flow sheets illustrating these modes of operation are given in Fig. 5.10. Several balances are required before all mass flows can be determined. In fermentation processes, it is better to recycle biomass from spent

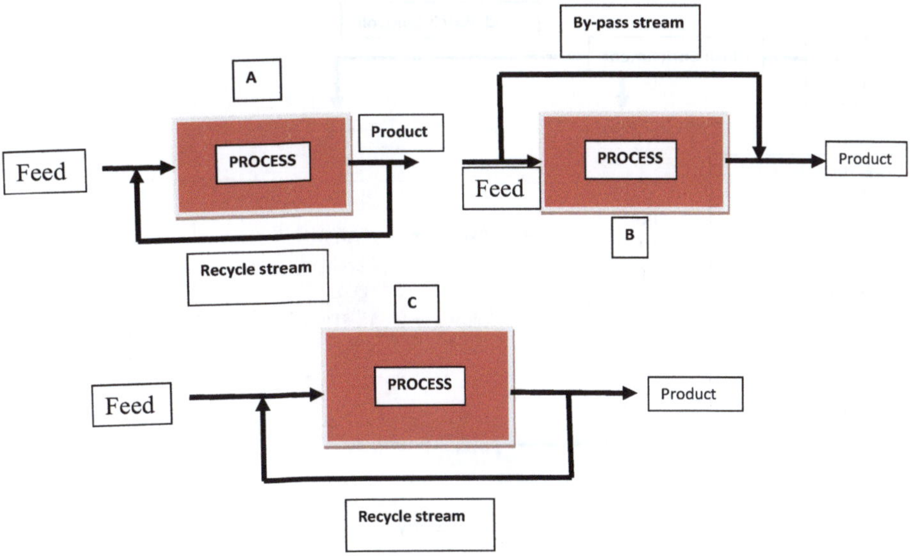

Fig. 5.10 Flow sheet for processes with (*A*) recycle, (*B*) by-pass, and (*C*) purge streams

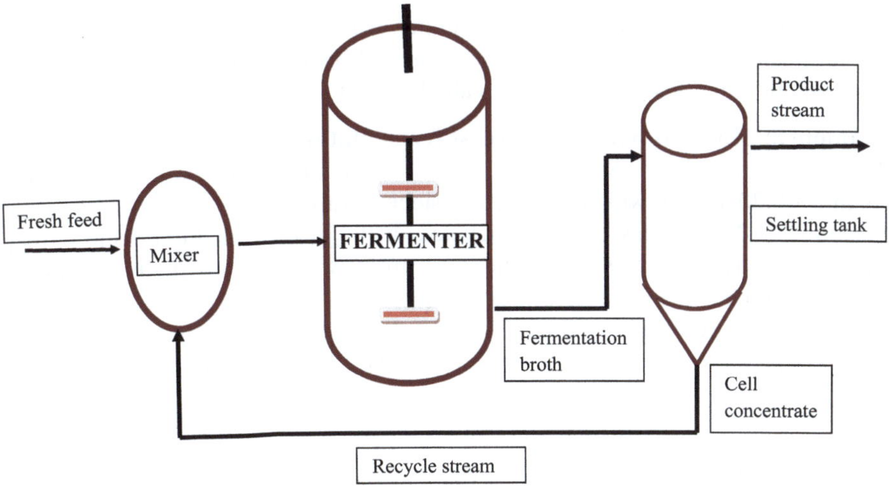

Fig. 5.11 Fermenter with cell recycling

fermentation broth. Cell recycling requires a separation device, such as a centrifuge or gravity settling tank (Fig. 5.11).

Figure 5.12 shows how at least four different system boundaries can be defined. System I represents the overall recycle process; only the fresh feed and final product streams cross this system boundary. In addition, separate material balances can be performed for each process unit (mixer, fermenter and settler). Other system

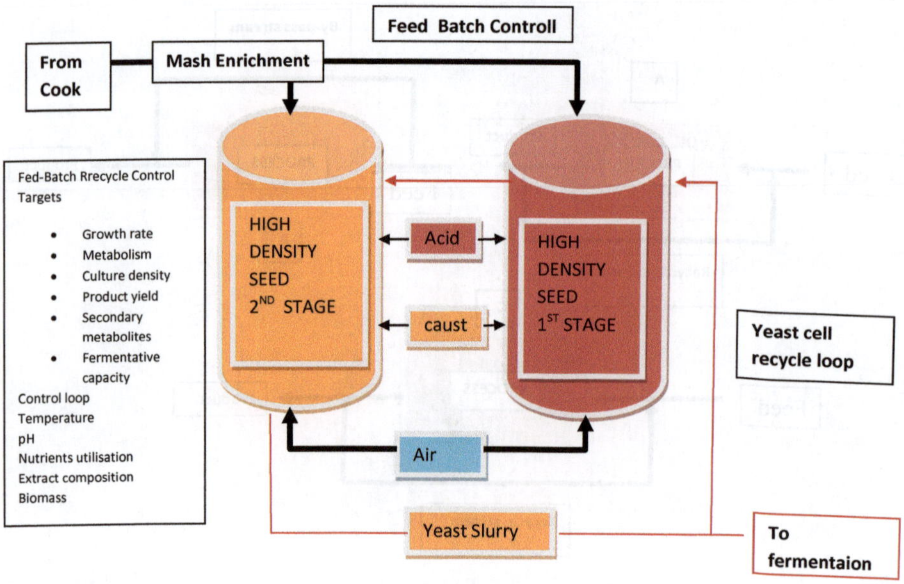

Fig. 5.12 High density yeast cultivation

boundaries can also be defined. For example, we could group together the mixer and fermenter, or settler and fermenter. Material balances with recycling involve carrying out individual mass-balance calculations for each designated system.

Example 8: Material balance for yeast fermentation

Baker's yeast is to be grown in a continuous fermentation system using a fermenter volume of 20 m^3 in which the flow residence time is 16 h. A 2% inoculum containing 1.2% of yeast cells is included in the growth medium. This is then passed to the fermenter, in which the yeast grows with a steady doubling time of 2.9 h. The broth leaving the fermenter then passes to a continuous centrifuge, which produces a yeast cream containing 7% yeast (97% of total yeast in the broth).

Calculate the rate of flow of the yeast cream and of the residual broth from the centrifuge.

Solution:

The volume of the fermenter is 20 m^3 and the residence time in this is 16 h, so the flow rate through the fermenter must be $20/16 = 1.25\ m^3\ h^{-1}$. Assuming the broth to have a density substantially equal to that of water (i.e., 1000 kg m^{-3}), the mass flow rate of broth is 1250 kg h^{-1}.

Yeast concentration in the liquid flowing to the fermenter can be given by:

(concentration in inoculum)/(dilution of inoculum) $= (1.2/100)/(100/2) = 2.4 \times 10^{-4}$ kg kg^{-1}.

The yeast mass doubles every 2.9 h, so in 2.9 h, 1 kg becomes 1×2^1 kg (one generation). In 16 h there are $16/2.9 = 5.6$ doubling times (generations) so 1 kg yeast grows to $1 \times 2^{5.6}$ kg $= 48.5$ kg.

Table 5.5 Material balance over the centrifuge hour

	Mass in (kg)		Mass out (kg)
Yeast-free broth	1235	Broth	1042
Yeast	15	Yeast in broth	0.45
		Yeast stream	208
		Yeast in stream	14.55
Total	1250	Total	1250

Yeast in broth flowing to fermenter $= 48.5 \times 2.4 \times 10^{-4}$ kg kg^{-1}.

Yeast leaving fermenter = initial concentration × growth × flow rate
$= 2.4 \times 10^{-4} \times 48.5 \times 1250$
$= 15$ kg h^{-1}.

Yeast-free broth flow leaving fermenter $=(1250 - 15) = 1235$ kg h^{-1}.

From the centrifuge flows a (yeast-rich) stream with 7% yeast, this being 97% of the total yeast. The yeast rich stream flow is $(15 \times 0.97) \times 100/7 = 208$ kg h^{-1} and the broth (yeast-lean) stream flow is $(1250 - 208) = 1042$ kg h^{-1}, which contains $(15 \times 0.03) = 0.45$ kg h^{-1} yeast. The yeast concentration in the residual broth is $0.45/1042 = 0.043\%$ per material balance over the centrifuge hour (see Table 5.5).

A materials balance, such as in this example for the manufacture of yeast, can be prepared in much greater detail if necessary and if the appropriate information is available. Not only broad constituents, such as the yeast, can be balanced but all other constituents must also balance.

5.3.11 Kinetic Study of Microbial Growth

Various structured and unstructured kinetic models have been reported in the scientific literature for fermentative production of acetic acid by bacteria. Unstructured, non-segregated kinetic models play an important role in monitoring and predicting the batch fermentation process [44–48]. Unstructured models are much easier to use and have been proven for the description of a wide range of experimental conditions and media. Kinetic and fed-batch models are described for the example below.

Example 9: Fed-batch process for ethanol production and cell recycling
The substrate used for alcoholic fermentation was formulated with only sugarcane juice with the following contents per liter: sucrose 133.01 g, glucose 16.79 g, and fructose 14.85 g, or in terms of total reducing sugar (TRS) 171.65 g. This is a typical feedstock for pilot-level ethanol production. The batch feeding was performed in 4 h (flow of 6.25 mL min^{-1}) up to the final volume of 1.5 L and was maintained for a further 2 h to ensure the uptake of accumulated sugar, even though all sugar was consumed. Temperature was maintained at 33 °C and pH at about 5. The

Table 5.6 Initial values and operational conditions for experiments on fed-batch fermentation for ethanol production with cell recycling

Initial values	Fermentation 1	Fermentation 2	Fermentation 3
X_i (kg/m^3)	40.55	44.83	44.65
S_i (kg/m^3)	17.95	15.52	4.22
P_i (kg/m^3)	9.84	14.56	11.13
Operational conditions	–	–	–
V_i (m^3)	0.5	0.5	0.5
S_A (kg/m^3)	171.7	171.7	171.7
F_A (m^3/h)	0.43	0.22	0.23
t_F (h)	2	3	3

fermented wine was centrifuged at 8000 rpm for 20 min in a Sorvall centrifuge and then the yeast was suspended with sterilized water and centrifuged again under the same conditions.

The centrifuged yeast biomass was carried back to the bioreactor for treatment with H_2SO_4 at pH 3.0 and aeration for 1 h. This treatment was performed before each fermentative cycle during yeast cell recycling. The fermentative cycle comprises fed-batch fermentation, cell treatment, and recycling. Two fermentative cycles were performed, that is, three fed-batch fermentation experiments, and two cell treatments and recycling.

Work out a spreadsheet model for the kinetics of the alcoholic fermentation of sugarcane juice using the above data for a fed-batch fermentation with cell recycling [49].

Initial conditions are given in Table 5.6.

Assumptions:

- The bioreactor content is well mixed and has a constant medium volume at certain initial conditions
- Temperature is constant
- pH is maintained at optimal pH of 5.00
- All reactants or nutrients are present in excess except for sugar substrate

Kinetic model

The state variables involved in this fermentation process were concentration of total cell mass X (kg m^{-3}), substrate S (kg m^{-3}), and ethanol P (kg m^{-3}). Experimental observations have shown that cell, substrate, and product inhibitions are significant for ethanol fermentation [46]. Equation (1) shows the equation for cell growth rate, r_x, which includes terms for such types of inhibition:

$$r_x = \mu_{\max}\frac{S}{K_3 + S}\exp(-k_i\ S)\left(\frac{X}{X_{\max}}\right)^m\left(1 + \frac{P}{P_{\max}}\right)X \qquad (1)$$

where:

$\mu_{\max}$ is maximum specific growth rate (h^{-1})

K_s is substrate saturation constant (kg m^{-3})

K_i is substrate inhibition parameter (m^{-3} kg)
X_{max} is cell concentration where the growth ceases (kg m^{-3})
P_{max} is ethanol concentration where the cell growth ceases (kg m^{-3})
m and n are empirical parameters.

A modified Luedking–Piret expression was used to account for the ethanol formation rate, r_p, as shown in Eq. (2). This rate depends on the specific growth rate and cell concentration (X). $Y_{p/x}$ (kg kg^{-1}) is the product yield based on cell growth, β_{mp} (kg kg^{-1} h^{-1}) is a parameter associated with maintenance, and $K_{\beta s1}$ (kg m^{-3}) is a saturation parameter.

$$Y_p = Y_{p/x} r_x + \frac{\beta_{mp}}{K_{\beta s1}} \times \quad (2)$$

The substrate consumption rate, r_s, is expressed as follows:

$$r_s = (r_x/Y_x) + \frac{\beta_{mp} S}{K_{\beta s2} + S} \times \quad (3)$$

where Y_x (kg kg^{-1}) denotes the limit cellular yield, β_{ms} (kg kg^{-1} h^{-1}) is a maintenance parameter, and $K_{\beta s2}$ (kg m^{-3}) is a saturation parameter.

Fed-Batch Model

Mechanistic models comprise mass balance differential equations, with microorganism growth, substrate consumption, and ethanol formation for a fed-batch reactor described as follows:

$$\text{Total cell}: \quad \frac{dx}{dt} = r_x - \frac{F_{AX}}{V} \quad (4)$$

$$\text{Substrate}: \quad \frac{ds}{dt} = \frac{F_A (S_A - S)}{V} Y_s \quad (5)$$

$$\text{Ethanol}: \quad \frac{dp}{dt} = r_p - \frac{F_A P}{V} \quad (6)$$

$$\text{Volume}: \quad \frac{dv}{dt} = F_A \quad (7)$$

The mass-balance differential equations were solved using the LSODE (Livermore solver for ordinary differential equations) [18, 50].

Parameter Estimation Method

The proposed method for estimation of kinetic parameters is depicted in Fig. 5.12. First the kinetic parameters are initialized (including fixed parameters and the influential parameters to be estimated) as well as the operational condition values for the fermentation process (feeding time t_F, feed stream flow rate F_A, and feed substrate concentration S_A). After this step, the proposed method is able to find optimum values for the parameters that produce the best fit between the experimental observations and the simulated response variables by minimizing cost functions.

$$E\theta = \sum_{n=1}^{np} \frac{(X_n - X_{e_n})^2}{Xe_{\max}^2} + \frac{(S_n - SX_{e_n})^2}{Se_{\max}^2} + \frac{(PX_n - PX_{e_n})^2}{Pe_{\max}^2} \tag{8}$$

$$E\theta = \sum_{n=1}^{np} \frac{(X_n - X_{e_n})^2}{\left(\frac{X_n - X_{e_n}}{2}\right)^2} + \frac{(S_n - S_{e_n})^2}{\left(\frac{S_n - S_{e_n}}{2}\right)^2} + \frac{(P_n - P_{e_n})^2}{\left(\frac{P_n - P_{e_n}}{2}\right)^2} \tag{9}$$

where θ, the vector of kinetic parameters, is constrained by bounds within a realistic range (i.e., biological means). X_{en}, S_{en}, and P_{en} are the experimental observations of cell, substrate and ethanol at the sampling time n. X_n, S_n, and P_n are the concentration of cell, substrate, and ethanol computed by the model at sampling time n. $Xe_{\max}$, $Se_{\max}$, and $Pe_{\max}$ are the maximum measured concentration.

5.4 Results

Fed-batch experiments with cell recycle were performed for estimation of kinetic parameters. The initial values of the state variables (X_i, S_i, and P_i) and the operational conditions of these experiments are as stated earlier (Table 5.6). From kinetic rates described in Eqs. (1)–(3), a set of 13 kinetic parameters can be obtained.

A sensitivity analysis approach applied to an analogous kinetic system concluded that a subset of those parameters differ within a range according to changes in operational conditions and fluctuations in the quality of raw materials [44]. In this sense, $\mu_{\max}$, $X_{\max}$, $P_{\max}$, Y_x, and $Y_{p/x}$ are known to be influential parameters in the system, and were estimated using the proposed methodology. In this study, β_{mp}, β_{ms}, $K_{\beta s1}$, and $K_{\beta s2}$ were also determined. The estimated numerical values are indicated in Table 5.7 . The remaining parameters were fixed at the previous values used in several studies [45], as $K_s = 4.1$ (kg m^{-3}), $K_i = 0.002$ m^3 kg^{-1}, $m = 1.0$, and $n = 1.5$.

The above data-based analysis shows that the model described the experimental data accurately, as evaluated by RSD (%). Also, in all cases, R^2 was close to unity, indicating a good fit of the model, as can be seen in Table 5.8. Results showed that it is possible to infer concentration accurately in a fed-batch fermentation with intensive recycling.

5.4.1 *Spreadsheet for a Typical Fermentation Process and Material Balance*

Yeasts can grow in the presence or absence of air. Anaerobic growth (growth in the absence of oxygen) is quite slow and inefficient. In this energy-yielding process,

Table 5.7 Estimated parameters values

Parameter	Estimated value
μ_{max}	0.33 h^{-1}
P_{max}	85.7 kg m^{-3}
X_{max}	56.2
$Y_{p/x}$	3.56 kg kg^{-1}
Y_x	0.0909 kg kg^{-1}
Y_x	0.229 kg kg^{-1}
β_{mp}	0.001 kg kg^{-1}
$K_{\beta s1}$	0.001 kg m^{-3}
$K_{\beta s}$	0.201 kg/kg h
$K_{\beta s2}$	0.001 kg m^{-3}

Table 5.8 Statistical criteria for characterizing the predictive quality of the fed-batch model

	X (kg/m^3)		S (kg/m^3)		P (kg/m^3)	
Fermentation	RSD (%)	R^2	RSD (%)	R^2	RSD (%)	R^2
No 1	6.25	0.99 35	35.24	0.98	10.54	0.98
No 2	3.73	0.98	11.06	0.98	15.60	0.95
No 3	6.59	0.99	99 35.	73 0.99	5.67	0.99

yeast converts organic molecules (such as sugar) into energy, carbon dioxide, or/and ethanol depending on the respiration pathway. Only a small portion of the sugar is used for cell maintenance and growth. In contrast, under aerobic conditions, in the presence of a sufficient quantity of dissolved oxygen, yeast grow by using most of the available sugar for growth and producing only negligible quantities of alcohol.

However, yeasts obtain more energy from aerobic respiration, but in the absence of oxygen they can continue to respire anaerobically, although not getting as much energy from the substrate. The complete breakdown of glucose to carbon dioxide and water in aerobic respiration yields much more energy than alcoholic fermentation: 36–38 ATP molecules versus only 2 ATP produced by anaerobic respiration. Ethanol molecules produced by alcoholic fermentation diffuse from yeast cells into the surrounding aqueous environment. Because ethanol is harmful to cellular membranes, yeast cells die if the ethanol concentrations reach a critical level.

Example 10: Sucrose fermentation

Sucrose is fermented into ethanol catalytically by yeast in the absence of air. Sucrose is a dimer of glucose and fructose molecules. In the first step of alcoholic fermentation, the enzyme invertase (also known as sucrase) cleaves the glycosidic linkage between glucose and fructose molecules (Fig. 5.13).

The chemical equations below summarize the fermentation of sucrose ($C_{12}H_{22}O_{11}$) into ethanol (C_2H_5OH). Alcoholic fermentation converts one mole of glucose into two moles of ethanol and two moles of carbon dioxide, producing two moles of ATP in the process.

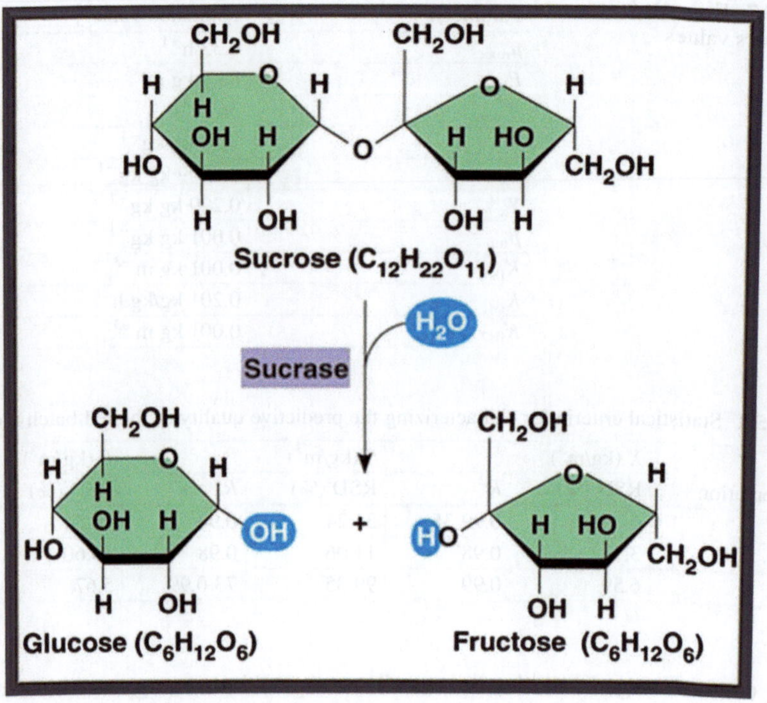

Fig. 5.13 The enzyme sucrase cleaves the glycosidic linkage between the glucose and fructose molecules of sucrose (Source: www.meetcolab.com)

Glycolysis:

$$\text{Glucose} + 2\text{NAD}^+ + 2\text{ADP} + 2\text{P}_i \rightarrow 2\text{Pyruvic acid} + 2\text{ATP} + 2\text{NADH} + 2\text{H}^+$$

Decarboxylation of pyruvate:

$$2\text{Pyruvic acid} \rightarrow 2\text{CO}_2$$

Reduction of acetaldehyde:

$$2\text{Acetaldehyde} + 2\text{NADH} + 2\text{H}^+ \rightarrow \text{Ethanol} + 2\text{NAD}^+$$

Net reaction:

$$\text{Glucose} + 2\text{ADP} + 2\text{P}_i \rightarrow 2\text{Ethanol} + 2\text{CO2} + 2\text{ATP}$$

5.5 Objectives

1. Find the overall stoichiometry of anaerobic fermentation catalyzed by yeasts using following methods:

- Nonlinear regression (guess for k_s and μ_m)
- The sum of least squares
- Lineweaver–Burk plot to determine the parameters μ_m and k_s

2. Determine the yield coefficient and project the minimum and maximum amounts of yeast cell mass, carbon dioxide, and ethanol produced

Experimental Procedure
Biochemistry analyzers and a spectrophotometer were used to measure and make calibration curves for sugar and yeast cell concentrations.

Reactant initial concentrations:

- Dextrose or sucrose 25 g L^{-1}
- Yeast 3 g L^{-1}
- Volume reactant solution 2 L

Initial conditions:

- 2 L solution
- 50 g sugar
- pH around 5.0
- Temperature around 28–30 °C

Assumptions:

- Bioreactor content is well mixed and has a constant medium volume at a certain initial conditions
- Temperature is constant
- pH maintained at optimal pH of 3
- All reactants or nutrients present in excess except for sugar substrate

Theory
An ideal fermentation process, in which the growing cells consume the substrate (sugars) and produce more cells, can be expressed according to the scheme in Fig. 5.14, where r_{sx} is the rate of substrate consumption, r_x is the rate of cell growth, s is substrate concentration, x is cell concentration, and P is ethanol concentration (in anaerobic case):

$$\gamma_x = \frac{dx}{dt}\gamma_x r_x = \frac{dx}{dt}$$

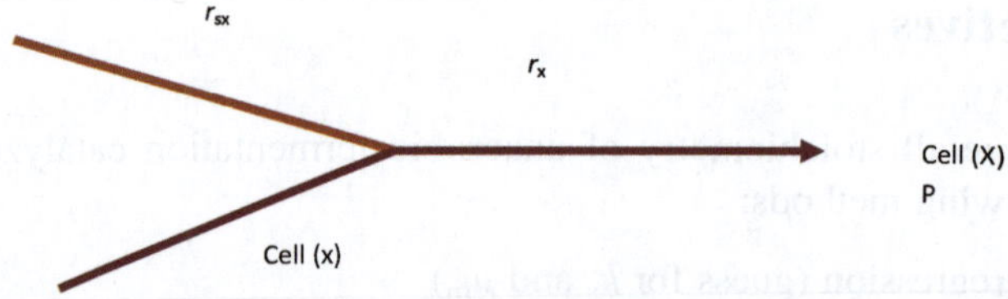

Fig. 5.14 Biomass production in an ideal fermentation process. r_{sx} rate of substrate consumption, r_x rate of cell growth, s substrate concentration, x cell concentration, P ethanol concentration (in anaerobic case)

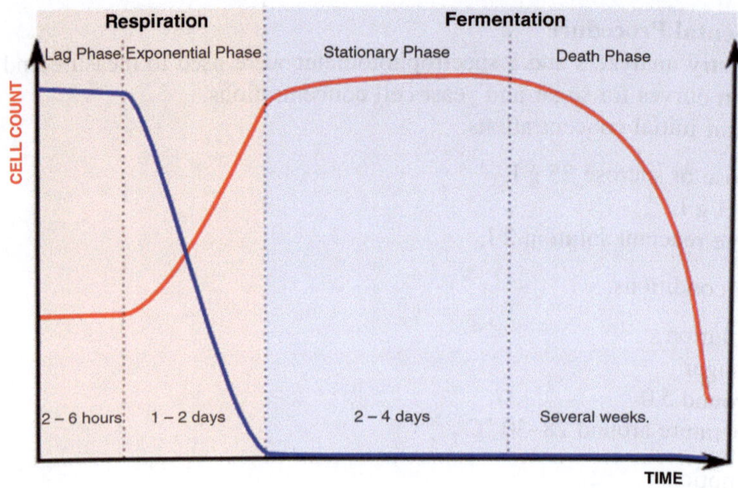

Fig. 5.15 Growth curve of yeast showing four stages

The plot in Fig. 5.15 shows the trend for yeast cell growth over time. Yeast growth occurs in four stages:

- Lag phase, during which yeast mature and acclimate to environment (no growth occurs).
- Exponential growth phase, during which the rate of reaction follows first-order kinetics.
- Deceleration phase, during which a large number of parameters, each with saturation effects, affect the kinetics of yeast growth (e.g., substrate and waste concentrations). The growth rate is ruled by the limiting substrate concentration (sugar). The final equation, often referred to as the Monod equation, looks very similar to the Michaelis–Menten equation.
- Stationary phase, during which no growth occurs as a result of the high waste concentration or complete substrate consumption.

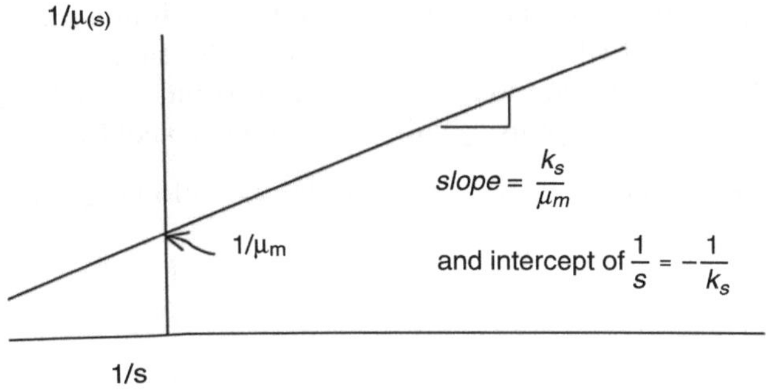

Fig. 5.16 Lineweaver–Burk plot

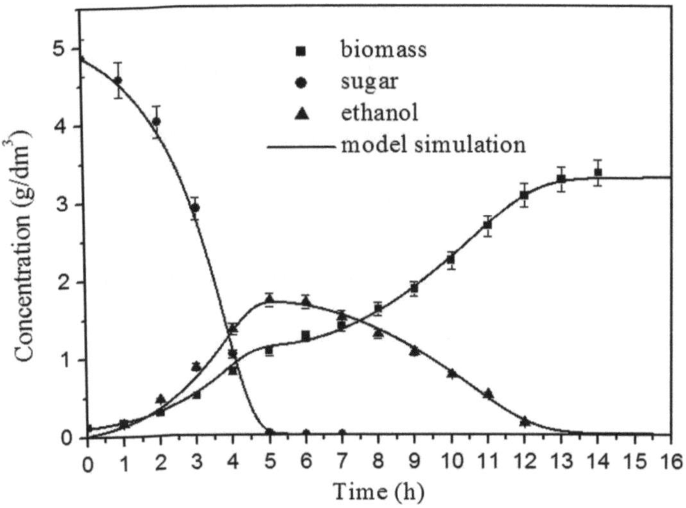

Fig. 5.17 Yeast biomass yield showing the consumption of different sources of carbon

The Lineweaver–Burk plot (Fig. 5.16) can expressed as follows:

$$\mu_s = \mu_m \left\{ \frac{S_s}{K_s + S_s} \right\} \frac{1}{\mu_s} = \frac{K_s + S}{\mu_m \; S} = \frac{K_s}{\mu_m} \frac{1}{S} + \frac{1}{\mu_m}$$

and nonlinear regression (Fig. 5.17) as:

$$\gamma_x = \frac{dx}{dt} = \mu_m \frac{S_s}{K_s + S} s$$

- Solve for $R_{\text{predicted}}$ (dx/dt) by calculating dx/dt from the polynomial equation fitted to the curve $x(t)$.

- Make initial guesses for k_s and μ_m where μ_m is the maximum specific growth rate that can be achieved when $S >> k_s$, where k_s is the saturation constant or the limiting substrate concentration S, at which μ_s is equal to half of μ_m.
- Minimize $\Sigma(R - R_{predicted})^2$ using solver function in Excel by varying k_s and μ_m.

The yield coefficient can be determined using the following expressions:

$$Y_{\frac{x}{s}} = \frac{dx}{ds} = \frac{X - X_0}{S_0 - S} \qquad Y_{\frac{p}{s}} = \frac{dp}{ds} = \frac{p - P_0}{S_0 - S}$$

Knowing $Y_{x/s}$ gives an idea of how much additional yeast cell mass, on average, is produced for a given amount of sugar substrate consumed. It also allows calculation of a lower boundary for the experimental stoichiometric coefficient γ and, therefore, calculation of ranges for ethanol and CO_2 production:

$$(\text{Yeast cells}) + C_6H_{12}O_6 \rightarrow \gamma(CO_2 + CH_3CH_2OH) + (\text{Yeast cells})$$

where, theoretically, $\gamma = 2$. Assume all yeast generation is attributable only to complete sugar consumption. Conservation of mass requires that the remaining products be equimolar amounts of CO_2 and ethanol.

The following errors can occur in Lineweaver–Burk parameters:

- Errors in k_s and μ_m relative to errors in slope and y intercept of linear fit
- Random error in y values: $S_{\frac{y}{x}} \sqrt{\frac{\sum\left(y_i - \hat{y}_i\right)^2}{(n-2)}}$
- Standard deviation of slope: $S_b = \frac{S_{y/x}}{\sqrt{\sum\left(x_i - \hat{x}\right)^2}}$
- Time deviation of y intercept: $S_\alpha = S_{y/x}\sqrt{\frac{\sum X_i^2}{n\sum\left(x_i - \hat{x}\right)^2}}$

References

1. David A and Landes S (1969) The unbound prometheus. Press Syndicate of the University of Cambridge. p. 104. ISBN 0-521-09418-6
2. Rosen William (2012) The most powerful idea in the world: a story of steam, industry and invention. University of Chicago Press. p. 185. ISBN 978-0226726342
3. Ahmad Y Hassan (1976) Taqi al-Din and Arabic mechanical engineering, p 34–35. Institute for the History of Arabic Science, University of Aleppo
4. Robert Henry Thurston (1903) A history of the growth of the steam-engine, D. Appleton and company, Google Print, p 15–16 (public domain)
5. Nag PK (2002) Power plant engineering. Tata McGraw-Hill. p. 432. ISBN 0-07-043599-5
6. Garcia Nicholas (2007) Mas alla de la Leyenda Negra. Valencia: Universidad de Valencia. pp. 443–454. ISBN 9788437067919
7. Clausius Rudolf (1850) On the Motive Power of Heat, and on the Laws which can be deduced from it for the Theory of Heat. Poggendorff's Annalen der Physik, LXXIX (Dover Reprint). ISBN 0-486-59065-8

8. William Thomson (1882) Mathematical and physical papers. 1. London, Cambridge: C.J. Clay, M.A. & Son, Cambridge University Press, p 232
9. Gibbs Willard J (1876) Transactions of the Connecticut Academy, III, pp 108–248, Oct. 1875–May 1876, and pp 343–524, May 1877–July 1878
10. Duhem PMM (1886) Le potentiel thermodynamique et ses applications à la mécanique chimique et à l'étude des phénomènes électriques. A. Hermann, Paris. http://gallica.bnf.fr/ark:/12148/bpt6k62445r
11. Lewis Gilbert N, Randall Merle (1923) Thermodynamics and the free energy of chemical substances. McGraw-Hill Book
12. Guggenheim EA (1933) Modern Thermodynamics by the Methods of J.W. Gibbs, Methuen, London
13. Guggenheim EA (1949/1967) Thermodynamics. An advanced treatment for chemists and physicists, 1st edition 1949, 5th edition 1967, North-Holland, Amsterdam
14. Ilya Prigogine I et al (1954) Chemical thermodynamics. Longmans, Green & Co., London. Includes classical non-equilibrium thermodynamics
15. Enrico Fermi (1956) Thermodynamics. Courier Dover Publications. pp. (ix). ISBN 048660361X. OCLC 230763036
16. Perrot Pierre (1998) A to Z of thermodynamics. Oxford University Press. ISBN 0-19-856552-6. OCLC 123283342
17. Clark John OE (2004) The essential dictionary of science. Barnes & Noble Books. ISBN 0-7607-4616-8. OCLC 58732844
18. Erickson LE et al (1978) Application of mass and energy balance regularities in fermentation. Biotechnol Bioeng 20:1595–1621
19. Gaudet G et al (1992) Futile cycle of glycogen in Fibrobacter succinogenes as shown by in situ 1H NMR and 13C NMR investigations. Eur J Biochem 107:155–162
20. Guiavarch E et al (2008) Application of a data reconciliation method to the stoichiometric analysis of Fibrobacter succinogenes growth. Appl Biochem Biotechnol 151:201–210
21. Guiavarch E et al (2010) Analysis of a continuous culture of Fibrobacter succinogenes S85 on a standardized glucose medium. Bioprocess Biosyst Eng 33:417–425
22. Himmelblau DM (1970) Process analysis by statistical methods. Wiley, New York
23. Humphrey AE (1974) Current developments in fermentation. Chem Eng 81:98–112
24. Alberty R (2004) A short history of the thermodynamics of enzyme-catalyzed reactions. J Biol Chem 279:27831–27836
25. Akihiko Ito and Takehisa Oikawa (2004) Global mapping of terrestrial primary productivity and light-use efficiency with a process-based model. In M. Shiyomi et al. Global Environmental Change in the Ocean and on Land (PDF), pp. 343–358
26. Alberty R (2004) A short history of the thermodynamics of enzyme-catalyzed reactions. J Biol Chem 279:27831. doi:10.1074/jbc.X400003200. PMID 15073189
27. Hungate RE (1950) The anaerobic mesophilic cellulolytic bacteria. Bacteriol Rev 14:1–49
28. Kristiansen B (1994) Integrated desugn of a fermentation plant. The production of baker's yeast. VCH (Ed.) Weinheim
29. Minkevich IG (1983) Mass-energy balance for microbial product synthesis—biochemical and cultural aspects. Biotechnol Bioeng 25:1267–1293
30. Nielsen J, Villadsen J, Liden G (2003) Bioreaction engineering principles, 2nd edn. Kluwer, Dordrecht
31. Nouaille R et al (2005) Oligosaccharide synthesis in Fibrobacter succinogenes S85 and its modulation by the substrate. FEBS J 272:2416–2427
32. Patton CJ, Crouch SR (1977) Spectrophotometric and kinetics investigation of the Berthelot reaction for the determination of ammonia. Anal Chem 49:464–469
33. Roels JA (1980) Application of macroscopic principles to microbial metabolism. Biotechnol Bioeng 22:2457–2514
34. Roels JA (1983) Energetics and kinetics in Biotechnology, 117. Elsevier Biomedical Press, Amsterdam

35. Solomon BO et al (1982) Maximum likelihood estimation of growth yields. Biotechnol Bioeng 24:633–649
36. Solomon BO et al (1984) Estimation of parameters where dependent observations are related by equality constraints. AICHE J 30:747–757
37. Summer JB, Howell SF (1935) A method for determination of saccharase activity. J Biochem Chem 108:51–54
38. Urrieta-Saltijeral JM et al (2001) Metabolic flux modeling as a tool to analyse the behavior of a genetically modified strain of Saccharomyces cerevisiae. Eng Manuf Biotechnol 4:143–156
39. Wang NS, Stephanopoulos G (1983) Application of macroscopic balances to the identification of gross measurement errors. Biotechnol Bioeng 25:2177–2208
40. Minkevich IG, Eroshin VK (1973) Folio Microbiol 18:376–385
41. Stainer RY, Adelberg EA, Ingraham J (1976) The microbial world, 4th edn. Prentice-Hall, New Jersey
42. Roel JA (1980) Application of macroscopic principles to microbial metabolism. Biotechnol Bioeng 22:2457–2514
43. Pauline Doran (1995) Material balance, bioprocess engineering and principles, Elsevier Science and Technology Books, 51–85, ISBN:122208552
44. Andrade RR et al (2009) Study of kinetic parameters in a mechanistic model for bioethanol production through a screening technique and optimization. Bioprocess Biosyst Eng 32:673–680
45. Andrade RR et al (2013) Kinetic of ethanol production from sugarcane bagasse enzymatic hydrolysate concentrated with molasses under cell recycle. Bioresour Technol 130:351–359
46. Rivera EC et al (2007) Development of adaptive modeling techniques to describe the temperature-dependent kinetics of biotechnological process. Biochem Eng J 36:157–166
47. Rivera EC (2010) Enzymatic hydrolysis of sugarcane bagasse for bioethanol production: determining optimal enzyme loading using neural networks. J Chem Technol Biotechnol 85:983–992
48. Shuler ML and Kargi F (1992) Bioprocess Engineering Basic Concept. Technology & Engineering, 479 pages; Prentice-Hall, New Jersey, 155–206
49. Rivera EC et al (2013) A procedure for estimation of fermentation kinetic parameters in fed-batch bioethanol production process with cell recycle. Chem Eng Trans 32:1369–1374
50. Radhakrishnan K and Hindmarsh A (1993) Description and use of LSODE, the Livermore solver for differential equations, NASA reference publication 1327

Index

B. Kumara Behera, A. Varma, *Microbial Biomass Process Technologies and Management*, DOI 10.1007/978-3-319-53913-3

T

U

V

MIX
Papier aus verantwortungsvollen Quellen
Paper from responsible sources
FSC® C105338

If you have any concerns about our products,
you can contact us on
ProductSafety@springernature.com

In case Publisher is established outside the EU,
the EU authorized representative is:
Springer Nature Customer Service Center GmbH
Europaplatz 3, 69115 Heidelberg, Germany

Printed by Libri Plureos GmbH
in Hamburg, Germany